Springer Series in Operations Research

Editors:
Peter Glynn   Stephen Robinson

Springer Series in Operations Research

John R. Birge   François Louveaux

Introduction to Stochastic Programming

With 38 Illustrations

 Springer

John R. Birge
McCormick School of Engineering
  and Applied Science
Northwestern University
Evanston, IL 60208
USA

François Louveaux
Facultés Universitaires Notre Dame
  de la Paix
Département des Méthodes Quantitatives
Rempart de la Vierge 8
B-5000 Namur
Belgium

Series Editors:
Peter Glynn
Department of Operations Research
Stanford University
Stanford, CA 94305
USA

Stephen Robinson
Department of Industrial Engineering
University of Wisconsin
Madison, WI 53786-1572
USA

Library of Congress Cataloging-in-Publication Data
Birge, John R.
    Introduction to stochastic programming / John R. Birge, François Louveaux.
      p. cm.
    Includes bibliographical references and index.
    ISBN 0-387-98217-5 (hardcover : alk. paper)
    1. Stochastic programming.   I. Louveaux, François.   II. Title.
    T57.79.B57 1997
    519.7—dc21   97-6931

Printed on acid-free paper.

Printed in the United States of America.   (SPI)

9 8 7 6 5 4

springer.com

To Pierrette and Marie

Preface

According to a French saying "Gérer, c'est prévoir," which we may translate as "(The art of) Managing is (in) foreseeing." Now, probability and statistics have long since taught us that the future cannot be perfectly forecast but instead should be considered random or uncertain. The aim of stochastic programming is precisely to find an optimal decision in problems involving uncertain data. In this terminology, *stochastic* is opposed to *deterministic* and means that some data are random, whereas programming refers to the fact that various parts of the problem can be modeled as linear or nonlinear mathematical programs. The field, also known as *optimization under uncertainty*, is developing rapidly with contributions from many disciplines such as operations research, economics, mathematics, probability, and statistics. The objective of this book is to provide a wide overview of stochastic programming, without requiring more than a basic background in these various disciplines.

Introduction to Stochastic Programming is intended as a first course for beginning graduate students or advanced undergraduate students in such fields as operations research, industrial engineering, business administration (in particular, finance or management science), and mathematics. Students should have some basic knowledge of linear programming, elementary analysis, and probability as given, for example, in an introductory book on operations research or management science or in a combination of an introduction to linear programming (optimization) and an introduction to probability theory.

Instructors may need to add some material on convex analysis depending on the choice of sections covered. We chose not to include such introductory material because students' backgrounds may vary widely and other texts include these concepts in detail. We did, however, include an introduction to random variables while modeling stochastic programs in Section 2.1 and short reviews of linear programming, duality, and nonlinear programming at the end of Chapter 2. This material is given as an indication of the prerequisites in the book to help instructors provide any missing background. In the Subject Index, the first reference to a concept is where it is defined or, for concepts specific to a single section, where a source is provided.

In our view, the objective of a first course based on this book is to help students build an intuition on how to model uncertainty into mathematical programs, which changes uncertainty brings into the decision process, what difficulties uncertainty may bring, and what problems are solvable. To begin this development, the first section in Chapter 1 provides a worked example of modeling a stochastic program. It introduces the basic concepts, without using any new or specific techniques. This first example can be complemented by any one of the other proposed cases of Chapter 1, in finance, in multistage capacity expansion, and in manufacturing. Based again on examples, Chapter 2 describes how a stochastic model is formally built. It also stresses the fact that several different models can be built, depending on the type of uncertainty and the time when decisions must be taken. This chapter links the various concepts to alternative fields of planning under uncertainty.

Any course should begin with the study of those two chapters. The sequel would then depend on the students' interests and backgrounds. A typical course would consist of elements of Chapter 3, Sections 4.1 to 4.5, Sections 5.1 to 5.3 and 5.7, and one or two more advanced sections of the instructor's choice. The final case study may serve as a conclusion. A class emphasizing modeling might focus on basic approximations in Chapter 9 and sampling in Chapter 10. A computational class would stress methods from Chapters 6 to 8. A more theoretical class might concentrate more deeply on Chapter 3 and the results from Chapters 9 to 11.

The book can also be used as an introduction for graduate students interested in stochastic programming as a research area. They will find a broad coverage of mathematical properties, models, and solution algorithms. Broad coverage cannot mean an in-depth study of all existing research. The reader will thus be referred to the original papers for details. Advanced sections may require multivariate calculus, probability measure theory, or an introduction to nonlinear or integer programming. Here again, the stress is clearly in building knowledge and intuition in the field. Mathematical results are given so long as they are either basic properties or helpful in developing efficient solution procedures. The importance of the various sections clearly reflects our own interests, which focus on results that may lead to practical applications of stochastic programming.

To conclude, we may use the following little story. An elderly person, celebrating her one hundredth birthday, was asked how she succeeded in reaching that age. She answered, "It's very simple. You just have to wait."

In comparison, stochastic programming may well look like a field of young impatient people who not only do not want to wait and see but who consider waiting to be suboptimal. We realize how much patience was needed from our friends and colleagues who encouraged us to write this book, which took us much longer than expected. To all of them, we are extremely thankful for their support. The authors also wish to thank the Fonds National de la Recherche Scientifique and the National Science Foundation for their financial support. Both authors are deeply grateful to the people who introduced us to the field, George Dantzig, Roger Wets, Jacques Drèze, and Guy de Ghellinck. Our special thanks go to our wives, Pierrette and Marie, to whom we dedicate this book.

Ann Arbor, Michigan　　　　　　　　　　　　　John R. Birge
Namur, Belgium　　　　　　　　　　　　　　François Louveaux

Contents

Notation

The following describes the major symbols and notations used in the text. To the greatest extent possible, we have attempted to keep unique meanings for each item. In those cases where an item has additional uses, they should be clear from context. We include here only notation used in more than one section. Additional notation may be needed within specific sections and is explained when used.

In general, vectors are assumed to be columns with transposes to indicate row vectors. This yields $c^T x$ to denote the inner product of two n-vectors, c and x. We reserve prime ($\prime$) for first derivatives with respect to time (e.g., $f' = df/dt$).

Vectors in primal programs are represented by lowercase Latin letters while matrices are uppercase. Dual variables and certain scalars are generally Greek letters. Superscripts indicate a stage while subscripts indicate components followed by realization index. Boldface indicates a random quantity. Expectations of random variables are indicated by a bar ($\bar{\xi}$), μ, or ($E(\boldsymbol{\xi})$). We also use the bar notation to denote sample means in Chapter 10.

Equations are numbered consecutively in the text by section and number within the section (e.g., (1.2) for Section 1, Equation 2). For references to chapters other than the current one, we use three indices: chapter, section, and equation, (e.g., (3.1.2) for Chapter 3, Section 1, Equation 2). Exercises are given at the end of each section and are referenced in the same manner as equations. All other items (figures, tables, declarations, examples) are labeled consecutively through the entire chapter with a single reference (e.g., Figure 1) if within the current chapter and chapter and number if in a different chapter (e.g., Figure 3.1 for Chapter 3, Figure 1).

Symbol	Definition
$+$	Superscript indicates the positive part of a real (i.e., $a^+ = \max(a,0)$) or unrestricted variable (e.g., $y = y^+ - y^-, y^+ \geq 0, y^- \geq 0$) and its objective coefficients (e.g., q^+), subscript as non-negative values in a set (e.g., $\Re_+$) or the right-limit ($F^+(t) = \lim_{s\downarrow t} F(s)$)
$-$	Superscript indicates the negative part of a real (i.e., $a^- = \max(-a,0)$) or unrestricted variable (e.g., $y = y^+ - y^-, y^+ \geq 0, y^- \geq 0$) and its objective coefficients (e.g., q^-) or the left-limit ($F^-(t) = \lim_{s\uparrow t} F(s)$)
$*$	Indicates an optimal value or solution (e.g., x^*)
$0 \wedge / \sim$	Indicate given nonoptimal values or solutions (e.g., $x^0, \hat{x}, x', \tilde{x}$)
a	Ancestor scenario, real value or vector
A	First-stage matrix (e.g., $Ax = b$), also used to indicate a subset, $A \in \mathcal{A} \subset \Omega$
$\mathcal{A}$	Collection of subsets
b	First-stage right-hand side (e.g., $Ax = b$)
B	Matrix, basis submatrix, Borel sets, or index set of a basis
$\mathcal{B}$	Collection of subsets (notably Borel sets)
c	First-stage objective ($c^T x$), t-th stage objective ($(c^t(\omega))^T x^t$) or real vectors
C	Matrix or index set of continuous variables
d	Right-hand side of a feasibility cut in the L-shaped method, a demand, or real vector
D	Left-hand side vector of a feasibility cut in the L-shaped method, a matrix, a set, or an index set of discrete variables
$\mathcal{D}$	Set of descendant scenarios
e	Exponential, right-hand side of an optimality cut in the L-shaped method, an extreme point, or the unit vector ($e^T = (1, \ldots, 1)$)
E	Mathematical expectation operator or left-hand side vector of an optimality cut in the L-shaped method
f	Function (usually in an objective ($f(x)$ or $f_i(x)$) or a density
F	Cumulative probability distribution
g	Function (usually in constraints ($g(x)$ or $g_j(x)$))
h	Right-hand side in second-stage ($Wy = h - Tx$), also $h^t(\omega)$ in multistage problems

Symbol	Definition
H	Number of stages (horizon) in multistage problems
i	Subscript index of functions (f_i) or vector elements (x_i, x_{ij})
I	Identity matrix or index set $(i \in I)$
j	Subscript index of functions (g_j) or vector elements (y_j, y_{ij})
J	Matrix or index set
k	Index of a realization of a random vector $(k = 1, \ldots, K)$
K	Feasibility sets (K_1, K_2) or total number of realizations of a discrete random vector
l	Index or a lower bound on a variable
L	The L-shaped method, objective value lower bound, or real value
m	Number of constraints (m_1, m_2) or number of elements $(i = 1, \ldots, m)$
n	Number of variables (n_1, n_2) or number of elements $(i = 1, \ldots, n)$
N	Set, normal cone, normal distribution, or number of random elements
O	Zero matrix
p	Probability of a random element (e.g., $p_k = P(\boldsymbol{\xi} = \xi_k)$) or matrix of probabilities
P	Probability of events (e.g., $P(\boldsymbol{\xi} \leq 0)$)
q	Second-stage objective vector $(q^T y)$
Q	Second-stage (multistage) value function with random argument $(Q(x, \boldsymbol{\xi})$ or $Q^t(x^t, \xi^t))$
$\mathcal{Q}$	Second-stage (multistage) expected value value (recourse) function $(\mathcal{Q}(x)$ or $\mathcal{Q}^t(x^t))$
r	Revenue or return in examples, real vector, or index
R	Matrix or set
$\Re$	Real numbers
s	Scenario or index
S	Set or matrix
t	Superscript stage or period index for multistage programs $(t = 1, \ldots, H)$, a real-valued parameter, or an index
T	Technology matrix $(Wy = h - Tx$ or $T^{t-1}(\omega)(x))$; as a superscript, the transpose of a matrix or vector
u	General vector, upper-bound vector, or expected shortage
U	Objective value upper bound

Symbol	Definition
v	Variable vector or expected surplus
V	Set, matrix or an operator
w	Second-stage decision vector in some examples
W	Recourse matrix ($Wy = h - Tx$)
x	First-stage decision vector or multistage decision vector (x^t)
X	First-stage feasible set ($x \in X$) or tth stage feasible set (X^t)
y	Second-stage decision vector
Y	Second-stage feasible set ($y \in Y$)
z	Objective value ($\min z = c^T x + \cdots$)
Z	Integers
α	Real value, vector, or probability level with probabilistic constraints
β	Real value or vector
γ	Real value or function
δ	Real value or function
ϵ	Real value
ζ	Random variable
η	Real value or random variable
θ	Lower bound on $\mathcal{Q}(x)$ in the L-shaped method
κ	Index
λ	Dual multiplier, parameter in a convex combination, or measure
μ	Expectation (used mostly in examples of densities) or a parameter for non-negative multiples
ν	Algorithm iteration index
$\boldsymbol{\xi}$	Random vector (often indexed by time, ξ^t) with realizations as ξ (without boldface)
Ξ	Support of the random vector $\boldsymbol{\xi}$
π	Dual multiplier
Π	Product or projection operator
ρ	Dual multiplier or discount factor
σ	Dual multiplier, standard deviation, or σ-field
Σ	Summation
τ	Possible right-hand side in bundles or index of time
ϕ	Function in computing the value of the stochastic solution or a measure
Φ	Function, cumulative distribution of standard normal

Symbol	Definition
$\emptyset$	Empty set
χ	Tender or offer from first to second period ($\chi = Tx$)
ψ	Second stage value function defined on tenders and with random argument, $\psi(\chi, \xi(\omega)))$
Ψ	Expected second stage value function defined on tenders, $\Psi(\chi))$
ω	Random event ($\omega \in \Omega$)
Ω	Set of all random events

Part I

Models

1
Introduction and Examples

This chapter presents stochastic programming examples from a variety of areas with wide application in stochastic programming. These examples are intended to help the reader build intuition on how to model uncertainty. They also reflect different structural aspects of the problems. In particular, we show the variety of stochastic programming models in terms of the objectives of the decision process, the constraints on those decisions, and their relationships to the random elements.

In each example, we investigate the value of the stochastic programming model over a similar deterministic problem. We show that even simple models can lead to significant savings. These results provide the motivation to lead us into the following chapters on stochastic programs, solution properties, and techniques.

In the first section, we consider a farmer who must decide on the amounts of various crops to plant. The yields of the crops vary according to the weather. From this example, we illustrate the basic foundation of stochastic programming and the advantage of the stochastic programming solution over deterministic approaches. We also introduce the classical news vendor (or newsboy) problem and give the fundamental properties of these problems' general class, called *two-stage stochastic linear programs with recourse*.

The second section contains an example in planning finances for a child's education. This example fits the situation in many discrete time control problems. Decisions occur at different points in time so that the problem can be viewed as having multiple stages of observations and actions.

The third section considers power system capacity expansion. Here, decisions are taken dynamically about additional capacity and about the allocation of capacity to meet demand. The resulting problem has multiple decision stages and a valuable property known as *block separable recourse* that allows efficient solution. The problem also provides a natural example of constraints on reliability within the area called *probabilistic* or *chance-constrained programming*.

The fourth example concerns the design of a simple axle. It includes market reaction to the design and performance characteristics of products made by a manufacturing system with variable performance. The essential characteristics of the maximum performance of the product illustrate a problem with fundamental nonlinearities incorporated directly into the stochastic program.

The final section of this chapter briefly describes several other major application areas of stochastic programs. The exercises at the end of the chapter develop modeling techniques. This chapter illustrates some of the range of stochastic programming applications but is not meant to be exhaustive. Applications in routing and location, for example, are discussed in Chapter 2.

1.1   A Farming Example and the News Vendor Problem

a. The farmer's problem

Consider a European farmer who specializes in raising grain, corn, and sugar beets on his 500 acres of land. During the winter, he wants to decide how much land to devote to each crop. (We refer to the farmer as "he" for convenience and not to imply anything about the gender of European farmers.)

The farmer knows that at least 200 tons (T) of wheat and 240 T of corn are needed for cattle feed. These amounts can be raised on the farm or bought from a wholesaler. Any production in excess of the feeding requirement would be sold. Selling prices are $170 and $150 per ton of wheat and corn, respectively. The purchase prices are 40% more than this due to the wholesaler's margin and transportation costs.

Another profitable crop is sugar beet, which sells at $36/T; however, the European Commission imposes a quota on sugar beet production. Any amount in excess of the quota can be sold only at $10/T. The farmer's quota for next year is 6000 T.

Based on past experience, the farmer knows that the mean yield on his land is roughly 2.5 T, 3 T, and 20 T per acre for wheat, corn, and sugar

TABLE 1. Data for farmer's problem.

	Wheat	Corn	Sugar Beets
Yield (T/acre)	2.5	3	20
Planting cost ($/acre)	150	230	260
Selling price ($/T)	170	150	36 under 6000 T
			10 above 6000 T
Purchase price ($/T)	238	210	–
Minimum require- ment (T)	200	240	–
Total available land: 500 acres			

beets, respectively. Table 1 summarizes these data and the planting costs for these crops.

To help the farmer make up his mind, we can set up the following model. Let

x_1 = acres of land devoted to wheat,
x_2 = acres of land devoted to corn,
x_3 = acres of land devoted to sugar beets,
w_1 = tons of wheat sold,
y_1 = tons of wheat purchased,
w_2 = tons of corn sold,
y_2 = tons of corn purchased,
w_3 = tons of sugar beets sold at the favorable price,
w_4 = tons of sugar beets sold at the lower price.

The problem reads as follows:

$$\min \quad 150x_1 + 230x_2 + 260x_3 + 238y_1 - 170w_1$$
$$+210y_2 - 150w_2 - 36w_3 - 10w_4 \qquad (1.1)$$
$$\text{s.t. } x_1 + x_2 + x_3 \le 500, \ 2.5\,x_1 + y_1 - w_1 \ge 200,$$
$$3\,x_2 + y_2 - w_2 \ge 240, w_3 + w_4 \le 20x_3, w_3 \le 6000,$$
$$x_1, x_2, x_3, y_1, y_2, w_1, w_2, w_3, w_4 \ge 0.$$

After solving (1.1) with his favorite linear program solver, the farmer obtains an optimal solution, as in Table 2.

This optimal solution is easy to understand. The farmer devotes enough land to sugar beets to reach the quota of 6000 T. He then devotes enough land to wheat and corn production to meet the feeding requirement. The rest of the land is devoted to wheat production. Some wheat can be sold.

To an extent, the optimal solution follows a very simple heuristic rule: to allocate land in order of decreasing profit per acre. In this example, the order is sugar beets at a favorable price, wheat, corn, and sugar beets at the lower price. This simple heuristic would, however, no longer be valid

TABLE 2. Optimal solution based on expected yields.

Culture	Wheat	Corn	Sugar Beets
Surface (acres)	120	80	300
Yield (T)	300	240	6000
Sales (T)	100	–	6000
Purchase (T)	–	–	–
Overall profit: $118,600			

if other constraints, such as labor requirements or crop rotation, would be included.

After thinking about this solution, the farmer becomes worried. He has indeed experienced quite different yields for the same crop over different years mainly because of changing weather conditions. Most crops need rain during the few weeks after seeding or planting, then sunshine is welcome for the rest of the growing period. Sunshine should, however, not turn into drought, which causes severe yield reductions. Dry weather is again beneficial during harvest. From all these factors, yields varying 20 to 25% above or below the mean yield are not unusual.

In the next sections, we study two possible representations of these variable yields. One approach using discrete, correlated random variables is described in Sections 1.1b and 1.1c. Another, using continuous uncorrelated random variables, is described in Section 1.1d.

b. A scenario representation

A first possibility is to assume some correlation among the yields of the different crops. A very simplified representation of this would be to assume, e.g., that years are good, fair, or bad for all crops, resulting in above average, average, or below average yields for all crops. To fix these ideas, "above" and "below" average indicate a yield 20% above or below the mean yield given in Table 1. For simplicity, we assume that weather conditions and yields for the farmer do not have a significant impact on prices.

The farmer wishes to know whether the optimal solution is sensitive to variations in yields. He decides to run two more optimizations based on above average and below average yields. Tables 3 and 4 give the optimal solutions he obtains in these cases.

Again, the solutions in Tables 3 and 4 seem quite natural. When yields are high, smaller surfaces are needed to raise the minimum requirements in wheat and corn and the sugar beet quota. The remaining land is devoted to wheat, whose extra production is sold. When yields are low, larger surfaces are needed to raise the minimum requirements and the sugar beet quota. In

TABLE 3. Optimal solution based on above average yields (+ 20%).

Culture	Wheat	Corn	Sugar Beets
Surface (acres)	183.33	66.67	250
Yield (T)	550	240	6000
Sales (T)	350	–	6000
Purchase (T)	–	–	–
Overall profit: $167,667			

TABLE 4. Optimal solution based on below average yields (−20%).

Culture	Wheat	Corn	Sugar Beets
Surface (acres)	100	25	375
Yield (T)	200	60	6000
Sales (T)	–	–	6000
Purchase (T)	–	180	–
Overall profit: $59,950			

fact, corn requirements cannot be satisfied with the production, and some corn must be bought.

The optimal solution is very sensitive to changes in yields. The optimal surfaces devoted to wheat range from 100 acres to 183.33 acres. Those devoted to corn range from 25 acres to 80 acres and those devoted to sugar beets from 250 acres to 375 acres. The overall profit ranges from $59,950 to $167,667.

Long-term weather forecasts would be very helpful here. Unfortunately, as even meteorologists agree, weather conditions cannot be accurately predicted six months ahead. The farmer must make up his mind without perfect information on yields.

The main issue here is clearly on sugar beet production. Planting large surfaces would make it certain to produce and sell the quota, but would also make it likely to sell some sugar beets at the unfavorable price. Planting small surfaces would make it likely to miss the opportunity to sell the full quota at the favorable price.

The farmer now realizes that he is unable to make a perfect decision that would be best in all circumstances. He would, therefore, want to assess the benefits and losses of each decision in each situation. Decisions on land assignment (x_1, x_2, x_3) have to be taken now, but sales and purchases $(w_i, \ i = 1, \ldots, 4, y_j, \ j = 1, 2)$ depend on the yields. It is useful to index those decisions by a scenario index $s = 1, 2, 3$ corresponding to above average, average, or below average yields, respectively. This creates a new set of variables of the form w_{is}, $i = 1, 2, 3, 4$, $s = 1, 2, 3$ and y_{js}, $j = 1, 2$, $s = 1, 2, 3$. As an example, w_{32} represents the amount of sugar beets sold at the favorable price if yields are average.

Assuming the farmer wants to maximize long-run profit, it is reasonable for him to seek a solution that maximizes his expected profit. (This assumption means that the farmer is neutral about risk. For a discussion of risk aversion and alternative utilities, see Chapter 2.) If the three scenarios have an equal probability of $1/3$, the farmer's problem reads as follows:

$$\min \quad 150x_1 + 230x_2 + 260x_3 - \frac{1}{3}(170w_{11} - 238y_{11} + 150w_{21}$$
$$-210y_{21} + 36w_{31} + 10w_{41})$$
$$-\frac{1}{3}(170w_{12} - 238y_{12} + 150w_{22}$$
$$-210y_{22} + 36w_{32} + 10w_{42})$$
$$-\frac{1}{3}(170w_{13} - 238y_{13} + 150w_{23}$$
$$-210y_{23} + 36w_{33} + 10w_{43})$$

s.t.

$$x_1 + x_2 + x_3 \le 500, \ 3x_1 + y_{11} - w_{11} \ge 200, 3.6x_2 + y_{21} - w_{21} \ge 240,$$
$$w_{31} + w_{41} \le 24x_3, \ w_{31} \le 6000, \ 2.5x_1 + y_{12} - w_{12} \ge 200,$$
$$3x_2 + y_{22} - w_{22} \ge 240, \ w_{32} + w_{42} \le 20x_3, \ w_{32} \le 6000,$$
$$2x_1 + y_{13} - w_{13} \ge 200, \ 2.4x_2 + y_{23} - w_{23} \ge 240, \ w_{33} + w_{43} \le 16x_3,$$
$$w_{33} \le 6000, \ x, y, w \ge 0. \tag{1.2}$$

Such a model of a stochastic decision program is known as the *extensive form* of the stochastic program because it explicitly describes the second-stage decision variables for all scenarios. The optimal solution of (1.2) is given in Table 5. The top line gives the planting areas, which must be determined before realizing the weather and crop yields. This decision is called the *first stage*. The other lines describe the yields, sales, and purchases in the three scenarios. They are called the *second stage*. The bottom line shows the overall expected profit.

The optimal solution can be understood as follows. The most profitable decision for sugar beet land allocation is the one that always avoids sales at the unfavorable price even if this implies that some portion of the quota is unused when yields are average or below average.

The area devoted to corn is such that it meets the feeding requirement when yields are average. This implies sales are possible when yields are above average and purchases are needed when yields are below average. Finally, the rest of the land is devoted to wheat. This area is large enough to cover the minimum requirement. Sales then always occur.

This solution illustrates that it is impossible, under uncertainty, to find a solution that is ideal under all circumstances. Selling some sugar beets at the unfavorable price or having some unused quota is a decision that would never take place with a perfect forecast. Such decisions can appear in a

TABLE 5. Optimal solution based on the stochastic model (1.2).

		Wheat	Corn	Sugar Beets
First Stage	Area (acres)	170	80	250
s = 1 Above	Yield (T)	510	288	6000
	Sales (T)	310	48	6000 (favor. price)
	Purchase (T)	–	–	–
s = 2 Average	Yield (T)	425	240	5000
	Sales (T)	225	–	5000 (favor. price)
	Purchase (T)	–	–	–
s = 3 Below	Yield (T)	340	192	4000
	Sales (T)	140	–	4000 (favor. price)
	Purchase (T)	–	48	–
Overall profit: $108,390				

stochastic model because decisions have to be balanced or hedged against the various scenarios.

The hedging effect has an important impact on the expected optimal profit. Suppose yields vary over years but are cyclical. A year with above average yields is always followed by a year with average yields and then a year with below average yields. The farmer would then take optimal solutions as given in Table 3, then Table 2, then Table 4, respectively. This would leave him with a profit of $167,667 the first year, $118,600 the second year, and $59,950 the third year. The mean profit over the three years (and in the long run) would be the mean of the three figures, namely $115,406 per year.

Now, assume again that yields vary over years, but on a random basis. If the farmer gets the information on the yields before planting, he will again choose the areas on the basis of the solution in Table 2, 3, or 4, depending on the information received. In the long run, if each yield is realized one third of the years, the farmer will get again an expected profit of $115,406 per year. This is the situation under perfect information.

As we know, the farmer unfortunately does not get prior information on the yields. So, the best he can do in the long run is to take the solution as given by Table 5. This leaves the farmer with an expected profit of $108,390. The difference between this figure and the value, $115,406, in the case of perfect information, namely $7016, represents what is called *the expected value of perfect information* (EVPI). This concept, along with others, will be studied in Chapter 4. At this introductory level, we may just say that it represents the loss of profit due to the presence of uncertainty.

Another approach the farmer may have is to assume expected yields and always to allocate the optimal planting surface according to these yields, as in Table 2. This approach represents the *expected value solution.* It is common in optimization but can have unfavorable consequences. Here, as shown in Exercise 1, using the expected value solution every year results in a long run annual profit of $107,240. The loss by not considering the random variations is the difference between this and the stochastic model profit from Table 5. This value, $108,390−107,240=$1,150, is the *value of the stochastic solution* (VSS), the possible gain from solving the stochastic model. Note that it is not equal to the expected value of perfect information, and, as we shall see in later models, may in fact be larger than the EVPI.

These two quantities give the motivation for stochastic programming in general and remain a key focus throughout this book. EVPI measures the value of knowing the future with certainty while VSS assesses the value of knowing and using distributions on future outcomes. Our emphasis will be on problems where no further information about the future is available so the VSS becomes more practically relevant. In some situations, however, more information might be available through more extensive forecasting, sampling, or exploration. In these cases, EVPI would be useful for deciding whether to undertake additional efforts.

c. *General model formulation*

We may also use this example to illustrate the general formulation of a stochastic problem. We have a set of decisions to be taken without full information on some random events. These decisions are called *first-stage decisions* and are usually represented by a vector x. In the farmer example, they are the decisions on how many acres to devote to each crop. Later, full information is received on the realization of some random vector $\boldsymbol{\xi}$. Then, second-stage or corrective actions $\mathbf{y}$ are taken. We use boldface notation here and throughout the book to denote that these vectors are random and to differentiate them from their realizations. We also sometimes use a functional form, such as $\xi(\omega)$ or $y(s)$, to show explicit dependence on an underlying element, ω or s.

In the farmer example, the random vector is the set of yields and the corrective actions are purchases and sales of products. In mathematical programming terms, this defines the so-called two-stage stochastic program with recourse of the form

$$\min \quad c^T x + E_{\boldsymbol{\xi}} Q(x, \boldsymbol{\xi}) \tag{1.3}$$
$$\text{s.t.} \quad Ax = b,$$
$$x \geq 0,$$

where $Q(x, \xi) = \min\{q^T y | Wy = h - Tx, y \geq 0\}$, ξ is the vector formed by the components of q^T, h^T, and T, and E_ξ denote mathematical expectation with respect to ξ. We assume here that W is fixed (*fixed recourse*). Reasons for this restriction are explained in Section 3.1.

In the farmer example, the random vector is a discrete variable with only three different values. Only the T matrix is random. A second-stage problem for one particular scenario s can thus be written as

$$Q(x, s) = \min\{238y_1 - 170w_1 + 210y_2 - 150w_2 - 36w_3 - 10w_4\} \quad (1.4)$$
$$\text{s.t. } t_1(s)x_1 + y_1 - w_1 \geq 200,$$
$$t_2(s)x_2 + y_2 - w_2 \geq 240,$$
$$w_3 + w_4 \leq t_3(s)x_3,$$
$$w_3 \leq 6000,$$
$$y_1, w_1 \geq 0,$$

where $t_i(s)$ represents the yield of crop i under scenario s (or state of nature s). To illustrate the link between the general formulation (1.3) and the example (1.4), observe that in (1.4) we may say that the random vector $\xi = (t_1, t_2, t_3)$ is formed by the three yields and that ξ can take on three different values, say ξ_1, ξ_2, and ξ_3, which represent $(t_1(1), t_2(1), t_3(1)), (t_1(2), t_2(2), t_3(2))$, and $(t_1(3), t_2(3), t_3(3))$, respectively.

An alternative interpretation would be to say that the random vector $\xi(s)$ in fact depends on the scenario s, which takes on three different values[1].

In this section, we have illustrated two possible representations of a stochastic program. The form (1.2) given earlier for the farmer's example is known as the extensive form. It is obtained by associating one decision vector in the second-stage to each possible realization of the random vector. The second form (1.3) or (1.4) is called the implicit representation of the stochastic program. A more condensed implicit representation is obtained by defining $\mathcal{Q}(x) = E_\xi Q(x, \xi)$ as the *value function* or *recourse function* so that (1.3) can be written as

$$\min \quad c^T x + \mathcal{Q}(x) \quad (1.5)$$
$$\text{s.t.} \quad Ax = b,$$
$$x \geq 0.$$

d. Continuous random variables

Contrary to the assumption made in Section 1.2, we may also assume that yields for the different crops are independent. In that case, we may as well consider a continuous random vector for the yields. To illustrate this, let us

---

[1]Note that the decisions y_1, y_2, w_1, w_2, w_3, and w_4 also depend on the scenario. This dependence is not always made explicit. It appears explicitly in (1.7) but not in (1.4).

assume that the yield for each crop i can be appropriately described by a uniform random variable, inside some range $[l_i, u_i]$ (see Appendix A.2). For the sake of comparison, we may take l_i to be 80% of the mean yield and u_i to be 120% of the mean yield so that the expectations for the yields will be the same as in Section 1.b. Again, the decisions on land allocation are first-stage decisions because they are taken before knowledge of the yields. Second-stage decisions are purchases and sales after the growing period. The second-stage formulation can again be described as $Q(x) = E_{\boldsymbol{\xi}} Q(x, \boldsymbol{\xi})$, where $Q(x, \boldsymbol{\xi})$ is the value of the second stage for a given realization of the random vector.

Now, in this particular example, the computation of $Q(x, \boldsymbol{\xi})$ can be separated among the three crops due to independence of the random vector. (Note that this separability property also holds in the discrete representation of Section 1.b.) We can then write:

$$E_{\boldsymbol{\xi}} Q(x, \boldsymbol{\xi}) = \sum_{i=1}^{3} E_{\boldsymbol{\xi}} Q_i(x_i, \boldsymbol{\xi}) = \sum_{i=1}^{3} Q_i(x_i), \qquad (1.6)$$

where $Q_i(x_i, \boldsymbol{\xi})$ is the optimal second-stage value of purchases and sales of crop i.

We are in fact in position to give an exact analytical expression for the second-stage value functions $Q_i(x_i)$, $i = 1, \ldots, 3$. We first consider sugar beet sales. For a given value $t_3(\boldsymbol{\xi})$ of the sugar beet yield, one obtains the following second-stage problem:

$$Q_3(x_3, \boldsymbol{\xi}) = \min -36w_3(\boldsymbol{\xi}) - 10w_4(\boldsymbol{\xi}) \qquad (1.7)$$
$$\text{s.t. } w_3(\boldsymbol{\xi}) + w_4(\boldsymbol{\xi}) \leq t_3(\boldsymbol{\xi})x_3,$$
$$w_3(\boldsymbol{\xi}) \leq 6000,$$
$$w_3(\boldsymbol{\xi}), w_4(\boldsymbol{\xi}) \geq 0.$$

The optimal decisions for this problem are clearly to sell as many sugar beets as possible at the favorable price, and to sell the possible remaining production at the unfavorable price, namely

$$w_3(\boldsymbol{\xi}) = \min[6000, t_3(\boldsymbol{\xi})x_3], \qquad (1.8)$$
$$w_4(\boldsymbol{\xi}) = \max[t_3(\boldsymbol{\xi})x_3 - 6000, 0].$$

This results in a second-stage value of

$$Q_3(x_3, \boldsymbol{\xi}) = -36\min[6000, t_3(\boldsymbol{\xi})x_3] - 10\max[t_3(\boldsymbol{\xi})x_3 - 6000, 0].$$

We first assume that the surface x_3 devoted to sugar beets will not be so large that quota would be exceeded for any possible yield or so small that production would always be less than the quota for any possible yield. In other words, we assume that the following relation holds:

$$l_3 x_3 \leq 6000 \leq u_3 x_3, \qquad (1.9)$$

where, as already defined, l_3 and u_3 are the bounds on the possible values of $t_3(\xi)$. Under this assumption, the expected value of the second stage for sugar beet sales is

$$\begin{aligned}
\mathcal{Q}_3(x_3) &= E_\xi Q_3(x_3, \xi_3) \\
&= -\int_{l_3}^{6000/x_3} 36tx_3 f(t)dt \\
&\quad - \int_{6000/x_3}^{u_3} (216000 + 10tx_3 - 60000)f(t)dt,
\end{aligned}$$

where $f(t)$ denotes the density of the random yield $t_3(\xi)$. Given the assumption that this density is uniform over the interval $[l_3, u_3]$, one obtains, after some computation, the following analytical expression

$$\mathcal{Q}_3(x_3) = -18\frac{(u_3^2 - l_3^2)x_3}{u_3 - l_3} + \frac{13(u_3 x_3 - 6000)^2}{x_3(u_3 - l_3)},$$

which can also be expressed as

$$\mathcal{Q}_3(x_3) = -36\bar{t}_3 x_3 + \frac{13(u_3 x_3 - 6000)^2}{x_3(u_3 - l_3)}, \tag{1.10}$$

where $\bar{t}_3$ denotes the expected yield for sugar beet production, which is $\frac{u_3 + l_3}{2}$ for a uniform density.

Note that assumption (1.9) is not really limiting. We can still compute the analytical expression of $\mathcal{Q}_3(x_3)$ for the other situations.

For example, if the surface x_3 is such that the production exceeds the quota for any possible yield ($l_3 x_3 > 6000$), then the optimal second-stage decisions are simply

$$w_3(\xi) = 6000,$$
$$w_4(\xi) = t_3(\xi)x_3 - 6000, \text{ for all } \xi.$$

The second-stage value for a given ξ is now

$$Q_3(x_3, \xi) = -216000 - 10(t_3(\xi)x_3 - 6000) = -156000 - 10t_3(\xi)x_3,$$

and the expected value is simply

$$\mathcal{Q}_3(x_3) = -156000 - 10\bar{t}_3 x_3. \tag{1.11}$$

Similarly, if the surface devoted to sugar beets is so small that for any yield the production is lower than the quota, the second-stage value function is

$$Q_3(x_3) = -36\bar{t}_3 x_3. \tag{1.12}$$

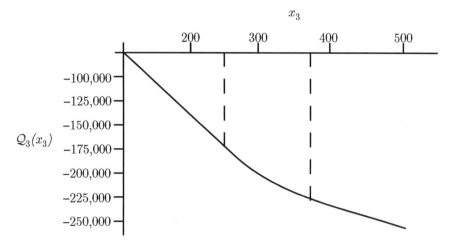

FIGURE 1. The expected recourse value for sugar beets as a function of acres planted.

We may therefore draw the graph of the function $Q_3(x_3)$ for all possible values of x_3 as in Figure 1. Note that with our assumption of $\bar{t}_3 = 20$, we would then have the limits on x_3 in (1.9) as $250 \le x_3 \le 375$.

We immediately see that the function has three different pieces. Two of these pieces are linear and one is nonlinear, but the function $Q_3(x_3)$ is continuous and convex. This property will be proved when we consider the generalization of this problem, known as the *news vendor, newsboy, or Christmas tree problem*. In fact, this property holds for a large class of second-stage problems, as will be seen in Chapter 3.

Similar computations can be done for the other two crops. For wheat, we obtain

$$Q_1(x_1) = \begin{cases} 47600 - 595x_1 & \text{for } x_1 \le 200/3, \\ 119\frac{(200-2x_1)^2}{x_1} - 85\frac{(200-3x_1)^2}{x_1} & \text{for } \frac{200}{3} \le x_1 \le 100, \\ 34000 - 425x_1 & \text{for } x_1 \ge 100, \end{cases}$$

and, for corn, we obtain

$$Q_2(x_2) = \begin{cases} 50400 - 630x_2 & \text{for } x_2 \le 200/3, \\ 87.5\frac{(240-2.4x_2)^2}{x_2} - 62.5\frac{(240-3.6x_2)^2}{x_2} & \text{for } 200/3 \le x_2 \le 100, \\ 36000 - 450x_2 & \text{for } x_2 \ge 100. \end{cases}$$

The global problem is therefore

$$\min \ 150x_1 + 230x_2 + 260x_3 + \mathcal{Q}_1(x_1) + \mathcal{Q}_2(x_2) + \mathcal{Q}_3(x_3)$$
$$\text{s.t.} \ \ x_1 + x_2 + x_3 \leq 500,$$
$$x_1, x_2, x_3 \geq 0.$$

Given that the three functions $\mathcal{Q}_i(x_i)$ are convex, continuous, and differentiable functions and the first-stage objective is linear, this problem is a convex program for which Karush-Kuhn-Tucker (K-K-T) conditions are necessary and sufficient for a global optimum. (This result is from nonlinear programming. For more on this result about optimality, see Section 2.9.) Denoting by λ the multiplier of the surface constraint and as before by c_i the first-stage objective coefficient of crop i, the K-K-T conditions require

$$x_i \left[c_i + \frac{\partial \mathcal{Q}_i(x_i)}{\partial x_i} + \lambda \right] = 0, \ \ c_i + \frac{\partial \mathcal{Q}_i(x_i)}{\partial x_i} + \lambda \geq 0, \ \ x_i \geq 0, \ \ \ i = 1, 2, 3;$$
$$\lambda[x_1 + x_2 + x_3 - 500] = 0, \ \ x_1 + x_2 + x_3 \leq 500, \ \ \lambda \geq 0.$$

Assume the optimal solution is such that $100 \leq x_1, \frac{200}{3} \leq x_2 \leq 100$, and $250 \leq x_3 \leq 375$ with $\lambda \neq 0$. Then the conditions read

$$\begin{cases} -275 + \lambda = 0, \\ -76 - \frac{1.44 \ 10^6}{x_2^2} + \lambda = 0, \\ 476 - \frac{5.85 \ 10^7}{x_3^2} + \lambda = 0, \\ x_1 + x_2 + x_3 = 500. \end{cases}$$

Solving this system of equations gives $\lambda = 275.00, x_1 = 135.83, x_2 = 85.07, x_3 = 279.10$, which satisfies all the required conditions and is therefore optimal. We observe that this solution is similar to the one obtained by using the scenario approach, although more surface is devoted to sugar beet and less to wheat than before. This similarity represents a characteristic robustness of a well-formed stochastic programming formulation. We shall consider it in more detail in our discussion of approximations in Chapter 9.

e. The news vendor problem

The previous section illustrates an example of a famous and basic problem in stochastic optimization, *the news vendor problem*. In this problem, a news vendor goes to the publisher every morning and buys x newspapers at a price of c per paper. This number is usually bounded above by some limit u, representing either the news vendor's purchase power or a limit set by the publisher to each vendor. The vendor then walks along the streets to sell as many newspapers as possible at the selling price q. Any unsold newspaper can be returned to the publisher at a return price r, with $r < c$.

We are asked to help the news vendor decide how many newspapers to buy every morning. Demand for newspapers varies over days and is described by a random variable $\boldsymbol{\xi}$.

It is assumed here that the news vendor cannot return to the publisher during the day to buy more newspapers. Other news vendors would have taken the remaining newspapers. Readers also only want the last edition.

To describe the news vendor's profit, we define y as the effective sales and w as the number of newspapers returned to the publisher at the end of the day. We may then formulate the problem as

$$\min \ cx + \mathcal{Q}(x)$$
$$0 \leq x \leq u,$$

where

$$\mathcal{Q}(x) = E_{\boldsymbol{\xi}} Q(x, \boldsymbol{\xi})$$

and

$$Q(x, \boldsymbol{\xi}) = \min \ -qy(\boldsymbol{\xi}) - rw(\boldsymbol{\xi})$$
$$\text{s.t.} \quad y(\boldsymbol{\xi}) \leq \boldsymbol{\xi},$$
$$y(\boldsymbol{\xi}) + w(\boldsymbol{\xi}) \leq x,$$
$$y(\boldsymbol{\xi}), w(\boldsymbol{\xi}) \geq 0,$$

where again $E_{\boldsymbol{\xi}}$ denotes the mathematical expectation with respect to $\boldsymbol{\xi}$.

In this notation, $-\mathcal{Q}(x)$ is the expected profit on sales and returns, while $-Q(x, \boldsymbol{\xi})$ is the profit on sales and returns if the demand is at level $\boldsymbol{\xi}$. The model illustrates the two-stage aspect of the news vendor problem. The buying decision has to be taken before any information is given on the demand. When demand is known in the so-called second stage, which represents the end of the sales period of a given edition, the profit can be computed. This is done using the following simple rule:

$$y^*(\boldsymbol{\xi}) = \min(\boldsymbol{\xi}, x),$$
$$w^*(\boldsymbol{\xi}) = \max(x - \boldsymbol{\xi}, 0).$$

Sales can never exceed the number of available newspapers or the demand. Returns occur only when demand is less than the number of newspapers available. The second-stage expected value function is simply

$$\mathcal{Q}(x) = E_{\boldsymbol{\xi}}[-q\min(\boldsymbol{\xi}, x) - r\max(x - \boldsymbol{\xi}, 0)].$$

As we will learn later, this function is convex and continuous. It is also differentiable when $\boldsymbol{\xi}$ is a continuous random vector. In that case, the optimal solution of the news vendor's problem is simply:

$$x = 0 \text{ if } c + \mathcal{Q}'(0) > 0,$$
$$x = u \text{ if } c + \mathcal{Q}'(u) < 0,$$
$$\text{a solution of } c + \mathcal{Q}'(x) = 0 \quad \text{otherwise,}$$

where $\mathcal{Q}'(x)$ denotes the first order derivative of $\mathcal{Q}(x)$ evaluated at x. By construction, $\mathcal{Q}(x)$ can be computed as

$$\mathcal{Q}(x) = \int_{-\infty}^{x} (-q\xi - r(x - \xi))dF(\xi) + \int_{x}^{\infty} -qx \, dF(\xi)$$
$$= -(q - r) \int_{-\infty}^{x} \xi \, dF(\xi) - rx \, F(x) - qx(1 - F(x)),$$

where $F(\xi)$ represents the cumulative probability distribution of ξ (see Section 2.1).

Integrating by parts, we observe that

$$\int_{-\infty}^{x} \xi \, dF(\xi) = xF(x) - \int_{-\infty}^{x} F(\xi)d\xi$$

under mild conditions on the distribution function $F(\xi)$. It follows that

$$\mathcal{Q}(x) = -qx + (q - r) \int_{-\infty}^{x} F(\xi)d\xi.$$

We may thus conclude that

$$\mathcal{Q}'(x) = -q + (q - r)F(x).$$

and therefore that the optimal solution is

$$\begin{cases} x^* = 0 & \text{if} & \frac{q-c}{q-r} < F(0), \\ x^* = u & \text{if} & \frac{q-c}{q-r} > F(u), \\ x^* = F^{-1}(\frac{q-c}{q-r}) & \text{otherwise,} \end{cases}$$

where $F^{-1}(\alpha)$ is the α-quantile of F (see Section 2.1). If F is continuous, $x = F^{-1}(\alpha)$ means $\alpha = F(x)$. Any reasonable representation of the demand would imply $F(0) = 0$ so that the solution is never $x^* = 0$.

As we shall see in Chapter 3, this problem is an example of a basic type of stochastic program called the *stochastic program with simple recourse*. The ideas of this section can be generalized to larger problems in this class of examples. Also observe that, as such, we only come to a partial answer, under the form of an expression for x^*. The vendor may still need to consult a statistician, who would provide an accurate cumulative distribution $F(\cdot)$. Only then will a precise figure be available for x^*.

Exercises

1. *Value of the stochastic solution*
   Assume the farmer allocates his land according to the solution of Table 2, i.e., 120 acres for wheat, 80 acres for corn, and 300 acres for sugar beets. Show that if yields are random (20% below average, average, and 20% above average for all crops with equal probability one third), his expected annual profit is $107,240. To do this observe that planting costs are certain but sales and purchases depend on the yield. In other words, fill in a table such as Table 5 but with the first-stage decisions given here.

2. *Price effect*
   When yields are good for the farmer, they are usually also good for many other farmers. The supply is thus increasing, which will lower the prices. As an example, we may consider prices going down by 10% for corn and wheat when yields are above average and going up by 10% when yields are below average. Formulate the model where these changes in prices affect both sales and purchases of corn and wheat. Assume sugar beet prices are not affected by yields.

3. *Binary first stage*
   Consider the case where the farmer possesses four fields of sizes 185, 145, 105, and 65 acres, respectively. Observe that the total of 500 acres is unchanged. Now, the fields are unfortunately located in different parts of the village. For reasons of efficiency the farmer wants to raise only one type of crop on each field. Formulate this model as a two-stage stochastic program with a first-stage program with binary variables.

4. *Integer second stage*
   Consider the case where sales and purchases of corn and wheat can only be obtained through contracts involving multiples of hundred tons. Formulate the model as a stochastic program with a mixed-integer second stage.

5. Consider any one of Exercises 2 to 4. Using standard mixed integer programming software, obtain an optimal solution of the extensive form of the stochastic program. Compute the expected value of perfect information and the value of the stochastic solution.

6. *Multistage program*
   It is typical in farming to implement crop rotation in order to maintain good soil quality. Sugar beets would, for example, appear in triennial crop rotation, which means they are planted on a given field only one out of three years. Formulate a multistage program to describe this situation. To keep things simple, describe the case when

sugar beets cannot be planted two successive years on the same field, and assume no such rule applies for wheat and corn.

(On a two-year basis, this exercise consists purely of formulation: with the basic data of the example, the solution is clearly to repeat the optimal solution in Table 5, i.e., to plant 170 acres of wheat, 80 acres of corn, and 250 acres of sugar beets. The problem becomes more relevant on a three-year basis. It is also relevant on a two-year basis with fields of given sizes as in Exercise 2.

In terms of formulation, it is sufficient to consider a three-stage model. The first stage consists of first-year planting. The second stage consists of first-year purchases and sales and second-year planting. The third-stage consists of second-year purchases and sales. Alternatively, a four-stage model can be built, separating first-year purchases and sales from second-year planting. Also discuss the question of discounting the revenues and expenses of the various stages.)

7. *Risk aversion*
   Economic theory tells us that, like many other people, the farmer would normally act as a risk-averse person. There are various ways to model risk aversion. One simple way is to plan for the worst case. More precisely, it consists of maximizing the profit under the worst situation. Note that for some models, it is not known in advance which scenario will turn out to induce the lowest profit. In our example, the worst situation corresponds to Scenario 3 (below average yields). Planning for the worst case implies the solution of Table 4 is optimal.

   (a) Compute the loss in expected profit if that solution is taken.

   (b) A median situation would be to require a reasonable profit under the worst case. Find the solution that maximizes the expected profit under the constraint that in the worst case the profit does not fall below $58,000. What is now the loss in expected profit?

   (c) Repeat part (b) with other values of minimal profit: $56,000, $54,000, $52,000, $50,000, and $48,000. Graph the curve of expected profit loss. Also compare the associated optimal decisions.

8. If prices are also random variables, the problem becomes more complicated. However, if prices and demands are independent random variables, show that the solution of the news vendor's problem is the one obtained before, where q and r are replaced by their expected values. Indicate under which conditions the same proposition is true for the farmer's problem.

9. In the news vendor's problem, we have assumed for simplicity that the random variable takes value from $-\infty$ to $+\infty$. Show that the

optimal decisions are insensitive to this assumption, so that if the random variables have a nonzero density on a limited interval then the optimal solutions are obtained by the same analytical expression.

10. Suppose $c = 10, q = 25, r = 5$, and demand is uniform on $[50, 150]$. Find the optimal solution of the news vendor problem. Also, find the optimal solution of the deterministic model obtained by assuming a demand of 100. What is the value of the stochastic solution?

1.2 Financial Planning and Control

Financial decision-making problems can often be modeled as stochastic programs. In fact, the essence of financial planning is the incorporation of risk into investment decisions. The area represents one of the largest application areas of stochastic programming. Many references can be found in, for example, Mulvey and Vladimirou [1989, 1991b, 1992], Ziemba and Vickson [1975], and Zenios [1992].

We consider a simple example that illustrates additional stochastic programming properties. As in the farming example of Section 1, this example involves randomness in the constraint matrix instead of the right-hand side elements. These random variables reflect uncertain investment yields.

This section's example also has the characteristic that decisions are highly dependent on past outcomes. In the following capacity expansion problem of Section 3, this is not the case. In Chapter 3, we define this difference by a block separable recourse property that is present in some capacity expansion and similar problems.

For the current problem, suppose we wish to provide for a child's college education Y years from now. We currently have $\$b$ to invest in any of I investments. After Y years, we will have a wealth that we would like to have exceed a tuition goal of $\$G$. We suppose that we can change investments every v years, so we have $H = Y/v$ investment periods. For our purposes here, we ignore transaction costs and taxes on income although these considerations would be important in reality. We also assume that all figures are in constant dollars.

In formulating the problem, we must first describe our objective in mathematical terms. We suppose that exceeding $\$G$ after Y years would be equivalent to our having an income of $q\%$ of the excess while not meeting the goal would lead to borrowing for a cost $r\%$ of the amount short. This gives us the concave utility function in Figure 2. Many other forms of nonlinear utility functions are, of course, possible. See Kallberg and Ziemba [1983] for a description of their relevance in financial planning.

The major uncertainty in this model is the return on each investment i within each period t. We describe this random variable as $\boldsymbol{\xi}(i, t) = \xi(i, t, \omega)$ where ω is some underlying random element. The decisions on investments

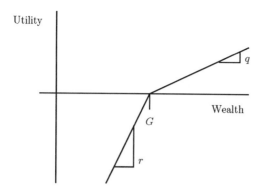

FIGURE 2. Utility function of wealth at year Y for a goal G.

will also be random. We describe these decisions as $\mathbf{x}(i,t) = x(i,t,\omega)$. From the randomness of the returns and investment decisions, our final wealth will also be a random variable.

A key point about this investment model is that we cannot completely observe the random element ω when we make all our decisions $x(i,t,\omega)$. We can only observe the returns that have already taken place. In stochastic programming, we say that we cannot *anticipate* every possible outcome so our decisions are *nonanticipative* of future outcomes. Before the first period, this restriction corresponds to saying that we must make fixed investments, $x(i,1)$, for all $\omega \in \Omega$, the space of all random elements or, more specifically, returns that could possibly occur.

To illustrate the effects of including stochastic outcomes as well as modeling effects from choosing the time horizon Y and the coarseness of the period approximations H, we use a simple example with two possible investment types, stocks ($i = 1$) and government securities (bonds) ($i = 2$). We begin by setting Y at 15 years and allow investment changes every five years so that $H = 3$.

We assume that, over the three decision periods, eight possible scenarios may occur. The scenarios correspond to independent and equal likelihoods of having (inflation-adjusted) returns of 1.25 for stocks and 1.14 for bonds or 1.06 for stocks and 1.12 for bonds over the five-year period. We indicate the scenarios by an index $s = 1, \ldots, 8$, which represents a collection of the outcomes ω that have common characteristics (such as returns) in a specific model. When we wish to allow more general interpretations of the outcomes, we use the base element ω. With the scenarios defined here, we assign probabilities for each s, $p(s) = 0.125$. The returns are $\xi(1,t,s) = 1.25, \xi(2,t,s) = 1.14$ for $t = 1, s = 1, \ldots, 4$, for $t = 2, s = 1,2,5,6$, and for $t = 3, s = 1,3,5,7$. In the other cases, $\xi(1,t,s) = 1.06, \xi(2,t,s) = 1.12$.

The eight scenarios are represented by the tree in Figure 3. The scenario tree divides into branches corresponding to different realizations of the random returns. Because Scenarios 1 to 4, for example, have the same

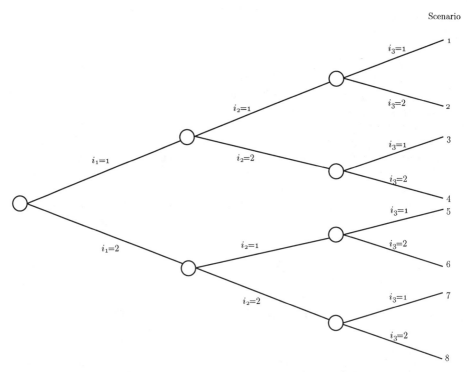

FIGURE 3. Tree of scenarios for three periods.

return for $t = 1$, they all follow the same first branch. Scenarios 1 and 2 then have the same second branch and finally divide completely in the last period. To show this more explicitly, we may refer to each scenario by the history of returns indexed by s_t for periods $t = 1, 2, 3$ as indicated on the tree in Figure 3. In this way, Scenario 1 may also be represented as $(s_1, s_2, s_3) = (1, 1, 1)$.

With the tree representation, we need only have a decision vector for each node of the tree. The decisions at $t = 1$ are just $x(1, 1)$ and $x(2, 1)$ for the amounts invested in stocks (1) and bonds (2) at the outset. For $t = 2$, we would have $x(i, 2, s_1)$ where $i = 1, 2$ for the type of investment and $s_1 = 1, 2$ for the first-period return outcome. Similarly, the decisions at $t = 3$ are $x(i, 3, s_1, s_2)$.

With these decision variables defined, we can formulate a mathematical program to maximize expected utility. Because the concave utility function in Figure 1 is piecewise linear, we just need to define deficit or shortage and excess or surplus variables, $w(i_1, i_2, i_3)$ and $y(i_1, i_2, i_3)$, and we can maintain a linear model. The objective is simply a probability- and penalty-weighted sum of these terms, which, in general, becomes:

$$\sum_{s_H} \cdots \sum_{s_1} p(s_1, \ldots, s_H)(-rw(s_1, \ldots, s_H) + qy(s_1, \ldots, s_H)).$$

The first-period constraint is simply to invest the initial wealth:

$$\sum_i x(i, 1) = b.$$

The constraints for periods $t = 2, \ldots, H - 1$ are, for each $s_1, \ldots, s_{t-1}$:

$$\sum_i -\xi(i, t - 1, s_1, \ldots, s_{t-1})x(i, t - 1, s_1, \ldots, s_{t-2})$$

$$+ \sum_i x(i, t, s_1, \ldots, s_{t-1}) = 0,$$

while the constraints for period H are:

$$\sum_i -\xi(i, H, s_1, \ldots, s_H)x(i, H, s_1, \ldots, s_{H-1}) - y(s_1, \ldots, s_H)$$

$$+ w(s_1, \ldots, s_H) = G.$$

Other constraints restrict the variables to be non-negative.

To specify the model in this example, we use initial wealth, $b = 55,000$; target value, $G = 80,000$; surplus reward, $q = 1$; and shortage penalty, $r = 4$. The result is a stochastic program in the following form where the units are thousands of dollars:

$$
\begin{aligned}
\max z = \sum_{s_1=1}^{2} \sum_{s_2=1}^{2} \sum_{s_3=1}^{2} & 0.125(y(s_1, s_2, s_3) - 4w(s_1, s_2, s_3)) \\
\text{s. t.} \qquad\qquad x(1,1) + x(2,1) &= 55, \\
-1.25x(1,1) - 1.14x(2,1) + x(1,2,1) + x(2,2,1) &= 0, \\
-1.06x(1,1) - 1.12x(2,1) + x(1,2,2) + x(2,2,2) &= 0, \\
-1.25x(1,2,1) - 1.14x(2,2,1) + x(1,3,1,1) + x(2,3,1,1) &= 0, \\
-1.06x(1,2,1) - 1.12x(2,2,1) + x(1,3,1,2) + x(2,3,1,2) &= 0, \\
-1.25x(1,2,2) - 1.14x(2,2,2) + x(1,3,2,1) + x(2,3,2,1) &= 0, \\
-1.06x(1,2,2) - 1.12x(2,2,2) + x(1,3,2,2) + x(2,3,2,2) &= 0, \\
1.25x(1,3,1,1) + 1.14x(2,3,1,1) - y(1,1,1) + w(1,1,1) &= 80, \\
1.06x(1,3,1,1) + 1.12x(2,3,1,1) - y(1,1,2) + w(1,1,2) &= 80, \\
1.25x(1,3,1,2) + 1.14x(2,3,1,2) - y(1,2,1) + w(1,2,1) &= 80, \\
1.06x(1,3,1,2) + 1.12x(2,3,1,2) - y(1,2,2) + w(1,2,2) &= 80, \\
1.25x(1,3,2,1) + 1.14x(2,3,2,1) - y(2,1,1) + w(2,1,1) &= 80, \\
1.06x(1,3,2,1) + 1.12x(2,3,2,1) - y(2,1,2) + w(2,1,2) &= 80, \\
1.25x(1,3,2,2) + 1.14x(2,3,2,2) - y(2,2,1) + w(2,2,1) &= 80, \\
1.06x(1,3,2,2) + 1.12x(2,3,2,2) - y(2,2,2) + w(2,2,2) &= 80, \\
x(i, t, s_1, \dots, s_{t-1}) \geq 0, y(s_1, s_2, s_3) \geq 0, w(s_1, s_2, s_3) &\geq 0, \\
\text{for all } i, t, s_1, s_2, s_3. &
\end{aligned}
$$

$$(2.1)$$

Solving the problem in (2.1) yields an optimal expected utility value of -1.52. We call this value, RP, for the expected *recourse problem* solution value. The optimal solution (in thousands of dollars) appears in Table 6.

In this solution, the initial investment is heavily in stock (\$41,500) with only \$13,500 in bonds. Notice the reaction to first-period outcomes, however. In the case of Scenarios 1 to 4, stocks are even more prominent, while Scenarios 5 to 8 reflect a more conservative government security portfolio. In the last period, notice how the investments are either completely in stocks or completely in bonds. This is a general trait of one-period decisions. It occurs here because in Scenarios 1 and 2, there is no risk of missing the target. In Scenarios 3 to 6, stock investments may cause one to miss the target, so they are avoided. In Scenarios 7 and 8, the only hope of reaching the target is through stocks.

We compare the results in Table 6 to a deterministic model in which all random returns are replaced by their expectation. For that model, because the expected return on stock is 1.155 in each period, while the expected return on bonds is only 1.13 in each period, the optimal investment plan places all funds in stocks in each period. If we implement this policy each period, but instead observed the random returns, we would have an expected utility called the *expected value* solution, or EV. In this case, we would realize an expected utility of $EV = -3.79$, while the stochastic program value is again $RP = -1.52$. The difference between these quantities is the value of the stochastic solution:

$$VSS = RP - EV = -1.52 - (-3.79) = 2.27.$$

TABLE 6. Optimal solution with three-period stochastic program.

Period, Scenario	Stock	Bonds
1,1-8	41.5	13.50
2,1-4	65.1	2.17
2,5-8	36.7	22.40
3,1-2	83.0	0.00
3,3-4	0.0	71.40
3,5-6	0.0	71.40
3,7-8	64.0	0.00
Scenario	Above G	Below G
1	24.80	0.0
2	8.87	0.0
3	1.42	0.0
4	0.00	0.0
5	1.42	0.0
6	0.00	0.0
7	0.00	0.0
8	0.00	12.2

This comparison gives us a measure of the utility value in using a decision from a stochastic program compared to a decision from a deterministic program. Another comparison of models is in terms of the probability of reaching the goal. Models with these types of objectives are called *chance-constrained programs* or *programs with probabilistic constraints* (see Charnes and Cooper [1959] and Prékopa [1973]). Notice that the stochastic program solution reaches the goal 87.5% of the time. The expected value deterministic model solution only reaches the goal 50% of the time. In this case, the value of the stochastic solution may be even more significant.

The formulation we gave in (2.1) can become quite cumbersome as the time horizon, H, increases and the decision tree of Figure 3 grows quite bushy. Another modeling approach to this type of multistage problem is to consider the full horizon scenarios, s, directly, without specifying the history of the process. We then substitute a scenario set S for the random elements Ω. Probabilities, $p(s)$, returns, $\xi(i,t,s)$, and investments, $x(i,t,s)$, become functions of the H-period scenarios and not just the history until period t.

The difficulty is that, when we have split up the scenarios, we may have lost nonanticipativity of the decisions because they would now include knowledge of the outcomes up to the end of the horizon. To enforce nonanticipativity, we add constraints explicitly in the formulation. First, the scenarios that correspond to the same set of past outcomes at each period form groups, $S^t_{s_1,\ldots,s_{t-1}}$, for scenarios at time t. Now, all actions up to time t must be the same within a group. We do this through an explicit constraint. The new general formulation of (2.1) becomes:

$$\max z = \sum_s p(s)(qy(s) - rw(s))$$

$$\text{s. t.} \sum_{i=1}^{I} x(i,1,s) = b, \forall s \in S,$$

$$\sum_{i=1}^{I} \xi(i,t,s)x(i,t-1,s) - \sum_{i=1}^{I} x(i,t,s) = 0, \forall s \in S,$$

$$t = 2, \ldots, H,$$

$$\sum_{i=1}^{I} \xi(i,H,s)x(i,H,s) - y(s) + w(s) = G,$$

$$\left(\sum_{s' \in S^t_{J(s,t)}} p(s')x(i,t,s') \right) - \left(\sum_{s' \in S^t_{J(s,t)}} p(s') \right) x(i,t,s) = 0,$$

$$\forall 1 \leq i \leq I, \forall 1 \leq t \leq H, \forall s \in S,$$

$$x(i,t,s) \geq 0, y(s) \geq 0, w(s) \geq 0,$$

$$\forall 1 \leq i \leq I, \forall 1 \leq t \leq H, \forall s \in S, \tag{2.2}$$

where $J(s,t) = \{s_1, \ldots, s_{t-1}\}$ such that $s \in S^t_{s_1,\ldots,s_{t-1}}$. Note that the last equality constraint indeed forces all decisions within the same group at time t to be the same. Formulation 2.2 has a special advantage for the problem here because these *nonanticipativity* constraints are the only constraints linking the separate scenarios. Without them, the problem would decompose into a separate problem for each s, maintaining the structure of that problem.

In modeling terms, this simple additional constraint makes it relatively easy to move from a deterministic model to a stochastic model of the same problem. This ease of conversion can be especially useful in modeling languages. For example, Figure 4 gives a complete AMPL (Fourer, Gay, and Kernighan [1993]) model of the problem in (2.2). In this language, *set*, *param*, and *var* are keywords for sets, parameters, and variables. The addition of the scenario indicators and nonanticipativity constraints (*nonanticip*) are the only additions to a deterministic model.

Given the ease of this modeling effort, standard optimization procedures can be simply applied to this problem. However, as we noted earlier, the number of scenarios can become extremely large. Standard methods may not be able to solve the problem in any reasonable amount of time, necessitating other techniques. The remaining chapters in this book focus on these other methods and on procedures for creating models that are amenable to those specialized techniques.

In financial problems, it is particularly worthwhile to try to exploit the underlying structure of the problem without the nonanticipativity constraints. This relaxed problem is in fact a generalized network that allows

```
# This problem describes a simple financial planning problem
# for financing college education
set investments; # different investment options
param initwealth; # initial holdings
param scenarios; # number of scenarios (total S)
# The following 0-1 array shows which scenarios are combined at period T
param scen_links{1..scenarios,1..scenarios,1..T};
param target; # target value G at time T
param H; # number of periods
param invest; # value of investing beyond target value
param penalty; # penalty for not meeting target
param return {investments,1..scenarios,1..T}; # return on each inv
param prob {1..scenarios}; # probability of each scenario
# variables
var amtinvest{investments,1..scenarios,1..T} >= 0; #actual amounts inv'd
var above_target{1..scenarios}>= 0; # amt above final target
var below_target{1..scenarios} >= 0; # amt below final target
# objective
maximize exp_value : sum{i in 1..scenarios} prob[i]*(invest*above_target[i]
- penalty*below_target[i]);
# constraints
subject to budget{i in 1..scenarios} :
sum{k in investments}(amtinvest[k,i,1]) = initwealth;#invest initial wealth
subject to nonanticip{k in investments,j in 1..scenarios,H in 1..T}:
(sum{i in 1..scenarios}scen_links[j,i,H]*prob[i]*amtinvest[k,i,H]) -
(sum{i in 1..scenarios}scen_links[j,i,H]*prob[i])*
amtinvest[k,j,H] = 0; # makes all investments nonanticipative
subject to balance{j in 1..scenarios, t in 1..T-1}:
(sum{k in investments}return[k,j,H]*amtinvest[k,j,H]) - sum{k in
investments} amtinvest[k,j,H+1] = 0; # reinvest each time period
subject to scenario_value{j in 1..scenarios}: (sum{k in
investments}return[k,j,H]*amtinvest[k,j,H]) - above_target[j] +
below_target[j] = target; # amounts not meeting target
```

FIGURE 4. AMPL format of financial planning model.

the use of efficient network optimization methods that cannot apply to the full problem in (2.2). We discuss this option more thoroughly in Chapters 5 and 7.

With either formulation (2.1) or (2.2), in completing the model, some decisions must be made about the possible set of outcomes or scenarios and the coarseness of the period structure, i.e., the number of periods H allowed for investments. We must also find probabilities to attach to outcomes within each of these periods. These probabilities are often approximations that can, as we shall see in Chapter 9, provide bounds on true values or on uncertain outcomes with incompletely known distributions. A key observation is that the important step is to include stochastic elements at least approximately and that deterministic solutions most often give misleading results.

In closing this section, note that the mathematical form of this problem actually represents a broad class of control problems (see, for example, Varaiya and Wets [1989]). In fact, it is basically equivalent to any control problem governed by a linear system of differential equations. We have merely taken a discrete time approach to this problem. This approach can be applied to the control of a wide variety of electrical, mechanical, chemical, and economic systems. We merely redefine state variables (now, wealth) in each time period and controls (investment levels). The random gain or loss is reflected in the return coefficients. Typically, these types of control problems would have nonlinear (e.g., *quadratic*) costs associated with the control in each time period. This presents no complication for our purposes, so we may include any of these problems as potential applications. In the last section, we will look at a fundamentally nonlinear problem in more detail.

Exercises

1. Suppose you consider just a five-year planning horizon. Choose an appropriate target and solve over this horizon with a single first-period decision.

2. Suppose that goal G is also a random parameter and could be $75,000 or $85,000 with equal probabilities. Formulate and solve this problem. Compare this solution to the solution for the problem with a known target.

1.3   Capacity Expansion

Capacity expansion models optimal choices of the timing and levels of investments to meet future demands of a given product. This problem has many applications. Here we illustrate the case of power plant expansion for electricity generation: we want to find optimal levels of investment in various types of power plants to meet future electricity demand.

We first present a *static deterministic analysis* of the electricity generation problem. *Static* means that decisions are taken only once. *Deterministic* means that the future is supposed to be fully and perfectly known.

Three properties of a given power plant i can be singled out in a static analysis: the investment cost r_i, the operating cost q_i, and the availability factor a_i, which indicates the percent of time the power plant can effectively be operated. Demand for electricity can be considered a single product, but the level of demand varies over time. Analysts usually represent the demand in terms of a so-called *load duration curve* that describes the demand over time in decreasing order of demand level (Figure 5). The curve gives the time, τ, that each demand level, D, is reached. Because here we are concerned with investments over the long run, the load duration curve we consider is taken over the life cycle of the plants.

The load duration curve can be approximated by a piecewise constant curve (Figure 6) with m segments. Let $d_1 = D_1, d_j = D_j - D_{j-1}, j = 2, \ldots, m$ represent the additional power demanded in the so-called *mode j* for a duration τ_j. To obtain a good approximation of the load curve, it is necessary to consider large values of m. In the static situation, the problem consists of finding the optimal investment for each mode j, i.e., to find the particular type of power plant i, $i = 1, \ldots, n$, that minimizes the total cost of effectively producing 1 MW of electricity during the time τ_j. It is given by

$$i(j) = \arg \min_{i=1,\ldots,n} \left\{ \frac{r_i + q_i \, \tau_j}{a_i} \right\}, \tag{3.1}$$

where n is the number of available technologies and arg min represents the index i for which the minimum is achieved.

The static model (3.1) captures one essential feature of the problem, namely, that base load demand (associated with large values of τ_j, i.e., small indices j) is covered by equipment with low operating costs (scaled by availability factor), while peak-load demand (associated with small values of τ_j, i.e., large indices j) is covered by equipment with low investment costs (also scaled by their availability factor). For the sake of completeness, peak-load equipment should also offer operational flexibility.

At least four elements justify considering a *dynamic* or *multistage model* for the electricity generation investment problem:

- the long-term evolution of equipment costs;

- the long-term evolution of the load curve;

- the appearance of new technologies;

- the obsolescence of currently available equipment.

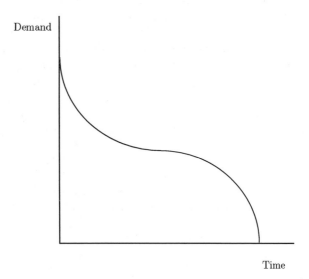

FIGURE 5. The load duration curve.

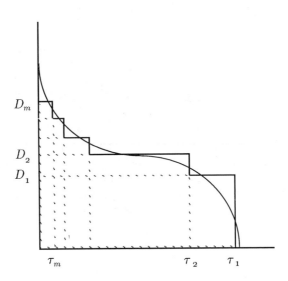

FIGURE 6. A piecewise constant approximation of the load duration curve.

The equipment costs are influenced by technological progress but also (and, for some, drastically) by the evolution of fuel costs.

Of significant importance in the evolution of demand is both the total energy demanded (the area under the load curve) and the peak-level D_m, which determines the total capacity that should be available to cover demand. The evolution of the load curve is determined by several factors, including the level of activity in industry, energy savings in general, and the electricity producers' rate policy.

The appearance of new technologies depends on the technical and commercial success of research and development while obsolescence of available equipment depends on past decisions and the technical lifetime of equipment. All the elements together imply that it is no longer optimal to invest only in view of the short-term ordering of equipment given by (3.1) but that a long-term optimal policy should be found.

The following multistage model can be proposed. Let

- $t = 1, \ldots, H$ index the periods or stages;

- $i = 1, \ldots, n$ index the available technologies;

- $j = 1, \ldots, m$ index the operating modes in the load duration curve.

Also define the following:

- a_i = availability factor of i;

- L_i = lifetime of i;

- g_i^t = existing capacity of i at time t, decided before $t = 1$;

- r_i^t = unit investment cost for i at time t (assuming a fixed plant life cycle for each type i of plant);

- q_i^t = unit production cost for i at time t;

- d_j^t = maximal power demanded in mode j at time t;

- τ_j^t = duration of mode j at time t.

Consider, finally, the set of decisions

- x_i^t = new capacity made available for technology i at time t;

- w_i^t = total capacity of i available at time t;

- y_{ij}^t = capacity of i effectively used at time t in mode j.

The electricity generation H-stage problem can be defined as

$$\min_{x,y,w} \sum_{t=1}^{H} \left(\sum_{i=1}^{n} r_i^t \cdot w_i^t + \sum_{i=1}^{n} \sum_{j=1}^{m} q_i^t \cdot \tau_j^t \cdot y_{ij}^t \right) \qquad (3.2)$$

subject to $w_i^t = w_i^{t-1} + x_i^t - x_i^{t-L_i}$, $i = 1, \ldots, n$, $t = 1, \ldots, H$, $\quad$ (3.3)

$$\sum_{i=1}^{n} y_{ij}^t = d_j^t, \quad j = 1, \ldots, m, \quad t = 1, \ldots, H, \quad (3.4)$$

$$\sum_{j=1}^{m} y_{ij}^t \leq a_i(g_i^t + w_i^t), \quad i = 1, \ldots, n, \quad t = 1, \ldots, H, \quad (3.5)$$

$$x, y, w \geq 0.$$

Decisions in each period t involve new capacities x_i^t made available in each technology and capacities y_{ij}^t operated in each mode for each technology.

Newly decided capacities increase the total capacity w_i^t made available, as given by (3.3), where the equipment's becoming obsolete after its lifetime is also considered. We assume $x_i^t = 0$ if $t \leq 0$, so equation (3.3) only involves newly decided capacities.

By (3.4), the optimal operation of equipment must be chosen to meet demand in all modes using available capacities, which by (3.5) depend on capacities g_i^t decided before $t = 1$, newly decided capacities x_i^t, and the availability factor.

The objective function (3.2) is the sum of the investment plus maintenance costs and operating costs. Compared to (3.1), availability factors enter constraints (3.5) and do not need to appear in the objective function. The operating costs are exactly the same and are based on operating decisions y_{ij}^t, while the investment annuities and maintenance costs r_i^t apply on the cumulative capacity w_i^t. Placing annuities on the cumulative capacity, instead of charging the full investment cost to the decision x_i^t, simplifies the treatment of end of horizon effects and is currently used in many power generation models. It is a special case of the salvage value approach and other period aggregations discussed in Section 11.2.

The same reasons that plead for the use of a multistage model motivate resorting to a *stochastic model*. The evolution of equipment costs, particularly fuel costs, the evolution of total demand, the date of appearance of new technologies, even the lifetime of existing equipment, can all be considered truly random. The main difference between the stochastic model and its deterministic counterpart is in the definition of the variables x_i^t and w_i^t. In particular, x_i^t now represents the new capacity of i decided at time t, which becomes available at time $x_i^{t+\Delta_i}$, where Δ_i is the construction delay for equipment i. In other words, to have extra capacity available at time t, it is necessary to decide at $t - \Delta_i$, when less information is available on the evolution of demand and equipment costs. This is especially important because it would be preferable to be able to wait until the last moment to take decisions that would have immediate impact.

Assume that each decision is now a random variable. Instead of writing an explicit dependence on the random element, ω, we use boldface notation to denote random variables. We then have:

- $\mathbf{x}_i^t$ = new capacity decided at time t for equipment i, $i = 1, \ldots, n$;

- $\mathbf{w}_i^t$ = total capacity of i available and in order at time t;

- $\boldsymbol{\xi}$ = the vector of random parameters at time t;

and all other variables as before. The stochastic model is then

$$\min E_{\boldsymbol{\xi}} \sum_{t=1}^{H} \left(\sum_{i=1}^{n} \mathbf{r}_i^t \mathbf{w}_i^t + \sum_{i=1}^{n} \sum_{j=1}^{m} \mathbf{q}_i^t \, \boldsymbol{\tau}_j^t \, \mathbf{y}_{ij}^t \right) \tag{3.6}$$

$$\text{s. t. } \mathbf{w}_i^t = \mathbf{w}_i^{t-1} + \mathbf{x}_i^t - \mathbf{x}_i^{t-L_i}, \ i = 1, \ldots, n, t = 1, \ldots, H, \tag{3.7}$$

$$\sum_{i=1}^{n} \mathbf{y}_{ij}^t = \mathbf{d}_j^t, \ j = 1, \ldots, m, t = 1, \ldots, H, \tag{3.8}$$

$$\sum_{j=1}^{m} \mathbf{y}_{ij}^t \leq a_i(g_i^t + \mathbf{w}_i^{t-\Delta_i}), \ i = 1, \ldots, n, t = 1, \ldots, H, \tag{3.9}$$

$$\mathbf{w}, \mathbf{x}, \mathbf{y} \geq 0,$$

where the expectation is taken with respect to the random vector $\boldsymbol{\xi} = (\boldsymbol{\xi}^2, \ldots, \boldsymbol{\xi}^H)$. Here, the elements forming $\boldsymbol{\xi}^t$ are the demands, $(\mathbf{d}_1^t, \ldots, \mathbf{d}_k^t)$, and the cost vectors, $(\mathbf{r}^t, \mathbf{q}^t)$. In some cases, $\boldsymbol{\xi}^t$ can also contain the lifetimes L_i, the delay factors Δ_i, and the availability factors a_i, depending on the elements deemed uncertain in the future.

Formulation (3.6)–(3.9) is a convenient representation of the stochastic program. At some point, however, this representation might seem a little confusing. For example, it seems that the expectation is taken only on the objective function, while the constraints contain random coefficients (such as $\mathbf{d}_j^t$ in the right-hand side of (3.8)).

Another important aspect is the fact that decisions taken at time t, $(\mathbf{w}^t, \mathbf{y}^t)$, are dependent on the particular realization of the random vector, $\boldsymbol{\xi}^t$, but cannot depend on future realizations of the random vector. This is clearly a desirable feature for a truly stochastic decision process. If demands in several periods are high, one would expect investors to increase capacity much more than if, for example, demands remain low.

Formally, if the decision variables $(\mathbf{w}^t, \mathbf{y}^t)$ were not dependent on $\boldsymbol{\xi}^t$, the objective function in (3.6) could be replaced by

$$\sum_t \sum_i \left(E_{\boldsymbol{\xi}} \mathbf{r}_i^t \, w_i^t + \sum_j E_{\boldsymbol{\xi}} \, \mathbf{q}_i^t \, \boldsymbol{\tau}_i^t \, y_{ij}^t \right)$$

$$= \sum_t \sum_i \left(\bar{r}_i^t \cdot w_i^t + \sum_j \overline{(q_i \tau_j)} y_{ij}^t \right),$$

where $\bar{r}_i^t = E_\xi r_i^t$ and $\overline{q_i \tau_j} = E_\xi(q_i^t \ \tau_j^t)$, making problem (3.6) to (3.9) deterministic. In the next section, we will make the dependence of the decision variables on the random vector explicit.

The formulation given earlier is convenient in its allowing for both continuous and discrete random variables. Theoretical properties such as continuity and convexity can be derived for both types of variables. Solution procedures, on the other hand, strongly differ.

Problem (3.6) to (3.9) is a multistage stochastic linear program with several random variables that actually has an additional property, called *block separable recourse*. This property stems from a separation that can be made between the aggregate-level decisions, $(\mathbf{x}^t, \mathbf{w}^t)$, and the detailed-level decisions, $\mathbf{y}^t$.

We will formally define block separability in Chapter 3, but we can make an observation about its effect here. Suppose future demands are always independent of the past. In this case, the decision on capacity to install in the future at some t only depends on available capacity and does not depend on the outcomes up to time t. The same $\mathbf{x}^t$ must then be optimal for any realization of ξ. The only remaining stochastic decision is in the operation-level vector, $\mathbf{y}^t$, which now depends separately on each period's capacity. The overall result is that a multiperiod problem now becomes a much less complex two-period problem.

As a simple example, consider the following problem that appears in Louveaux and Smeers [1988]. In this case, the resulting two period model has three operating modes, $n = 4$ technologies, $\Delta_i = 1$ period of construction delay, full availabilities, $a \equiv 1$, and no existing equipment, $g \equiv 0$. The only random variable is $\mathbf{d}_1 = \xi$. The other demands are $d_2 = 3$ and $d_3 = 2$. The investment costs are $r^1 = (10, 7, 16, 6)^T$ with production costs $q^2 = (4, 4.5, 3.2, 5.5)^T$ and load durations $\tau^2 = (10, 6, 1)^T$. We also add a budget constraint to keep all investment below 120. The resulting two-period stochastic program is:

$$\min 10x_1^1 + 7x_2^1 + 16x_3^1 + 6x_4^1 + E_\xi\left[\sum_{j=1}^{3} \tau_j^2(4\mathbf{y}_{1j}^2 + 4.5\mathbf{y}_{2j}^2\right.$$
$$\left. + 3.2\mathbf{y}_{3j}^2 + 5.5\mathbf{y}_{4j}^2)\right]$$
$$\text{s. t. } 10x_1^1 + 7x_2^1 + 16x_3^1 + 6x_4^1 \le 120,$$
$$-x_i^1 + \sum_{j=1}^{3} \mathbf{y}_{ij}^2 \le 0, i = 1, \dots, 4,$$
$$\sum_{i=1}^{y} \mathbf{y}_{i1}^2 = \xi,$$
$$\sum_{i=1}^{y} \mathbf{y}_{ij}^2 = d_j^2, j = 2, 3,$$
$$x_1^1 \ge 0, x_2^1 \ge 0, x_3^1 \ge 0, x_4^1 \ge 0, \mathbf{y}_{ij}^2 \ge 0, \ i = 1, \dots, 4, j = 1, 2, 3. \quad (3.10)$$

Assuming that ξ takes on the values 3, 5, and 7 with probabilities 0.3, 0.4, and 0.3, respectively, an optimal stochastic programming solution to (3.10) includes $x^{1*} = (2.67, 4.00, 3.33, 2.00)^T$ with an optimal objective value of 381.85. We can again consider the expected value solution, which would substitute $\xi \equiv 5$ in (3.10). An optimal solution here (again not unique) is $\bar{x}^1 = (0.00, 3.00, 5.00, 2.00)^T$. The objective value, if this single event occurs, is 365. However, if we use this solution in the stochastic problem, then with probability 0.3, demand cannot be met. This would yield an infinite value of the stochastic solution.

Infinite values probably do not make sense in practice because an action can be taken somehow to avoid total system collapse. The power company could buy from neighboring utilities, for example, but the cost would be much higher than any company operating cost. An alternative technology (internal or external to the company) that is always available at high cost is called a *backstop* technology. If we assume, for example, in problem (3.10) that some other technology is always available, without any required investment costs at a unit operating cost of 100, then the expected value solution would be feasible and have an expected stochastic program value of 427.82. In this case, the value of the stochastic solution becomes $427.82 - 381.85 = 45.97$.

In many power problems, focus is on the reliability of the system or the system's ability to meet demand. This reliability is often described as expressing a minimum probability for meeting demand using the non-backstop technologies. If these technologies are $1, \ldots, n-1$, then the reliability restriction (in the two-period situation where capacity decisions need not be random) is:

$$P[\sum_{i=1}^{n-1} a_i(g_i^t + w_i^t) \geq \sum_{j=1}^{m} \mathbf{d}_j^t] \geq \alpha, \forall t, \qquad (3.11)$$

where $0 < \alpha \leq 1$. Inequality (3.11) is called a *chance* or *probabilistic constraint* in stochastic programming. In production problems, these constraints are often called *fill rate* or *service rate constraints*. They place restrictions on decisions so that constraint violations are not too frequent. Hence, we would often have α quite close to 1.

If the only probabilistic constraints are of the form in (3.11), then we simply want the cumulative available capacity at time t to be at least the α quantile of the cumulative demand in all modes at time t. We then obtain a *deterministic equivalent* constraint to (3.11) of the following form:

$$\sum_{i=1}^{n-1} a_i(g_i^t + w_i^t) \geq (F^t)^{-1}(\alpha), \forall t, \qquad (3.12)$$

where F^t is the (assumed continuous) distribution function of $\sum_{j=1}^{m} \mathbf{d}_j^t$ and $F^{-1}(\alpha)$ is the α-quantile of F. Constraints of the form in (3.12) can then be added to (3.6) to (3.9) or, indeed, to the deterministic problem in (3.2) to (3.5), where expected values replace the random variables.

By adding these chance constraint equivalents, many of the problems of deterministic formulations can be avoided. For example, if we choose $\alpha = 0.7$ for the problem in (3.10), then adding a constraint of the form in (3.12) would not change the deterministic expected value solution. However, we would get a different result if we set $\alpha = 1.0$. In this case, constraint (3.12) for the given data becomes simply:

$$\sum_{i=1}^{4} w_i^1 \geq 12. \tag{3.13}$$

Adding (3.13) to the expected value problem results in an optimal solution with $w^{1*} = (0.833, 3.00, 4.17, 4.00)^T$. The expected value of using this solution in the stochastic program is 383.99, or only 2.14 more than the optimal value in (3.10).

In general, probabilistic constraints are represented by deterministic equivalents and are often included in stochastic programs. We discuss some of the theory of these constraints in Chapter 3. Our emphasis in this book is, however, on optimizing the expected value of continuous utility functions, such as the costs in this capacity expansion problem. We, therefore, concentrate on recourse problems and assume that probabilistic constraints are represented by deterministic equivalents within our formulations.

This problem illustrates a multistage decision problem and the addition of probabilistic constraints. The structure of the problem, however, allows for a two-stage equivalent problem. In this way, the capacity expansion problem provides a bridge between the two-stage example of Section 1 and the multistage problem of Section 2.

This problem also has a natural interpretation with discrete decision variables. For most producing units, only a limited number of possible sizes exists. Typical sizes for high-temperature nuclear reactors would be 1000 MW and 1300 MW, so that capacity decisions could only be taken as integer multiples of these values.

Exercises

1. The detailed-level decisions can be found quite easily according to an *order of merit* rule. In this case, one begins with mode 1 and uses the least expensive equipment until its capacity is exhausted or demand is satisfied. One continues to exhaust capacity or satisfy demand in order of increasing unit operating cost and mode. Show that this procedure is indeed optimal for determining the y_{ij}^t values.

2. Prove that, in the case of no serial correlation (ξ^t and ξ^{t+1} stochastically independent), an optimal solution has the same value for $\mathbf{w}^t$

and $\mathbf{x}^t$ for all $\boldsymbol{\xi}$. Give an example where this does not occur with serial correlation.

3. For the example in (3.10), suppose we add a reliability constraint of the form in (3.13) to the expected value problem, but we use a right-hand side of 11 instead of 12. What is the stochastic program expected value of this solution?

1.4 Design for Manufacturing Quality

This section illustrates a common engineering problem that we model as a stochastic program. The problem demonstrates nonlinear functions in stochastic programming and provides further evidence of the importance of the stochastic solution.

Consider a designer deciding various product specifications to achieve some measure of product cost and performance. The specifications may not, however, completely determine the characteristics of each manufactured product. Key characteristics of the product are often random. For example, every item includes variations due to machining or other processing. Each consumer also does not use the product in the same way. Cost and performance characteristics thus become random variables.

Deterministic methods may yield costly results that are only discovered after production has begun. From this experience, designing for quality and consideration of variable outcomes has become an increasingly important aspect of modern manufacturing (see, for example, Taguchi et al. [1989]). In industry, the methods of Taguchi have been widely used (see also Taguchi [1986]). Taguchi methods can, in fact, be seen as examples of stochastic programming, although they are often not described this way.

In this section, we wish to give a small example of the uses of stochastic programming in manufacturing design and to show how the general stochastic programming approach can be applied. We note that we base our analysis on actual performance measures, whereas the Taguchi methods generally attach surrogate costs to deviations from nominal parameter values.

We consider the design of a simple axle assembly for a bicycle cart. The axle has the general appearance in Figure 7.

The designer must determine the specified length w and design diameter ξ of the axle. We use inches to measure these quantities and assume that other dimensions are fixed. Together, these quantities determine the performance characteristics of the product. The goal is to determine a combination that gives the greatest expected profit.

The initial costs are for manufacturing the components. We assume that a single process is used for the two components. No alternative technologies are available, although, in practice, several processes might be available.

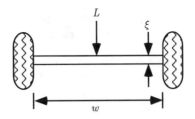

FIGURE 7. An axle of length w and diameter ξ with a central load L.

When the axle is produced, the actual dimensions are not exactly those that are specified. For this example, we suppose that the length w can be produced exactly but that the diameter ξ is a random variable, $\xi(x)$, that depends on a specified mean value, x, that represents, for example, the setting on a machine. We assume a triangular distribution for $\xi(x)$ on $[0.9x, 1.1x]$. This distribution has a density,

$$
f_x(\xi) = \begin{cases} (100/x^2)(\xi - 0.9x) & \text{if } 0.9x \le \xi < x, \\ (100/x^2)(1.1x - \xi) & \text{if } x \le \xi \le 1.1x, \\ 0 & \text{otherwise.} \end{cases} \tag{4.1}
$$

The decision is then to determine w and x, subject to certain limits, $w \le w^{\max}$ and $x \le x^{\max}$, in order to maximize expected profits. For revenues, we assume that if the product is profitable, we sell as many as we can produce. This amount is fixed by labor and equipment regardless of the size of the axle. We, therefore, only wish to determine the maximum selling price that generates enough demand for all production. From marketing studies, we determine that this maximum selling price depends on the length and is expressed as

$$
r(1 - e^{-0.1w}), \tag{4.2}
$$

where r is the maximum possible for any such product.

Our production costs for labor and equipment are assumed fixed, so only material cost is variable. This cost is proportional to the mean values of the specified dimensions because material is acquired before the actual machining process. Suppose c is the cost of a single axle material unit. The total manufacturing cost for an item is then

$$
c\left(\frac{w\pi x^2}{4}\right). \tag{4.3}
$$

In this simplified model, we assume that no quantity discounts apply in the production process.

Other costs are incurred after the product is made due to warranty claims and potential future sales losses from product defects. These costs are often called *quality losses*. In stochastic programming terms, these are the recourse costs. Here, the product may perform poorly if the axle becomes bent or broken due to excess stress or deflection. The stress limit, assuming a steel axle and 100-pound maximum central load, is

$$\frac{w}{\xi^3} \leq 39.27. \tag{4.4}$$

For deflection, we use a maximum 2000-rpm speed (equivalent to a speed of 60 km/hour for a typical 15-centimeter wheel) to obtain:

$$\frac{w^3}{\xi^4} \leq 63,169. \tag{4.5}$$

When either of these constraints is violated, the axle deforms. The expected cost for not meeting these constraints is assumed proportional to the square of the violation. We express it as

$$Q(w, x, \xi) = \min_{y}\{qy^2 \text{ s. t. } \frac{w}{\xi^3} - y \leq 39.27, \frac{w^3}{\xi^4} - 300y \leq 63,169\}, \tag{4.6}$$

where y is, therefore, the maximum of stress violation and (to maintain similar units) $\frac{1}{300}$ of the deflection violation.

The expected cost, given w and x, is

$$\mathcal{Q}(w, x) = \int_{\xi} Q(w, x, \xi) f_x(\xi) d\xi, \tag{4.7}$$

which can be written as:

$$\mathcal{Q}(w, x) = q \int_{.9x}^{1.1x} (100/x^2) \min\{\xi - .9x, 1.1x - \xi\}$$
$$[\max\{0, (\frac{w}{\xi^3}) - 39.27, (\frac{w^3}{300\xi^4}) - 210.56\}]^2 d\xi. \tag{4.8}$$

The overall problem is to find:

max (total revenue per item − manufacturing cost per item

− expected future cost per item). \tag{4.9}

Mathematically, we write this as:

$$\max z(w, x) = r(1 - e^{-0.1w}) - c(\frac{w\pi x^2}{4}) - \mathcal{Q}(w, x)$$
$$\text{s. t. } 0 \leq w \leq w^{\max}, 0 \leq x \leq x^{\max}. \tag{4.10}$$

In stochastic programming terms, this formulation gives the deterministic equivalent problem to the stochastic program for minimizing the current value for the design decision plus future reactions to deviations in the axle diameter. Standard optimization procedures can be used to solve this problem. Assuming maximum values of $w^{\max} = 36$, $x^{\max} = 1.25$, a maximum sales price of \$10 ($r = 10$), a material cost of \$0.025 per cubic inch ($c = .025$), and a unit penalty $q = 1$, an optimal solution is found at $w^* = 33.6$, $x^* = 1.038$, and $z^* = z(w^*, x^*) = 8.94$. The graphs of z as a function of w for $x = x^*$ and as a function of x for $w = w^*$ appear in Figures 8 and 9. In this solution, the stress constraint is only violated when $.9x = 0.934 \leq \xi \leq 0.949 = (w/39.27)^{1/3}$.

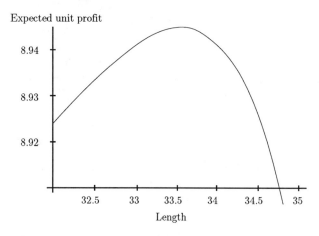

FIGURE 8. The expected unit profit as a function of length with a diameter of 1.038 inches.

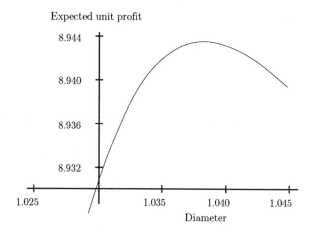

FIGURE 9. The expected unit profit as a function of diameter with a length of 33.6 inches.

We again consider the expected value problem where random variables are replaced with their means to obtain a deterministic problem. For this problem, we would obtain:

$$\max z(w, x, \bar{\xi}) = r(1 - e^{-0.1w}) - c\left(\frac{w\pi x^2}{4}\right)$$

$$-[\max\{0, \left(\frac{w}{x^3}\right) - 39.27, \left(\frac{w^3}{300x^4}\right) - 210.56\}]^2$$

$$\text{s. t. } 0 \le w \le w^{\max}, 0 \le x \le x^{\max}. \tag{4.11}$$

Using the same data as earlier, an optimal solution to (4.11) is $\bar{w}(\bar{\xi}) = 35.0719$, $\bar{x}(\bar{\xi}) = 0.963$, and $z(\bar{w}, \bar{x}, \bar{\xi}) = 9.07$.

At first glance, it appears that this solution obtains a better expected profit than the stochastic problem solution. However, as we shall see in Chapter 9 on approximations, this deterministic problem paints an overly optimistic picture of the actual situation. The deterministic objective is (in the case of concave maximization) always an *overestimate* of the actual expected profit. In this case, the true *expected* value of the deterministic solution is $z(\bar{w}, \bar{x}) = -26.79$. This problem then has a value of the stochastic solution equal to the difference between the expected value of the stochastic solution and the expected value of the deterministic solution, $z^* - z(\bar{w}, \bar{x}) = 35.73$. In other words, solving the stochastic program yields an expected profit increase over solving the deterministic problem.

This problem is another example of how stochastic programming can be used. The problem has nonlinear functions and a simple recourse structure. We will discuss further computational methods for problems of this type in Chapter 6. In other problems, decisions may also be taken after the observation of the outcome. For example, we could inspect and then decide whether to sell the product (Exercise 3). This often leads to tolerance settings and is the focus of much of quality control.

The general stochastic program provides a framework for uniting design and quality control. Many loss functions can be used to measure performance degradation to help improve designs in their initial stages. These functions may include the stress and performance penalties described earlier, the Taguchi-type quadratic loss, or methods based on reliability characterizations.

Most traditional approaches assume some form for the distribution as we have done here. This situation rarely matches practice, however. Approximations can nevertheless be used that obtain bounds on the actual solution value so that robust decisions may be made without complete distributional information. This topic will be discussed further in Chapter 9.

Exercises

1. For the example given, what is the probability of exceeding the stress constraint for an axle designed according to the stochastic program optimal specifications?

2. Again, for the example given, what is the probability of exceeding the stress constraint for an axle designed according to the deterministic program's (4.11) optimal specifications?

3. Suppose that every axle can be tested before being shipped at a cost of s per test. The test completely determines the dimensions of the product and thus informs the producer of the risk of failure. Formulate the new problem with testing.

1.5 Other Applications

In this chapter, we discussed a few examples of stochastic programming applications. The examples were chosen because of their frequency in stochastic programming application as well as to illustrate various aspects of stochastic programming models in terms of number of stages, continuous or discrete variables, separable or nonseparable recourse, probabilistic constraints, and linear or nonlinear constraint and objective functions.

Several other application areas deserve some recognition but were not discussed yet. A particular example is in airline planning. One of the first applications of stochastic programming was a decision on the allocation of aircraft to routes (*fleet assignment*) by Ferguson and Dantzig [1956]. In this problem, penalties were incurred for lost passengers. The problem becomes a simple recourse problem in stochastic programming terms that they solved using a variant of the standard transportation simplex method.

Production planning is another major area that was not in our examples. This area also has been the subject of stochastic programming models for many years. The original chance-constrained stochastic programming model of Charnes, Cooper, and Symonds [1958], for example, considered the production of heating oil with constraints on meeting sales and not exceeding capacity. More recent examples include the study by Escudero et al. [1993] for IBM procurement policies.

Water resource modeling has also received widespread application. A good example of this area is the paper by Prékopa and Szántai [1976], where they discuss regulation of Lake Balaton's water level and show how stochastic programming could have avoided floods that occurred before such planning methods were available. Approaches to pollution and the environmental area of water resource planning are also common. An example discussion appears in Somlyódy and Wets [1988].

Energy planning has been the focus of many stochastic programming studies. We note in particular Manne's [1974] analysis of the U.S. decision on whether to invest in breeder reactors. The more recent work of Manne and Richels [1992] on buying insurance against the greenhouse effect is also an excellent example of how stochastic programming can model uncertain future situations so that informed public policy decisions may be made.

Stochastic programming has been applied in many other areas. Of particular note is the forestry planning model in Gassmann ([1989]) and the hospital staffing problem in Kao and Queyranne ([1985]). We also include two exercises in stochastic programming in sports. Many other references appear in King's survey (King [1988b]) in the volume by Ermoliev and Wets [1988]. Many more applications are open to stochastic programming, especially with the powerful techniques now available. In the remainder of this book, we will explore those methods, their properties, and the general classes of problems they solve.

Exercises

These exercises all contain a stochastic programming problem that can be solved using standard linear, nonlinear and integer programming software. For each problem, you should develop the model, solve the stochastic program, solve the expected value problem, and find the value of the stochastic solution.

1. Northam Airlines is trying to decide how to partition a new plane for its Chicago–Detroit route. The plane can seat 200 economy class passengers. A section can be partitioned off for first class seats but each of these seats takes the space of 2 economy class seats. A business class section can also be included, but each of these seats takes as much space as 1.5 economy class seats. The profit on a first class ticket is, however, three times the profit of an economy ticket. A business class ticket has a profit of two times an economy ticket's profit. Once the plane is partitioned into these seating classes, it cannot be changed. Northam knows, however, that the plane will not always be full in each section. They have decided that three scenarios will occur with about the same frequency: (1) weekday morning and evening traffic, (2) weekend traffic, and (3) weekday midday traffic. Under Scenario 1, they think they can sell as many as 20 first class tickets, 50 business class tickets, and 200 economy tickets. Under Scenario 2, these figures are 10, 25, and 175. Under Scenario 3, they are 5, 10, and 150. You can assume they cannot sell more tickets than seats in each of the sections. (In reality, the company may allow overbooking, but then it faces the problem of passengers with reservations who do not appear for the flight (*no-shows*). The problem of determining how many passengers to accept is part of the field called *yield management*. For one approach to this problem, see Brumelle and McGill [1993]. This subject is explored further in Exercise 1 of Section 2.7.)

2. Tomatoes Inc. (TI) produces tomato paste, ketchup, and salsa from four resources: labor, tomatoes, sugar, and spices. Each box of the tomato paste requires 0.5 labor hours, 1.0 crate of tomatoes, no sugar,

and 0.25 can of spice. A ketchup box requires 0.8 labor hours, 0.5 crate of tomatoes, 0.5 sacks of sugar, and 1.0 can of spice. A salsa box requires 1.0 labor hour, 0.5 crate of tomatoes, 1.0 sack of sugar, and 3.0 cans of spice.

The company is deciding production for the next three periods. It is restricted to using 200 hours of labor, 250 crates of tomatoes, 300 sacks of sugar, and 100 cans of spices in each period at regular rates. The company can, however, pay for additional resources at a cost of 2.0 per labor hour, 0.5 per tomato crate, 1.0 per sugar sack, and 1.0 per spice can. The regular production costs for each product are 1.0 for tomato paste, 1.5 for ketchup, and 2.5 for salsa.

Demand is not known with certainty until after the products are made in each period. TI forecasts that in each period two possibilities are equally likely, corresponding to a good or bad economy. In the good case, 200 boxes of tomato paste, 40 boxes of ketchup, and 20 boxes of salsa can be sold. In the bad case, these values are reduced to 100, 30, and 5, respectively. Any surplus production is stored at costs of 0.5, 0.25, and 0.2 per box for tomato paste, ketchup, and salsa, respectively. TI also considers unmet demand important and assigns costs of 2.0, 3.0, and 6.0 per box for tomato paste, ketchup, and salsa, respectively, for any demand that is not met in each period.

3. The Clear Lake Dam controls the water level in Clear Lake, a well-known resort in Dreamland. The Dam Commission is trying to decide how much water to release in each of the next four months. The Lake is currently 150 mm below flood stage. The dam is capable of lowering the water level 200 mm each month, but additional precipitation and evaporation affect the dam. The weather near Clear Lake is highly variable. The Dam Commission has divided the months into two two-month blocks of similar weather. The months within each block have the same probabilities for weather, which are assumed independent of one another. In each month of the first block, they assign a probability of 1/2 to having a natural 100-mm increase in water levels and probabilities of 1/4 to having a 50-mm decrease or a 250-mm increase in water levels. All these figures correspond to natural changes in water level without dam releases. In each month of the second block, they assign a probability of 1/2 to having a natural 150-mm increase in water levels and probabilities of 1/4 to having a 50-mm increase or a 350-mm increase in water levels. If a flood occurs, then damage is assessed at $10,000 per mm above flood level. A water level too low leads to costly importation of water. These costs are $5000 per mm less than 250 mm below flood stage. The commission first considers an overall goal of minimizing expected costs. They also consider minimizing the probability of violating the maximum and minimum water

levels. (This makes the problem a special form of chance-constrained model.) Consider both objectives.

4. The Energy Ministry of a medium-size country is trying to decide on expenditures for new resources that can be used to meet energy demand in the next decade. There are currently two major resources to meet energy demand. These resources are, however, exhaustible. Resource 1 has a cost of 5 per unit of demand met and a total current availability equal to 25 cumulative units of demand. Resource 2 has a cost of 10 per unit of demand met and a total current availability of 10 demand units. An additional resource from outside the country is always available at a cost of 16.7 per unit of demand met.

Some investment is considered in each of Resources 1 and 2 to discover new supplies and build capital. Resource 1 is, however, elusive. A unit of investment in new sources of Resource 1 yields only 0.1 demand unit of Resource 1 with probability 0.5 and yields 1 demand unit with probability 0.5. For Resource 2, investment is well known. Each unit of investment yields a demand unit equivalent of Resource 2. Cumulative demand in the current decade is projected to be 10, while demand in the next decade will be 25.

The ministry wants to minimize expected costs of meeting demands in the current and following decade assuming that the results of Resource 1 investment will only be known when the current decade ends. Next-decade costs are discounted to 60% of their future real values (which should not change).

5. Pacific Pulp and Paper is deciding how to manage their main forest. They have trees at a variety of ages, which we will break into Classes 1 to 4. Currently, they have 8000 acres in Class 1, 10,000 acres in Class 2, 20,000 in Class 3, and 60,000 in Class 4. Each class corresponds to about 25 years of growth. The company would like to determine how to harvest in each of the next four 25-year periods to maximize expected revenue from the forest. They also foresee the company's continuing after a century, so they place a constraint of having 40,000 acres in Class 4 at the end of the planning horizon.

Each class of timber has a different yield. Class 1 has no yield, Class 2 yields 250 cubic feet per acre, Class 3 yields 510 cubic feet per acre, and Class 4 yields 700 cubic feet per acre. Without fires, the number of acres in Class i (for $i = 2, 3$) in one period is equal to the amount in Class $i - 1$ from the previous period minus the amount harvested from Class $i - 1$ in the previous period. Class 1 at period t consists of the total amount harvested in the previous period $t - 1$, while Class 4 includes all remaining Class 4 land plus the increment from Class 3.

While weather effects do not vary greatly over 25-year periods, fire damage can be quite variable. Assume that in each 25-year block, the probability is 1/3 that 15% of all timber stands are destroyed and that the probability is 2/3 that 5% is lost. Suppose that discount rates are completely overcome by increasing timber value so that all harvests in the 100-year period have the same current value. Revenue is then proportional to the total wood yield.

6. A hospital emergency room is trying to plan holiday weekend staffing for a Saturday, Sunday, and Monday. Regular-time nurses can work any two days of the weekend at a rate of $300 per day. In general, a nurse can handle 10 patients during a shift. The demand is not known, however. If more patients arrive than the capacity of the regular-time nurses, they must work overtime at an average cost of $50 per patient overload. The Saturday demand also gives a good indicator of Sunday–Monday demand. More nurses can be called in for Sunday–Monday duty after Saturday demand is observed. The cost is $400 per day, however, in this case. The hospital would like to minimize the expected cost of meeting demand.

Suppose that the following scenarios of 3 day demand are all equally likely: (100, 90, 20), (100, 110, 120), (100, 100, 110), (90, 100, 110), (90, 80, 110), (90, 90,100), (80, 90, 100), (80, 70, 100), (80, 80, 90).

7. After winning the pole at Monza, you are trying to determine the quickest way to get through the first right-hand turn, which begins 200 meters from the start and is 30 meters wide. You are through the turn at 100 meters past the beginning of the next stretch (see Figure 10). As in the figure, you will attempt to stay 10 meters inside the barrier on the starting stretch and accelerate as fast as possible until point d_1. At this distance, you will start braking as hard as possible and take the turn at the current velocity reached at some point d_2. (Assume a circular turn with radius equal to the square of velocity divided by maximum lateral acceleration.) Obviously, you do not want to go off the course.

The problem is that you can never be exactly sure of the car and track speed until you start braking at point d_1. At that point, you can tell whether the track is fast, medium, or slow, and you can then determine the point d_2 where you enter the turn. You suppose that the three kinds of track/car combinations are equally likely. If fast, you accelerate at 27 m/sec^2, decelerate at 45 m/sec^2, and have a maximum lateral acceleration of 1.8 g (= 17.5 m/sec^2). For medium, these values are 24, 42, and 16; for slow, the values are 20, 35, and 14. You want to minimize the expected time through this section. You also assume that if you follow an optimal strategy, other competitors

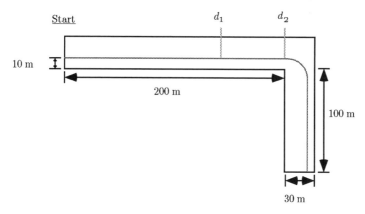

FIGURE 10. Opening straight and turn for Problem 7.

will not throw you out of the race (although you may not be sure of that).

8. In training for the Olympic decathlon, you are trying to choose your takeoff point for the long jump to maximize your expected official jump. Unfortunately, when you aim at a certain spot, you have a 50/50 chance of actually taking off 10 cm beyond that point. If that violates the official takeoff line, you foul and lose that jump opportunity. Assume that you have three chances and that your longest jump counts as your official finish.

You then want to determine your aiming strategy for each jump. Assume that your actual takeoff is independent from jump to jump. Initially you are equally likely to hit a 7.4- or 7.6-meter jump from your actual takeoff point. If you hit a long first jump, then you have a 2/3 chance of another 7.6-meter jump and 1/3 chance of jumping 7.4 meters. The probabilities are reversed if you jumped 7.4 meters the first time. You always seem to hit the third jump the same as the second.

First, find a strategy to maximize the expected official jump. Then, maximize decathlon points from the following Table 7.

TABLE 7. Decathlon Points for Problem 8.

Distance	Points	Distance	Points
7.30	886	7.46	925
7.31	888	7.47	927
7.32	891	7.48	930
7.33	893	7.49	932
7.34	896	7.50	935
7.35	898	7.51	937
7.36	900	7.52	940
7.37	903	7.53	942
7.38	905	7.54	945
7.39	908	7.55	947
7.40	910	7.56	950
7.41	913	7.57	952
7.42	915	7.58	955
7.43	918	7.59	957
7.44	920	7.60	960
7.45	922	7.61	962

2
Uncertainty and Modeling Issues

In the previous chapter, we gave several examples of stochastic programming models. These formulations fit into different categories of stochastic programs in terms of the characteristics of the model. This chapter presents those basic characteristics by describing the fundamentals of any modeling effort and some of the standard forms detailed in later chapters.

Before beginning general model descriptions, however, we first describe the probability concepts that we will assume in the rest of the book. Familiarity with these concepts is essential in understanding the structure of a stochastic program. This presentation is made simple enough to be understood by readers unfamiliar with the field and, thus, leaves aside some questions related to measure theory. Sections 2 through 7 build on these fundamentals and give the general forms in various categories. Section 8 gives some background on the relationship of stochastic programming to other areas of decision making under uncertainty. Section 9 briefly reviews the main optimization concepts used in the book.

2.1 Probability Spaces and Random Variables

Several parameters of a problem can be considered uncertain and are thus represented as random variables. Production and distribution costs typically depend on fuel costs, which are random. Future demands depend on uncertain market conditions. Crop returns depend on uncertain weather conditions.

Uncertainty is represented in terms of random experiments with outcomes denoted by ω. The set of all outcomes is represented by Ω. In a transport and distribution problem, the outcomes range from political conditions in the Middle East to general trade situations, while the random variable of interest may be the fuel cost. The relevant set of outcomes is clearly problem-dependent. Also, it is usually not very important to be able to define those outcomes accurately because the focus is mainly on their impact on some (random) variables.

The outcomes may be combined into subsets of Ω called *events*. We denote by $\mathcal{A}$ a collection of random events. As an example, if Ω contains the six possible results of the throw of a dice, $\mathcal{A}$ also contains combined outcomes such as an odd number, a result smaller than or equal to four, etc. If Ω contains weather conditions for a single day, $\mathcal{A}$ also contains combined events such as "a day without rain," which might be the union of a sunny day, a partly cloudy day, a cloudy day without showers, etc.

Finally, to each event $A \in \mathcal{A}$ is associated a value $P(A)$, called a *probability*, such that $0 \leq P(A) \leq 1, P(\emptyset) = 0, P(\Omega) = 1$ and $P(A_1 \cup A_2) = P(A_1) + P(A_2)$ if $A_1 \cap A_2 = \emptyset$.

The triplet $(\Omega, \mathcal{A}, P)$ is called a *probability space* that must satisfy a number of conditions (see, e.g., Chung [1974]). It is possible to define several random variables associated with a probability space, namely, all variables that are influenced by the random events in $\mathcal{A}$. If one takes as elements of Ω events ranging from the political situation in the Middle East to the general trade situations, they allow us to describe random variables such as the fuel costs and the interest rates and inflation rates in some Western countries. If the elements of Ω are the weather conditions from April to September, they influence random variables such as the production of corn, the sales of umbrellas and ice cream, or even the exam results of undergraduate students.

In terms of stochastic programming, there exists one situation where the description of random variables is closely related to Ω: in some cases indeed, the elements $\omega \in \Omega$ are used to describe a few *states of the world* or *scenarios*. All random elements then jointly depend on these finitely many scenarios. Such a situation frequently occurs in strategic models where the knowledge of the possible outcomes in the future is obtained through experts' judgments and only a few scenarios are considered in detail. In many situations, however, it is extremely difficult and pointless to construct Ω and $\mathcal{A}$; the knowledge of the random variables is sufficient.

For a particular random variable ξ, we define its cumulative distribution $F_\xi(x) = P(\xi \leq x)$, or more precisely $F_\xi(x) = P(\{\omega | \xi \leq x\})$. Two major cases are then considered. A discrete random variable takes a finite or countable number of different values. It is best described by its probability distribution, which is the list of possible values, $\xi^k, k \in K$, with associated probabilities,

$$f(\xi^k) = P(\boldsymbol{\xi} = \xi^k) \text{ s.t.} \sum_{k \in K} f(\xi^k) = 1.$$

Continuous random variables can often be described through a so-called *density* function $f(\xi)$. The probability of ξ being in an interval $[a, b]$ is obtained as

$$P(a \leq \boldsymbol{\xi} \leq b) = \int_a^b f(\xi)d\xi,$$

or equivalently

$$P(a \leq \boldsymbol{\xi} \leq b) = \int_a^b dF(\boldsymbol{\xi}),$$

where $F(.)$ is the cumulative distribution as earlier. Contrary to the discrete case, the probability of a single value $P(\boldsymbol{\xi} = a)$ is always zero for a continuous random variable. The distribution $F(.)$ must be such that $\int_{-\infty}^{\infty} dF(\boldsymbol{\xi}) = 1$.

The *expectation* of a random variable is computed as $\mu = \sum_{k \in K} \xi^k f(\xi^k)$ or $\mu = \int_{-\infty}^{\infty} \xi dF(\boldsymbol{\xi})$ in the discrete and continuous cases, respectively. The *variance* of a random variable is $E[(\boldsymbol{\xi} - \mu)^2]$. The expectation of $\boldsymbol{\xi}^r$ is called the rth *moment* of $\boldsymbol{\xi}$ and is denoted $\bar{\xi}^{(r)} = E[\boldsymbol{\xi}^r]$. A point η is called the α-quantile of $\boldsymbol{\xi}$ if and only if for $0 < \alpha < 1$, $\eta = \min\{x | F(x) \geq \alpha\}$.

The appendix lists the distributions used in the textbook and their expectations and variances. The concepts of probability distribution, density, and expectation easily extend to the case of multiple random variables. Some of the sections in the book use probability measure theory which generalizes these concepts. These sections contain a warning to readers unfamiliar with this field.

2.2 Deterministic Linear Programs

A deterministic linear program consists of finding a solution to

$$\min z = c^T x$$
$$\text{s.t. } Ax = b,$$
$$x \geq 0,$$

where x is an $(n \times 1)$ vector of decisions and $c, A,$ and b are known data of sizes $(n \times 1), (m \times n),$ and $(m \times 1)$, respectively. The value $z = c^T x$ corresponds to the objective function, while $\{x | Ax = b, x \geq 0\}$ defines the set of feasible solutions. An optimum x^* is a feasible solution such that $c^T x \geq c^T x^*$ for any feasible x. Linear programs typically search for a minimal-cost solution under some requirements (demand) to be met or for a maximum profit solution under limited resources. There exists a wide variety of applications, routinely solved in the industry. As introductory references, we cite Chvátal [1980], Dantzig [1963], and Murty [1983]. We assume the reader is familiar with linear programming and has some knowledge of basic duality theory as in these textbooks. A short review is given in Section 2.9.

2.3   Decisions and Stages

Stochastic linear programs are linear programs in which some problem data may be considered uncertain. *Recourse programs* are those in which some decisions or recourse actions can be taken after uncertainty is disclosed. To be more precise, data uncertainty means that some of the problem data can be represented as random variables. An accurate probabilistic description of the random variables is assumed available, under the form of the probability distributions, densities or, more generally, probability measures. As usual, the particular values the various random variables will take are only known after the random experiment, i.e., the vector $\xi = \xi(\omega)$ is only known after the experiment.

The set of decisions is then divided into two groups:

- A number of decisions have to be taken before the experiment. All these decisions are called *first-stage decisions* and the period when these decisions are taken is called the *first stage*.

- A number of decisions can be taken after the experiment. They are called *second-stage decisions*. The corresponding period is called the *second stage*.

First-stage decisions are represented by the vector x, while second-stage decisions are represented by the vector y or $y(\omega)$ or even $y(\omega, x)$ if one wishes to stress that second-stage decisions differ as functions of the outcome of the random experiment and of the first-stage decision. The sequence of events and decisions is thus summarized as

$$x \rightarrow \xi(\omega) \rightarrow y(\omega, x).$$

Observe here that the definitions of first and second stages are only related to before and after the random experiment and may in fact contain sequences of decisions and events. In the farming example of Section 1.1, the first stage corresponds to planting and occurs during the whole spring. Second-stage decisions consist of sales and purchases. Selling extra corn would probably occur very soon after the crop while buying missing corn will take place as late as possible.

A more extreme example is the following. A traveling salesperson receives one item every day. She visits clients hoping to sell that item. She returns home when a buyer is found or when all clients are visited. Clients buy or do not buy in a random fashion. The decision is not influenced by the previous days' decisions. The salesperson wishes to determine the order in which to visit clients, in such a way as to be at home as early as possible (seems reasonable, does it not?). Time spent involves the traveling time plus some service time at each visited client.

To make things simple, once the sequence of clients to be visited is fixed, it is not changed. Clearly the first stage consists of fixing the sequence and traveling to the first client. The second stage is of variable duration depending on the successive clients buying the item or not. Now, consider the following example. There are two clients with probability of buying 0.3 and 0.8, respectively and traveling times (including service) as in the graph of Figure 1.

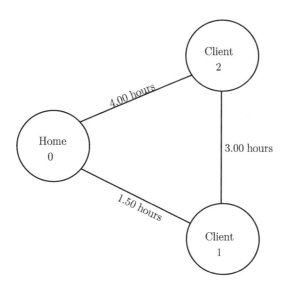

FIGURE 1. Traveling salesperson example.

Assume the day starts at 8 A.M. If the sequence is (1,2), the first stage goes from 8 to 9:30. The second stage starts at 9:30 and finishes either at 11 A.M. if 1 buys or 4:30 P.M. otherwise. If the sequence is (2,1), the first stage goes from 8 to 12:00, the second stage starts at 12:00 and finishes either at 4:00 P.M. or at 4:30 P.M. Thus, the first stage if sequence (2,1) is chosen may sometimes end after the second stage is finished when (1,2) is chosen if Client 1 buys the item.

2.4 Two-Stage Program with Fixed Recourse

The classical two-stage stochastic linear program with fixed recourse (originated by Dantzig [1955] and Beale [1955]) is the problem of finding

$$\min z = c^T x + E_{\xi}[\min q(\omega)^T y(\omega)] \tag{4.1}$$
$$\text{s.t. } Ax = b, \tag{4.2}$$
$$T(\omega)x + Wy(\omega) = h(\omega), \tag{4.3}$$
$$x \geq 0 , y(\omega) \geq 0. \tag{4.4}$$

As in the previous section, a distinction is made between the first stage and the second stage. The first-stage decisions are represented by the $n_1 \times 1$ vector x. Corresponding to x are the first-stage vectors and matrices c, b, and A, of sizes $n_1 \times 1, m_1 \times 1$, and $m_1 \times n_1$, respectively. In the second stage, a number of random events $\omega \in \Omega$ may realize. For a given realization ω, the second-stage problem data $q(\omega), h(\omega)$ and $T(\omega)$ become known, where $q(\omega)$ is $n_2 \times 1, h(\omega)$ is $m_2 \times 1$, and $T(\omega)$ is $m_2 \times n_1$.

Each component of q, T, and h is thus a possible random variable. Let $T_{i.}(\omega)$ be the ith row of $T(\omega)$. Piecing together the stochastic components of the second-stage data, we obtain a vector $\xi^T(\omega) = (q(\omega)^T, h(\omega)^T, T_{1.}(\omega), \cdots, T_{m_2.}(\omega))$, with potentially up to $N = n_2 + m_2 + (m_2 \times n_1)$ components. As indicated before, a single random event ω (or state of the world) influences several random variables, here, all components of ξ.

Let also $\Xi \subset \Re^N$ be the *support* of ξ, that is, the smallest closed subset in $\Re^N$ such that $P(\Xi) = 1$. As just said, when the random event ω is realized, the second-stage problem data, q, h, and T, become known. Then, the second-stage decision $y(\omega)$ or $(y(\omega, x))$ must be taken. The dependence of y on ω is of a completely different nature from the dependence of q or other parameters on ω. It is not functional but simply indicates that the decisions y are typically not the same under different realizations of ω. They are chosen so that the constraints (4.3) and (4.4) hold *almost surely* (denoted *a.s.*), i.e., for all $\omega \in \Omega$ except perhaps for sets with zero probability. We assume random constraints to hold in this way throughout this book unless a specific probability is given for satisfying constraints.

The objective function of (4.1) contains a deterministic term $c^T x$ and the expectation of the second-stage objective $q(\omega)^T y(\omega)$ taken over all realizations of the random event ω. This second-stage term is the more difficult one because, for each ω, the value $y(\omega)$ is the solution of a linear program. To stress this fact, one sometimes uses the notion of a deterministic equivalent program. For a given realization ω, let

$$Q(x, \xi(\omega)) = \min_{y}\{q(\omega)^T y | Wy = h(\omega) - T(\omega)x, y \geq 0\} \tag{4.5}$$

be the second-stage value function. Then, define the expected second-stage value function

$$Q(x) = E_\xi Q(x, \xi(\omega)) \qquad (4.6)$$

and the *deterministic equivalent program* (DEP)

$$\min z = c^T x + Q(x) \qquad (4.7)$$
$$\text{s.t. } Ax = b, \qquad (4.8)$$
$$x \geq 0.$$

This representation of a stochastic program clearly illustrates that the major difference from a deterministic formulation is in the second-stage value function. If that function is given, then a stochastic program is just an ordinary nonlinear program.

Formulation (4.1)–(4.4) is the simplest form of a stochastic two-stage program. Extensions are easily modeled. For example, if first-stage or second-stage decisions are to be integers, constraint (4.4) can be replaced by a more general form:

$$x \in X, y(w) \in Y,$$

where $X = Z_+^{n_1}$ and $Y = Z_+^{n_2}$. Similarly, nonlinear first-stage and second-stage objectives or constraints can easily be incorporated.

Examples of recourse formulation and interpretations

The definition of first stage versus second stage is not only problem dependent but also context dependent. We illustrate different examples of recourse formulations for one class of problems: *the location problem*.

Let $i = 1, \cdots, m$ index clients having demand d_i for a given commodity. The firm can open a facility (such as a plant or a warehouse) in potential sites $j = 1, \ldots, n$. Each client can be supplied from an open facility where the commodity is made available (i.e., produced or stored). The problem of the firm is to choose the number of facilities to open, their locations, and market areas to maximize profit or minimize costs.

Let us first present the deterministic version of the so-called simple plant location or uncapacitated facility location problem. Let x_j be a binary variable equal to one if facility j is open and zero otherwise. Let c_j be the fixed cost for opening and operating facility j and let v_j be the variable operating cost of facility j. Let y_{ij} be the fraction of the demand of client i served from facility j and t_{ij} be the unit transportation cost from j to i.

All costs and profits should be taken in conformable units, typically on a yearly equivalent basis. Let r_i denote the unit price charged to client i and $q_{ij} = (r_i - v_j - t_{ij})d_i$ be the total revenue obtained when all of client i's demand is satisfied from facility j. Then the simple plant location problem or uncapacitated facility location problem (UFLP) reads as follows:

$$\text{UFLP: } \max_{x,y} \; z(x,y) = -\sum_{j=1}^{n} c_j x_j + \sum_{i=1}^{m}\sum_{j=1}^{n} q_{ij} y_{ij} \tag{4.9}$$

$$\text{s.t.} \sum_{j=1}^{n} y_{ij} \le 1, i = 1, \cdots, m, \tag{4.10}$$

$$0 \le y_{ij} \le x_j, i = 1, \ldots, m, j = 1, \ldots, n, \tag{4.11}$$

$$x_j \in \{0,1\}, j = 1, \ldots, n. \tag{4.12}$$

Constraints (4.10) ensure that the sum of fractions of clients i's demand served cannot exceed one. Constraints (4.11) ensure that clients are served only through open plants.

It is customary to present the uncapacitated facility location in a different canonical form that minimizes the sum of the fixed costs of opening facilities and of the transportation costs plus possibly the variable operating costs. (There are several ways to arrive at this canonical representation. One is to assume that unit prices are much larger than unit costs in such a way that demand is always fully satisfied.) This presentation more clearly stresses the link between the deterministic and stochastic cases.

In the UFLP, a trade-off is sought between opening more plants, which results in higher fixed costs and lower transportation costs and opening fewer plants with the opposite effect. Whenever the optimal solution is known, the size of an open facility is computed as the sum of demands it serves. (In the deterministic case, it is always optimal to have each y_{ij} equal to either zero or one.) The market areas of each facility are then well-defined.

The notation x_j for the location variables and y_{ij} for the distribution variables is common in location theory and is thus not meant here as first stage and second stage, respectively, although in some of the models it is indeed the case.

Several parameters of the problem may be uncertain and may thus have to be represented by random variables. Production and distribution costs may vary over time. Future demands for the product may be uncertain.

As indicated in the introduction of the section, we will now discuss various situations of recourse. It is customary to consider that the location decisions x_j are first-stage decisions because it takes some time to implement decisions such as moving or building a plant or warehouse. The main modeling issue is on the distribution decisions. The firm may have full control on the distribution, for example, when the clients are shops owned by the firm. It may then choose the distribution pattern after conducting some random experiments. In other cases, the firm may have contracts that fix which plants serve which clients, or the firm may wish fixed distribution patterns in view of improved efficiency because drivers would have better knowledge of the regions traveled.

a. Fixed distribution pattern, fixed demand, r_i, v_j, t_{ij} stochastic

Assume the only uncertainties are in production and distribution costs
and prices charged to the client. Assume also that the distribution pattern
is fixed in advance, i.e., is considered first stage. The second stage then
just serves as a measure of the cost of distribution. We now show that
the problem is in fact a deterministic problem in which the total revenue
$q_{ij} = (r_i - v_j - t_{ij})d_i$ can be replaced by its expectation. To do this, we
formally introduce extra second-stage variables w_{ij}, with the constraint
$w_{ij}(\omega) = y_{ij}$ for all ω. We obtain

$$\max \quad -\sum_{j=1}^{n} c_j x_j + E_\xi \sum_{i=1}^{m} \sum_{j=1}^{n} q_{ij}(\omega) w_{ij}(\omega)$$

s.t. (4.10), (4.11), (4.12), and

$$w_{ij}(\omega) = y_{ij}, \quad i = 1,\ldots,m \, , \, j = 1,\ldots,n \quad \text{for all } \omega. \tag{4.13}$$

By (4.13), the second-stage objective function can be replaced by

$$E_\xi \sum_{i=1}^{m} \sum_{j=1}^{n} q_{ij}(\omega) y_{ij}$$

or

$$\sum_{i=1}^{n} \sum_{j=1}^{n} E_\xi q_{ij}(\omega) y_{ij},$$

because y_{ij} is fixed and summations and expectation can be interchanged.
The problem is thus the deterministic problem

$$\max \quad -\sum_{j=1}^{n} c_j x_j + \sum_{i=1}^{m} \sum_{j=1}^{n} (E_\xi q_{ij}(\omega)) y_{ij}$$

s.t. (4.10), (4.11), (4.12).

Although there exists uncertainty about the distribution costs and revenues, the only possible action is to plan in view of the expected costs.

b. Fixed distribution pattern, uncertain demand

Assume now that demand is uncertain, but, for some of the reasons cited
earlier, the distribution pattern is fixed in the first stage. Depending on the
context, the distribution costs and revenues (v_j, t_{ij}, r_i) may or may not be
uncertain.

We define y_{ij} = quantity transported from j to i, a quantity no longer
defined as a function of the demand d_i, because demand is now stochastic.
For simplicity, we assume that a penalty q_i^+ is paid per unit of demand d_i

which cannot be satisfied from all quantities transported to i (they might have to be obtained from other sources) and a penalty q_i^- is paid per unit on the products delivered to i in excess of d_i (the cost of inventory, for example). We thus introduce second-stage variables: $w_i^-(\omega)$ = amount of extra products delivered to i in state ω; $w_i^+(\omega)$ = amount of unsatisfied demand to i in state ω.

The formulation becomes

$$\max \quad -\sum_{j=1}^{n} c_j x_j + \sum_{i=1}^{m}\sum_{j=1}^{n} (E_\xi(-v_j - t_{ij}))y_{ij} + E_\xi[-\sum_{i=1}^{m} q_i^+ w_i^+(\omega) \quad (4.14)$$

$$-\sum_{i=1}^{m} q_i^- w_i^-(\omega)] + E_\xi \sum_{i=1}^{m} r_i d_i$$

$$\text{s.t.} \sum_{i=1}^{m} y_{ij} \leq M x_j, \ j = 1, \ldots, n, \quad (4.15)$$

$$w_i^+(\omega) - w_i^-(\omega) = d_i(\omega) - \sum_{j=1}^{n} y_{ij}, i = 1, \ldots, m, \quad (4.16)$$

$$x_j \in \{0,1\}, 0 \leq y_{ij}, w_i^+(\omega) \geq 0, w_i^-(\omega) \geq 0, i = 1, \ldots, m; j = 1, \ldots, n. (4.17)$$

This model is a location extension of the transportation model of Williams [1963]. The objective function contains the investment costs for opening plants, the expected production and distribution costs, the expected penalties for extra or insufficient demands, and the expected revenue. This last term is constant because it is assumed that all demands must be satisfied by either direct delivery or some other means reflected in the penalty for unmet demand. The problem only makes sense if q_i^+ is large enough, for example, larger than $E_\xi(v_j + t_{ij})$ for all j, although weaker conditions may sometimes suffice. Constraint (4.15) guarantees that distribution only occurs from open plants, i.e., plants such that $x_j = 1$. The constant M represents the maximum possible size of a plant.

Observe that here the variables y_{ij} are first-stage variables. Also observe that in the second stage, the constraints (4.16,4.17) have a very simple form, as $w_i^+(\omega) = \mathbf{d}_i - \sum_{j=1}^{n} y_{ij}$ if this quantity is non-negative and $w_i^-(\omega) = \sum_{j=1}^{n} y_{ij} - \mathbf{d}_i$ otherwise. This is an example of a *second stage with simple recourse*.

Also note that in Cases a and b, the size or capacity of plant j is simply obtained as the sum of the quantity transported from j, namely, $\sum_{i=1}^{m} d_i y_{ij}$ in Case a and $\sum_{i=1}^{m} y_{ij}$ in Case b.

c. Uncertain demand, variable distribution pattern

We now consider the case where the distribution pattern can be adjusted to the realization of the random event. This might be the case when uncertainty corresponds to long-term scenarios, of which only one is realized.

Then the distribution pattern can be adapted to this particular realization. This also implies that the sizes of the plants cannot be defined as the sum of the quantity distributed, because those quantities depend on the random event. We thus define as before:

$$x_j = \begin{cases} 1 & \text{if plant } j \text{ is open,} \\ 0 & \text{otherwise.} \end{cases}$$

We now let y_{ij} depend on ω with $y_{ij}(\omega) =$ fraction of demand $d_i(\omega)$ served from j and define new variables $w_j =$ size (capacity) of plant j, with unit investment cost g_j.

The model now reads

$$\max \quad -\sum_{j=1}^{n} c_j x_j - \sum_{j=1}^{n} g_j w_j + E_{\xi} \max \sum_{i=1}^{m} \sum_{j=1}^{n} q_{ij}(\omega) y_{ij}(\omega) \qquad (4.18)$$

$$\text{s.t.} \quad x_j \in \{0, 1\}, \ w_j \geq 0, j = 1, \ldots, n, \qquad (4.19)$$

$$\sum_{j=1}^{n} y_{ij}(\omega) \leq 1, i = 1, \ldots, m, \qquad (4.20)$$

$$\sum_{i=1}^{m} d_i(\omega) y_{ij}(\omega) \leq w_j, j = 1, \ldots, n, \qquad (4.21)$$

$$0 \leq y_{ij}(\omega) \leq x_j, i = 1, \ldots, m, j = 1, \ldots, n, \qquad (4.22)$$

where $q_{ij}(\omega) = (r_i - v_j - t_{ij}) d_i(\omega)$ now includes the demand $d_i(\omega)$.

Constraint (4.20) indicates that no more than 100% of i's demand can be served, but that the possibility exists that not all demand is served. Constraint (4.21) imposes that the quantity distributed from plant j does not exceed the capacity w_j decided in the first stage. For the sake of clarity, one could impose a constraint $w_j \leq M x_j$, but this is implied by (4.21) and (4.22). For a discussion of algorithmic solutions of this problem, see Louveaux and Peeters [1992].

d. Stages versus periods; Two-stage versus multistage

In this section, we highlight again the difference in a stochastic program between *stages* and *periods* of times. Consider the case of a distribution firm that makes its plans for the next 36 months. It may formulate a model such as (4.18)–(4.22). The location of warehouses would be first-stage decisions, while the distribution problem would be second-stage decisions. The duration of the first stage would be something like six months (depending on the type of warehouse) and the second stage would run over the 30 remaining months. Although we may think of a problem over 36 periods, a two-stage model is totally relevant. In this case, the only moment where the number of periods is important is when the precise values of the objective coefficients are computed.

In this example, a multistage model becomes necessary if the distribution firm foresees additional periods where it is ready to change the location of the warehouses. In this example, suppose the firm decides that the opening of new warehouses can be decided after one year. A three-stage model can be constructed. The first stage would consist of decisions on warehouses to be built now. The second stage would consist of the distribution patterns between months 7 and 18 as well and new openings decided in month 12. The third stage would consist of distribution patterns between months 19 and 36.

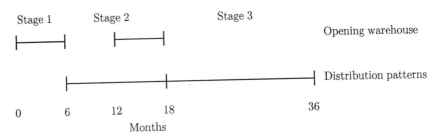

FIGURE 2. Three-stage model decisions and times.

Let x^1 and $x^2(\omega_2)$ be the binary vectors representing opening warehouses in stages 1 and 2, respectively. Let $y^2(\omega_2)$ and $y^3(\omega_3)$ be the vectors representing the distribution decisions in stages 2 and 3, respectively, where ω_2 and ω_3 are the states of the world in stages 2 and 3. Assuming each warehouse can only have a fixed size M, the following model can be built:

$$
\max \quad -\sum_{j=1}^{n} c_j x_j^1 + E_{\boldsymbol{\xi}_2} \max\{\sum_{i=1}^{m}\sum_{j=1}^{n} q_{ij}^2(\omega_2)y_{ij}^2(\omega_2) - \sum_{j=1}^{n} c_j^2(\omega_2)x_j^2(\omega_2)
$$

$$
+ E_{\boldsymbol{\xi}_3|\boldsymbol{\xi}_2} \max[\sum_{i=1}^{m}\sum_{j=1}^{n} q_{ij}^3(\omega_3)y_{ij}^3(\omega_3)]\}
$$

$$
\text{s.t.} \quad \sum_{j=1}^{n} y_{ij}^2(\omega_2) \leq 1, i = 1,\ldots,m,
$$

$$
\sum_{i=1}^{m} d_i(\omega_2)y_{ij}^2(\omega_2) \leq M x_j^1, j = 1,\ldots,n,
$$

$$
\sum_{j=1}^{n} y_{ij}^3(\omega_3) \leq 1, i = 1,\ldots,m,
$$

$$
\sum_{i=1}^{m} d_i(\omega_3)y_{ij}^3(\omega_3) \leq M(x_j^1 + x_j^2(\omega_2)), j = 1,\ldots,n,
$$

$$
x_j^1 + x_j^2(\omega_2) \leq 1, j = 1,\ldots,n,
$$

$$
x_j^1, x_j^2(\omega_2) \in \{0,1\}, j = 1,\ldots,n,
$$

$$
y_{ij}^2(\omega_2), y_{ij}^3(\omega_3) \geq 0, i = 1,\ldots,m, j = 1,\ldots,n.
$$

Multistage programs will be further studied in Section 3.5.

2.5 Random Variables and Risk Aversion

In our view, one can often classify random events and random variables in two major categories. In the first category, we would place uncertainties that recur frequently on a short-term basis. As an example, uncertainty may correspond to daily or weekly demands. This normally leads to a model similar to the one in Section 4, Case b (4.b), where allocation cannot be adjusted every time period. It follows that the expectation in the second stage somehow represents a mean over possible values of the random variables, of which many will occur. Thus, the expectation takes into account realizations that might not occur and many realizations that will occur. To fix ideas here, if in Model 4.b the units in the objective function are in a yearly basis and the randomness involves daily or weekly demands, one may expect that the value of the objective of stochastic model will closely match the realized total yearly revenue.

As one interesting example of a real-world application of a location model of this first category, we may recommend the paper by Psaraftis, Tharakan, and Ceder [1986]. It deals with the optimal location and size of equipment to fight oil spills. Occurrence and sizes of spills are random. The sizes of the spills are represented by a discrete random variable taking three possible values, corresponding to small, medium, or large spills. Sadly enough, spills are sufficiently frequent that the expectation may be considered close enough to the mean cost, as just described. Occurrence of spills at a given site is also random. It is described by a Poisson process. By making the assumption of nonconcomitant occurrence of spills, all equipment is made available for each spill, which simplifies the second-stage descriptions compared to (4.14)–(4.17).

In the second category, we would place uncertainties that can be represented as scenarios, of which basically only one or a small number are realized. This would be the case in long-term models where scenarios represent the general trend of the variables. As already indicated, this is the spirit in which Model c is built. In the second stage, among all scenarios over which expectation is taken, only one is realized. The objective function may then be considered a poor representation of risk aversion, which is typically assumed in decision making (if we exclude gambling).

Starting from the Von Neumann and Morgenstern [1944] theory of utility, this field of modeling preferences has been developed by economics. Models such as the mean-variance approach of Markowitz [1959] have been widely used. Other methods have been proposed based on mixes of mean-variance and other approaches (see, e.g, Ben-Tal and Teboulle [1986]). From a theoretical point of view, considering a nonlinear utility function transforms the problems into stochastic nonlinear programs. This topic is covered in Chapter 6. In practice, however, it seems difficult to solve large-scale non-linear stochastic programs, so that the choice has very often been either to include risk aversion in a somewhat small second-stage description or to maintain a linear utility function with a more detailed second-stage description.

One interesting alternative to nonlinear utility models is to include risk aversion in a linear utility model under the form of a linear constraint, called *downside risk* (Eppen, Martin, and Schrage [1989]). The problem there is to determine the type and level of production capacity at each of several locations. Plants produce various types of cars and may be open, closed, or retooled. The demand for each type of car in the medium term is random. The decisions about the locations and configurations of plants have to be made before the actual demands are known.

Scenarios are based on pessimistic, neutral, or optimistic realizations of demands. A scenario consists of a sequence of realizations for the next five years. The stochastic model maximizes the present value of expected discounted cash flows. The linear constraint on risk is as follows: the down-side risk of a given scenario is the amount by which profit falls below some given target value. It is thus zero for larger profits. The expected downside risk is simply the expectation of the downside risk over all scenarios. The constraint is thus that the expected downside risk must fall below some level.

To give an idea of how this works, consider a two-stage model similar to (4.1)–(4.4) but in terms of profit maximization, by

$$\max z = c^T x + E_\xi[\max q^T(\omega)y(\omega)]$$

s.t. (4.2)–(4.4).
Then define the target level g on profit. The downside risk $u(\xi)$ is thus defined by two constraints:

$$u(\xi(\omega)) \geq g - q^T(\omega)y(\omega) \qquad (5.1)$$
$$u(\xi(\omega)) \geq 0. \qquad (5.2)$$

The constraint on expected downside risk is

$$E_\xi u(\xi) \leq l, \qquad (5.3)$$

where l is some given level. For a problem with a discrete random vector ξ, constraint (5.3) is linear. Observe that (5.3) is in fact a first-stage constraint as it runs over all scenarios. It can be used directly in the extensive form. It can also be used indirectly in a sequential manner, by imposing such a constraint only when needed. This can be done in a way similar to the induced constraints for feasibility that we will study in Chapter 5.

2.6   Implicit Representation of the Second Stage

This book is mainly concerned with stochastic programs of the form (4.1)–(4.4), assuming that an adequate and computationally tractable representation of the recourse problem exists. This is not always the case. Two possibilities then exist that still permit some treatment of the problem:

- A closed form expression is available for the expected value function $\mathcal{Q}(x)$.

- For a given first-stage decision x, the expected value function $\mathcal{Q}(x)$ is computable.

These possibilities are described in the following sections.

a.   A closed form expression is available for $\mathcal{Q}(x)$

We may illustrate this case by the Stochastic Queue Median model (SQM) first proposed by Berman, Larson, and Chiu [1985] from which we take the following in a simplified form. The problem consists of locating an emergency unit (such as an ambulance). When a call arrives, there is a certain probability that the ambulance is already busy handling an earlier demand for ambulance service. In that event, the new service demand is either referred to a backup ambulance service or entered into a queue of other waiting "customers." Here, the first-stage decision consists of finding a location for the ambulance. The second stage consists of the day-to-day response of the system to the random demands. Assuming a first-in, first-out decision rule, decisions in the second stage are somehow automatic. On the other hand, the quality of response, measured, e.g., by the expected service time, depends on the first-stage decision. Indeed, when responding to a call, an ambulance typically goes to the scene and returns to the home location before responding to the next call. The time when it is unavailable for another call is clearly a function of the home location.

Let λ be the total demand rate, $\lambda \geq 0$. Let p_i be the probability that a demand originates from demand region i, with $\sum_{i=1}^{m} p_i = 1$. Let also $t(i, x)$ denote the travel time between location x and call i. On-scene service time is omitted for simplicity. Given facility location x, the expected response time is the sum of the mean-in-queue delay $w(x)$ and the expected travel time $\bar{t}(x)$,

$$\mathcal{Q}(x) = w(x) + \bar{t}(x), \qquad (6.1)$$

where

$$w(x) = \begin{cases} \frac{\lambda \bar{t}^{(2)}(x)}{2(1 - \lambda \bar{t}(x))} & \text{if } \lambda \bar{t}(x) < 1, \\ 0 & \text{otherwise,} \end{cases} \qquad (6.2)$$

$$\bar{t}(x) = \sum_{i=1}^{m} p_i t(i, x), \qquad (6.3)$$

and

$$\bar{t}^{(2)}(x) = \sum_{i=1}^{m} p_i t^2(i, x) .$$

(6.4)

The global problem is then of the form:

$$\min_{x \in X} \mathcal{Q}(x),$$

(6.5)

where the first-stage objective function is usually taken equal to zero and X represents the set of possible locations, which typically consists of a network.

It should be clear that no possibility exists to adequately describe the exact sequence of decisions and events in the so-called second stage and that the expected recourse $\mathcal{Q}(x)$ represents the result of a computation assuming the system is in steady state.

b. For a given x, $\mathcal{Q}(x)$ is computable

The deterministic traveling salesperson problem (TSP) consists of finding a Hamiltonian tour of least cost or distance. Following a Hamiltonian tour means that the traveling salesperson starts from her home location, visits all customers, (say $i = 1, \cdots, m$) exactly, and returns to the home location.

Now, assume each customer has a probability p_i of being present. A full optimization that would allow the salesperson to decide the next customer to visit at each step would be a difficult multistage stochastic program. A simpler two-stage model, known as *a priori optimization* is as follows: in the first-stage, an a priori Hamiltonian tour is designed. In the second stage, the a priori tour is followed by skipping the absent customers. The problem is to find the tour with minimal expected cost (Jaillet [1988]).

The exact representation of such a second-stage recourse problem as a mathematical program with binary decision variables might be possible in theory but would be so cumbersome that it would be of no practical value. On the other hand, the expected length of the tour (and thus $\mathcal{Q}(x)$) is easily computed when the tour (x) is given. Other examples of a priori optimization can be found in Bertsimas, Jaillet, and Odoni [1990].

2.7   Probabilistic Programming

In probabilistic programming, some of the constraints or the objective are expressed in terms of probabilistic statements about first-stage decisions. The description of second-stage or recourse actions is thus avoided. This is particularly useful when the cost and benefits of second-stage decisions are difficult to assess.

2.7 Probabilistic Programming 65

Consider the following covering location problem. Let $j = 1, \cdots, n$ be the potential locations with, as usual, $x_j = 1$ if site j is open and 0 otherwise, and c_j the investment cost. Let $i = 1, \cdots, m$ be the clients. Client i is served if there exists an open site within distance t_i. The distance between i and j is t_{ij}. Define $N_i = \{j | t_{ij} < t_i\}$ as the set of eligible sites for client i. The deterministic covering problem is

$$\min \sum_{j=1}^{n} c_j x_j \tag{7.1}$$

$$\text{s.t.} \sum_{j \in N_i} x_j \geq 1, i = 1, \cdots, m, \tag{7.2}$$

$$x_j \in \{0, 1\}, j = 1, \cdots, n. \tag{7.3}$$

Taking again the case of an ambulance service, one site may be covering more than one region or demand area. When a call is placed, the emergency units may be busy serving another call. Let q be the probability that no emergency unit is available at site j. For simplicity, assume this probability is the same for every site (see Toregas et al. [1971]). Then, the deterministic covering constraint (7.2) may be replaced by the requirement that P (at least one emergency unit from an open eligible site is available) $\geq \alpha$ where α is some confidence level, typically 90 or 95%. Here, the probability that none of the eligible sites has an available emergency unit is q to the power $\sum_{j \in N_i} x_j$, so that the probabilistic constraint is

$$1 - q^{\sum_{j \in N_i} x_j} \geq \alpha \, , \, i = 1, \cdots, m \tag{7.4}$$

or

$$q^{\sum_{j \in N_i} x_j} \leq 1 - \alpha.$$

Taking the logarithm on both sides, one obtains

$$\sum_{j \in N_i} x_j \geq b \tag{7.5}$$

with

$$b = \lceil \frac{\ln(1 - \alpha)}{\ln q} \rceil, \tag{7.6}$$

where $\lceil a \rceil$ denotes the smallest integer greater than or equal to a. Thus, the probabilistic constraint (7.4) has a linear deterministic equivalent (7.5). This is the desired situation with probabilistic constraints. Very often, the deterministic equivalents correspond to nonlinear constraints and the question is whether they define a convex feasible region. This will be studied in Section 3.2.

Exercises

1. Consider Exercise 1 of Section 1.5.

   (a) Show that this is a two-stage stochastic program with first-stage integer decision variables. Observe that for a random variable with integer realizations, the second-stage variables can be assumed continuous, because the optimal second-stage decisions are automatically integer.

      Assume that Northam revises its seating policy every year. Is a multistage program needed?

   (b) Assume that the data in Exercise 1 correspond to the demand for seat reservations. Assume that there is a 50% probability that all clients with a reservation effectively show up and that 10 or 20% no-shows occur with equal probability. Model this situation as a three-stage program, with first-stage decisions as before, second-stage decisions corresponding to the number of accepted reservations, and third-stage decisions corresponding to effective seat occupation. Show that the third stage is a simple recourse program with a reward for each occupied seat and a penalty for each denied reservation.

   (c) Consider now the situation where the number of seats has been fixed to 12, 24, and 140 for the first class, business class, and economy class, respectively. Assume the top management estimates the reward of an occupied seat to be 4, 2, and 1 in the first class, business class, and economy class, respectively, and the penalty for a denied reservation is 1.5 times the reward. Model the corresponding problem as a recourse program. Find the optimal acceptance policy with the data of Exercise 1 in Section 1.5 and no-shows as in (b) of the current exercise. To simplify, assume that passengers with a denied reservation are not seated in a higher class even if a seat is available there.

2. Let x represent the first-stage production of a given good. Let ξ be the demand for the same good. A typical second stage would consist of selling as much as possible, namely, $\min(\xi, x)$. Obtain a closed form expression for the recourse function $E_{\xi}[\min(\xi, x)]$ in the following cases of ξ:

   (a) Poisson distribution,

   (b) A normal distribution.

3. Consider an airplane with x seats. Assume passengers with reservations show up with probability 0.90, independently of each other.

(a) Let $x = 40$. If 42 passengers receive a reservation, what is the probability that at least one is denied seat.

(b) Let $x = 50$. How many reservations can be accepted under the constraint that the probability of seating all passengers who arrive for the flight is greater than 90% ?

2.8 Relationship to Other Decision-Making Models

The stochastic programming models considered in this section illustrate the general form of a stochastic program. While this form can apply to virtually all decision-making problems with unknown parameters, certain characteristics typify stochastic programs and form the major emphasis of this book. In general, stochastic programs are generalizations of deterministic mathematical programs in which some uncontrollable data are not known with certainty. The key features are typically many decision variables with many potential values, discrete time periods for decisions, the use of expectation functionals for objectives, and known (or partially known) distributions. The relative importance of these features contrasts with similar areas, such as statistical decision theory, decision analysis, dynamic programming, Markov decision processes, and stochastic control. In the following subsections, we consider these other areas of study and highlight the different emphases.

a. Statistical decision theory and decision analysis

Wald [1950] developed much of the foundation of optimal statistical decision theory (see also DeGroot [1970] and Berger [1985]). The basic motivation was to determine best levels of variables that affect the outcome of an experiment. With variables x in some set X, random outcomes, $\omega \in \Omega$, an associated distribution, $F(\omega)$, and a reward or loss associated with the experiment under outcome ω of $r(x, \omega)$, the basic problem is to find $x \in X$ to

$$\max E_\omega[r(x, \omega)|F] = \max \int_\omega r(x, \omega) dF(\omega). \tag{8.1}$$

The problem in (8.1) is also the fundamental form of stochastic programming. The major differences in emphases between the fields stem from underlying assumptions about the relative importance of different aspects of the problem.

In stochastic programming, one generally assumes that difficulties in finding the form of the function r and changes in the distribution F as a function of actions are small in comparison to finding the expectations with known distributions and an optimal value x with all other information known. The emphasis is on finding a solution after a suitable problem statement in the form (8.1) has been found. For example, in the simple farming example in Section 1.1, the number of possible planting configurations (even allowing only whole-acre lots) is enormous. Enumerating the possibilities would be hopeless. Stochastic programming avoids such inefficiencies through an optimization process.

We might suppose that the fields or crop varieties are new and that the farmer has little direct information about yields. In this case, the yield distribution would probably start as some prior belief but would be modified as time went on. This modification and possible effects of varying crop rotations to obtain information are the emphases from statistical decision theory. If we assumed that only limited variation in planting size (such as 50-acre blocks) was possible, then the combinatorial nature of the problem would look less severe. Enumeration might then be possible without any particular optimization process. If enumeration were not possible, the farmer might still update the distributions and objectives and use stochastic programming procedures to determine next year's crops based on the updated information.

In terms of (8.1), statistical decision theory places a heavy emphasis on changes in F to some updated distribution $\hat{F}_x$ that depends on a partial choice of x and some observations of ω. The implied assumption is that this part of the analysis dominates any solution procedure, as when X is a small finite set that can be enumerated easily.

Decision analysis (see, e.g., Raiffa [1968]) can be viewed as a particular part of optimal statistical decision theory. The key emphases are often on acquiring information about possible outcomes, on evaluating the utility associated with various outcomes, and on defining a limited set of possible actions (usually in the form of a decision tree). For example, consider the capacity expansion problem in Section 1.3. We considered a wide number of alternative technology levels and production decisions. In that model, we assumed that demand in each period was independent of the demand in the previous period. This characteristic gave the block separability property that can allow efficient solutions for large problems.

A decision analytic model might apply to the situation where an electric utility's demand depends greatly on whether a given industry locates in the region. The decision problem might then be broken into separate stochastic programs depending on whether the new industry demand materializes and whether the utility starts on new plants before knowing the industry decision. In this framework, the utility first decides whether to start its own projects. The utility then observes whether the new industry expands into the region and faces the stochastic program form from Section 1.4 with four possible input scenarios about the available capacity when the industry's location decision is known (see Figure 3).

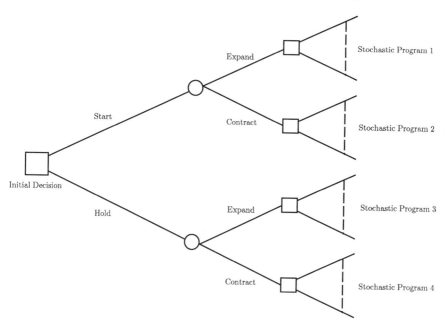

FIGURE 3. Decision tree for utility with stochastic programs on leaves.

The two stochastic programs given each initial decision allow for the evaluation of expected utility given the two possible outcomes and two possible initial decisions. The actual initial decision taken on current capacity expansion would then be made by taking expectations over these two outcomes.

Separation into distinct possible outcomes and decisions and the realization of different distributions depending on the industry decision give this model a decision analysis framework. In general, a decision analytic approach would probably also consider multiple attributes of the capacity decisions (for example, social costs for a given location) and would concentrate on the value of risk in the objective. It would probably also entail consideration of methods for obtaining information about the industry's decision and contingent decisions based on the outcomes of these investigations. Of course, these considerations can all be included in a stochastic program, but they are not typically the major components of a stochastic programming analysis.

b. Dynamic programming and Markov decision processes

Much of the literature on stochastic optimization considers dynamic programming and Markov decision processes (see, e.g., Heyman and Sobel [1984], Bellman [1957], Ross [1983], and Kall and Wallace [1994] for a discussion relating to stochastic programming). In these models, one searches for optimal actions to take at generally discrete points in time. The actions are influenced by random outcomes and carry one from some state at some stage t to another state at stage $t + 1$. The emphasis in these models is typically in identifying finite (or, at least, low-dimensional) state and action spaces and in assuming some Markovian structure (so that actions and outcomes only depend on the current state).

With this characterization, the typical approach is to form a backward recursion resulting in an optimal decision associated with each state at each stage. With large state spaces, this approach becomes quite computationally cumbersome although it does form the basis of many stochastic programming computation schemes as given in Chapter 7. Another approach is to consider an infinite horizon and use discounting to establish a stationary policy (see Howard [1960] and Blackwell [1965]) so that one need only find an optimal decision associated with a state for any stage.

A typical example of this is in investment. Suppose that instead of saving for a specific time period in the example of Section 1.2, you wish to maximize a discounted expected utility of wealth in all future periods. In this case, the state of the system is the amount of wealth. The decision or action is to determine what amount of the wealth to invest in stock and bonds. We could discretize to varying wealth levels and then form a problem as follows:

$$\max \sum_{t=1}^{\infty} \rho^t E[q\mathbf{y}(t) - r\mathbf{w}(t)] \qquad (8.2)$$

$$\text{s. t. } x(1,1) + x(2,1) = b,$$

$$\boldsymbol{\xi}(1,t)\mathbf{x}(1,t) + \boldsymbol{\xi}(2,t)\mathbf{x}(2,t) - \mathbf{y}(t) + \mathbf{w}(t) = G,$$

$$\boldsymbol{\xi}(1,t)\mathbf{x}(1,t) + \boldsymbol{\xi}(2,t)\mathbf{x}(2,t) = \mathbf{x}(1,t+1) + \mathbf{x}(2,t+1),$$

$$\mathbf{x}(i,t), \mathbf{y}(t), \mathbf{w}(t) \geq 0, \mathbf{x} \in \mathcal{N},$$

where $\mathcal{N}$ is the space of nonanticipative decisions and ρ is some discount factor. This approach could lead to finding a stationary solution to

$$z(b) = \max_{x(1)+x(2)=b} \{E[-q(G - \boldsymbol{\xi}(1)x(1) - \boldsymbol{\xi}(2)x(2))^-$$

$$-r(G - \boldsymbol{\xi}(1)x(1) - \boldsymbol{\xi}(2)x(2))^+ + \rho E[z(\boldsymbol{\xi}(1)x(1) + \boldsymbol{\xi}(2)x(2))]]\}. \qquad (8.3)$$

Again, problem (8.2) fits the general stochastic programming form, but particular solutions as in (8.3) are more typical of Markov decision processes. These are not excluded in stochastic programs, but stochastic programs generally do not include the Markovian assumptions necessary to derive (8.3).

c. Comparison to optimal stochastic control

Stochastic control models are often similar to stochastic programming models. The differences are mainly due to problem dimension (stochastic programs would generally have higher dimension), emphases on control rules in stochastic control, and more restrictive constraint assumptions in stochastic control. In many cases, the distinction is, however, not at all clear.

As an example, suppose a more general formulation of the financial model in Section 1.2. There, we considered a specific form of the objective function, but we could also use other forms. For example, suppose the objective was generally stated as minimizing some cost $r_t(\mathbf{x}(t), \mathbf{u}(t))$ in each time period t, where $\mathbf{u}(t)$ are the controls $u((i, j), t, s)$ that correspond to actual transactions of exchanging asset i into asset j in period t under scenario s. In this case, problem (1.2.2) becomes:

$$\min z = \sum_s p(s)(\sum_{t=1}^{H} r_t(x(t, s), u(t, s), s))$$
$$\text{s. t. } x(0, s) = b,$$
$$x(t, s) + \xi(s)^T u(t, s) = x(t + 1, s), t = 0, \ldots, H,$$
$$x(s), u(s) \text{ nonanticipative}, \qquad (8.4)$$

where $\xi(s)^T$ represents returns on investments minus transaction costs. Additional constraints may be incorporated into the objective of (8.4) through penalty terms.

Problem (8.4) is fairly typical of a discrete time control problem governed by a linear system. The general emphasis in control approaches to such problems is for *linear, quadratic, Gaussian* (LQG) models (see, for example, Kushner [1971], Fleming and Rishel [1975], and Dempster [1980]), where we have a linear system as earlier, but where the randomness is Gaussian in each period (for example, ξ is known but the state equation for $x(t + 1, s)$ includes a Gaussian term), and r_t is quadratic. In these models, one may also have difficulty observing x so that an additional observation variable $y(t)$ may be present.

The LQG problem leads to Kalman filtering solutions (see, for example, Kalman [1969]). Various extensions of this approach are also possible, but the major emphasis remains on developing controls with specific decision rules to link observations directly into estimations of the state and controls. In stochastic programming models, general constraints (such as non-negative state variables) are emphasized. In this case, most simple decision rules forms (such as when u is a linear function of state) fail to obtain satisfactory solutions (see, for example, Gartska and Wets [1974]). For this reason, stochastic programming procedures tend to search for more general solution characteristics.

Stochastic control procedures may, of course, apply but stochastic programming tends to consider more general forms of interperiod relationships and state space constraints. Other types of control formulations, such as robust control, may also be considered specific forms of a stochastic program that are amenable to specific techniques to find control policies with given characteristics.

Continuous time stochastic models (see, e.g., Harrison [1985]) are also possible but generally require more simplified models than those considered in stochastic programming. Again, continuous time formulations are consistent with stochastic programs but have not been the main emphasis of research or the examples in this book. In certain examples again, they may be quite relevant (see, for example, Harrison and Wein [1990] for an excellent application in manufacturing) in defining fundamental solution characteristics, such as the optimality of control limit policies.

In all these control problems, the main emphasis is on characterizing solutions of some form of the dynamic programming Bellman-Hamilton-Jacobi equation or application of Pontryagin's maximum principle. Stochastic programs tend to view all decisions from beginning to end as part of the procedure. The dependence of the current decision on future outcomes and the transient nature of solutions are key elements. Section 3.5 provides some further explanation by describing these characteristics in terms of general optimality conditions.

c. Summary

Stochastic programming is simply another name for the study of optimal decision making under uncertainty. The term *stochastic programming* emphasizes a link to mathematical programming and algorithmic optimization procedures. These considerations dominate work in stochastic programming and distinguish stochastic programming from other fields of study. In this book, we follow this paradigm of concentrating on representation and characterizations of optimal decisions and on developing procedures to follow in determining optimal or approximately optimal decisions. This development begins in the next chapter with basic properties of stochastic program solution sets and optimal values.

Exercises

1. Consider the design problem in Section 1.4. Suppose the design decision does not completely specify x in (1.4.1), but the designer only knows that if a value $\hat{x}$ is specified then $x \in [.99\hat{x}, 1.01\hat{x}]$. Suppose a uniform distribution for x is assumed initially on this interval. How would the formulation in Section 1.4 be modified to account for information as new parts are produced?

2. From the example in Section 1.2, suppose that a goal in each period is to realize a 16% return with penalties $q = 1$ and $r = 4$ as before. Formulate the problem as in (8.2).

2.9   Short Reviews

a. Linear Programming

Consider a linear program (L.P.) of the form

$$\max\{c^T x | Ax = b, x \geq 0\}, \tag{9.1}$$

where A is an $m \times n$ matrix, x and c are $n \times 1$ vectors, and b is an $m \times 1$ vector. If needed, any inequality constraint can be transformed into an equality by the addition of *slack variables*:

$$a_i.x \leq b_i \text{ becomes } a_i.x + s_i = b_i,$$

where s_i is the slack variable of row i and $a_i.$ is the ith row of matrix A.

A *solution* to (9.1) is a vector x that satisfies $Ax = b$. A *feasible solution* is a solution x with $x \geq 0$. An *optimal solution* x^* is a feasible solution such that $c^T x^* \geq c^T x$ for all feasible solutions x. A *basis* is a choice of n linearly independent columns of A. Associated to a basis is a submatrix B of the corresponding columns, so that, after a suitable rearrangement, A can be partitioned into $A = [B, N]$. Associated with a basis is a *basic solution*, $x_B = B^{-1}b$, $x_N = 0$, and $z = c_B^T B^{-1}b$, where $[x_B, x_N]$ and $[c_B, c_N]$ are partitions of x and c following the basic and nonbasic columns. We use B^{-1} to denote the inverse of B, which is known to exist because B has linearly independent columns and is square.

In geometric terms, basic solutions correspond to *extreme points* of the polyhedron, $\{x | Ax = b, x \geq 0\}$. A basis is feasible (optimal) if its associated basic solution is feasible (optimal). The conditions for feasibility are $B^{-1}b \geq 0$. The conditions for optimality are that in addition to feasibility, the inequalities, $c_N^T - c_B^T B^{-1} N \leq 0$, hold.

Linear programs are routinely solved by widely distributed, easy-to-use linear program solvers. Access to such a solver would be useful for some exercises in this book. For a better understanding, some examples and exercises also use manual solutions of linear programs.

Finding an optimal solution is equivalent to finding an optimal *dictionary*, a definition of individual variables in terms of the other variables. In the *simplex algorithm*, starting from a feasible dictionary, the next one is obtained by selecting an *entering variable* (any nonbasic variable whose increase leads to an increase in the objective value), then finding a *leaving variable* (the first to become negative as the entering variable increases), then realizing a *pivot* substituting the entering for the leaving variable in the dictionary. An optimal solution is reached when no entering variable can be found.

A linear program is *unbounded* if an entering variable exists for which no leaving variable can be found. In some cases, a feasible initial dictionary is not available at once. Then, *phase one* of the simplex method consists of finding such an initial dictionary. A number of artificial variables are introduced to make the dictionary feasible. The phase one procedure minimizes the sum of artificials using the simplex method. If a solution with a sum of artificials equal to zero exists, then the original problem is feasible and *phase two* continues with the true objective function. If the optimal solution of the phase one problem is nonzero, then the original problem is *infeasible*.

As an example, consider the following linear program:

$$
\begin{array}{llll}
\max & -x_1 & +3x_2 & \\
\text{s. t.} & 2x_1 & +x_2 & \geq 5, \\
& x_1 & +x_2 & \leq 3, \\
& x_1, & x_2 & \geq 0.
\end{array}
$$

Adding slack variables s_1 and s_2, the two constraints read

$$
\begin{array}{llll}
2x_1 & +x_2 & -s_1 & = 5, \\
x_1 & +x_2 & +s_2 & = 3.
\end{array}
$$

The natural choice for the initial basis is (s_1, s_2). This basis is infeasible as s_1 would obtain the value -5. An *artificial variable* (a_1) is added to row one to form:

$$2x_1 + x_2 - s_1 + a_1 = 5.$$

The phase one problem consists of minimizing a_1, i.e., finding $-\max -a_1$. Let $z = -a_1$ be the phase one objective, which after substituting for a_1 gives the initial dictionary in phase one:

$$
\begin{array}{lllll}
z & = -5 & +2x_1 & +x_2 & -s_1, \\
a_1 & = 5 & -2x_1 & -x_2 & +s_1, \\
s_2 & = 3 & -x_1 & -x_2, &
\end{array}
$$

corresponding to the initial basis (a_1, s_2). Entering candidates are x_1 and x_2 as they both increase the objective value. Choosing x_1, the leaving variable is a_1 (because it becomes zero for $x_1 = 2.5$ while s_2 becomes zero only for $x_1 = 3$). Substituting x_1 for a_1, the second dictionary becomes:

$$
\begin{array}{lllll}
z & = & & & -a_1, \\
x_1 & = 2.5 & -0.5x_2 & +0.5s_1 & -0.5a_1, \\
s_2 & = 0.5 & -0.5x_2 & -0.5s_1 & +0.5a_1.
\end{array}
$$

This dictionary is an optimal dictionary for phase one. (No nonbasic variable would possibly increase x.) This means the original problem is feasible. (In fact, the basis (x_1, s_2) is feasible with solution $x_1 = 2.5, x_2 = 0.0$).)

We now turn to phase two. We replace the phase one objective with the original objective:

$$z = -x_1 + 3x_2 = -2.5 + 3.5x_2 - 0.5s_1.$$

By removing the artificial variable a_1 (as it is not needed anymore), we obtain the following first dictionary in phase two:

$$
\begin{aligned}
z &= -2.5 &&+3.5x_2 &&-0.5s_1, \\
x_1 &= 2.5 &&-0.5x_2 &&+0.5s_1, \\
s_2 &= 0.5 &&-0.5x_2 &&-0.5s_1.
\end{aligned}
$$

The next entering variable is x_2 with leaving variable s_2. After substitution, we obtain the final dictionary:

$$
\begin{aligned}
z &= 1 &&-4s_1 &&-7s_2, \\
x_1 &= 2 &&+s_1 &&+s_2, \\
x_2 &= 1 &&-s_1 &&-2s_2,
\end{aligned}
$$

which is optimal because no nonbasic variable is a valid entering variable. The optimal solution is $x^* = (2,1)^T$ with $z^* = 1$.

b. Duality for linear programs

The *dual* of the so-called primal problem (9.1) is:

$$
\min\{\pi^T b \,|\, \pi^T A \geq c^T, \pi \text{ unrestricted}\}. \tag{9.2}
$$

Variables π are called *dual variables*. One such variable is associated with each constraint of the primal. When the primal constraint is an equality, the dual variable is *free* (unrestricted in sign). Dual variables are sometimes called *shadow prices* or *multipliers* (as in nonlinear programming). The dual variable π_i may sometimes be interpreted as the *marginal value* associated with resource b_i.

If the dual is unbounded, then the primal is infeasible. Similarly, if the primal is unbounded, then the dual is infeasible. Both problems can also be simultaneously infeasible.

If x is primal feasible and π is dual feasible, then $c^T x \leq \pi^T b$. The primal has an optimal solution x^* if and only if the dual has an optimal solution π^*. In that case, $c^T x^* = (\pi^*)^T b$ and the primal and dual solutions satisfy the *complementary slackness* conditions:

$$
(a_{i\cdot})x^* = b_i \text{ or } \pi_i^* = 0 \text{ or both, for any } i = 1,\dots,m,
$$

$$
(\pi^*)^T a_{\cdot j} = c_j \text{ or } x_j^* = 0 \text{ or both, for any } j = 1,\dots,n,
$$

where $a_{\cdot j}$ is the jth column of A and, as before, $a_{i\cdot}$ is the ith row of A.

An alternative presentation is to say that $s_i^* \pi_i^* = 0$, where s_i is the slack variable of the ith constraint, i.e., either the slack or the dual variable associated with a constraint is zero, and similarly for the second condition. Thus, the optimal solution of the dual can be recovered from the optimal solution for the primal, and vice versa.

The optimality conditions can also be interpreted to say that either there exists some *improving direction*, w, from a current feasible solution, $\hat{x}$, so that $c^T w > 0$, $w_j \geq 0$ for all $j \in N$, $N = \{j | \hat{x}_j = 0\}$, and $a_i.w \leq 0$ for all $i \in I$, $I = \{i | a_i.\hat{x} = b_i\}$ or there exists some $\pi \geq 0$ such that $\sum_{i \in I} \pi_i a_{ij} \leq c_j$ for all $j \in N$, $\sum_{i \in I} \pi_i a_{ij} = c_j$ for all $j \notin N$, but both cannot occur. This result is equivalent to the *Farkas lemma*, which gives alternative systems with or without solutions.

The *dual simplex method* replicates on the primal solution what the iterations of the simplex method would be on the dual problem: it first finds the leaving variable (one that is strictly negative) then the entering variable (the first one that would become positive in the objective line). The dual simplex is particularly useful when a solution is already available to the original primal problem and some extra constraint or bound is added to the problem. The reader is referred to Chvátal [1980, pp. 152–157] for a detailed presentation.

Other material not covered in this section is meant to be restrictive to a given topic area. The next section discusses more of the mathematical properties of solutions and functions.

c. Nonlinear programming and convex analysis

When objectives and constraints may contain nonlinear functions, the optimization problem becomes a *nonlinear program*. The nonlinear program analogous to (9.1) has the form

$$\min\{f(x)|g(x) \leq 0, h(x) = 0\}, \tag{9.3}$$

where $x \in \Re^n$, $f : \Re^n \rightarrow \Re$, $g : \Re^n \rightarrow \Re^m$, and $h : \Re^n \rightarrow \Re^l$. We may also assume that the range of f may include ∞ to allow the objective to include constraints directly through an *indicator function*:

$$\delta(x|X) = \begin{cases} 0 & \text{if } g(x) \leq 0, h(x) = 0, \\ +\infty & \text{otherwise,} \end{cases}$$

where X is the set of x satisfying the constraints in (9.3), i.e., the *feasible region*.

In this book, the feasible region is usually a *convex set* so that X contains any *convex combination*,

$$\sum_{i=1}^{s} \lambda^i x^i, \sum_{i=1}^{s} \lambda^i = 1, \lambda^i \geq 0, i = 1, \ldots, s,$$

of points, x^i, $i = 1, \ldots, s$, that are in the feasible region. Extreme points of the region are points that cannot be expressed as a convex combination of two distinct points also in the region. The set of all convex combinations of a given set of points is its *convex hull*.

The feasible region is also most generally *closed* so that it contains all limits of infinite sequences of points in the region. The region is also generally *connected*, so that, for any x^1 and x^2 in the region, there exists some path of points in the feasible region connecting x^1 to x^2 by a function, $\eta : [0, 1] \to \Re^n$ that is continuous with $\eta(0) = x^1$ and $\eta(1) = x^2$. For certain results, we may also assume the region is *bounded* so that a *ball* of radius M, $\{x | \|x\| \leq M\}$, contains the entire set of feasible points. Otherwise, the region is *unbounded*. Note that a region may be unbounded while the optimal value in (9.1) or (9.3) is still bounded. In this case, the region often contains a *cone*, i.e., a set S such that if $x \in S$, then $\lambda x \in S$ for all $\lambda \geq 0$. When the region is both closed and bounded, then it is *compact*.

The set of equality constraints, $h(x) = 0$, is often *affine*, i.e., they can be expressed as linear combinations of the components of x and some constant. In this case, each constraint, $h_i(x) = 0$, is a *hyperplane*, $a_i.x - b_i = 0$, as in the linear program constraints. In this case, $h(x) = 0$, defines an *affine space*, a *translation* of the *parallel* subspace, $Ax = 0$. The affine space *dimension* is the same as its parallel subspace, i.e., the maximum number of linearly independent vectors in the subspace.

With nonlinear constraints and inequalities, the region may not be an affine space, but we often consider the lowest-dimension affine space containing them, i.e., the *affine hull* of the region. The affine hull is useful in optimality conditions because it distinguishes *interior* points that can be the center of a ball entirely within the region from the *relative interior* (ri), which can be the center of a ball whose intersection with the affine hull is entirely within the region. When a point is not in a feasible region, we often take its *projection* into the region using an operator, Π. If the region is X, then the projection of x onto X is $\Pi(x) = argmin\{\|w - x\| | w \in X\}$.

In this book, we generally assume that the objective function f is a *convex function*, i.e., such that

$$f(\lambda x^1 + (1 - \lambda)x^2) \leq \lambda f(x^1) + (1 - \lambda)f(x^2),$$

$0 \leq \lambda \leq 1$. If f also is never $-\infty$ and is not $+\infty$ everywhere, then f is a *proper convex function*. The region where f is finite is called the *effective domain* of f $(domf)$. We can also define convex functions in terms of the *epigraph* of f, $epi(f) = \{(x, \beta) | \beta \geq f(x)\}$. In this case, f is convex if and only if its epigraph is convex. If $-f$ is convex, then f is *concave*.

Often, we assume that f has *directional derivatives*, $f'(x; w)$, that are defined as:

$$f'(x; w) = \lim_{\lambda \downarrow 0} \frac{f(x + \lambda w) - f(x)}{\lambda}.$$

When these limits exist and do not vary in all directions, then f is *differentiable*, i.e., there exists a *gradient*, ∇f, such that

$$\nabla f^T w = f'(x; w)$$

for all directions $w \in \Re^n$. We sometimes distinguish this standard form of differentiability from stricter forms as *Gâteaux* or *G-differentiability*. The stricter forms impose more conditions on the directional derivative such as uniform convergence over compact sets (*Hadamard* derivatives).

We also consider *Lipschitz* continuous or *Lipschitzian* functions such that $|f(x) - f(w)| \leq M\|x - w\|$ for any x and w and some $M < \infty$. If this property holds for all x and w in a set X, then f is Lipschitzian *relative to X*. When this property only holds *locally*, i.e., for $\|w - x\| \leq \epsilon$ for some $\epsilon > 0$, then f is *locally Lipschitz* at x.

Among differentiable functions, we often use *quadratic* functions that have a *Hessian* matrix of second derivatives, D, and can be written as

$$f(x) = c^T x + \frac{1}{2} x^T D x.$$

Many functions are not, however, differentiable. In this case, we express optimality in terms of *subgradients* at a point x, or vectors, η, such that

$$f(w) \geq f(x) + \eta^T (w - x)$$

for all w. In this case, $\{(x, \beta)|\beta = f(x) + \eta^T(w - x)\}$ is a *supporting hyperplane* of f at x. The set of subgradients at a point x is the *subdifferential* of f at x, written $\partial f(x)$.

Other useful properties include that f is *piecewise linear*, i.e., such that $f(x)$ is linear over regions defined by linear inequalities. When f is *separable* so that $f(x) = \sum_{i=1}^{n} f_i(x_i)$, then other advantages are possible in computation.

Given f convex and a convex feasible region in (9.3), we can define conditions that an optimal solution x^* and associated multipliers (π^*, ρ^*) must satisfy. In general, these conditions require some form of *regularity* condition. A common form is that there exists some $\hat{x}$ such that $g(\hat{x}) < 0$ and h is affine. This is generally called the *Slater condition*.

Given a regularity condition of this type, if the constraints in (9.3) define a feasible region, then x^* is optimal if and only if the *Karush-Kuhn-Tucker* conditions hold so that $x^* \in X$ and there exists $\pi^* \geq 0, \rho^*$ such that

$$\nabla f(x^*) + (\pi^*)^T \nabla g(x^*) + (\rho^*)^T \nabla h(x^*) = 0, \nabla g(x^*)^T \pi^* = 0. \quad (9.4)$$

Optimality can also be expressed in terms of the *Lagrangian*:

$$\mathcal{L}(x, \pi, \rho) = f(x) + \pi^T g(x) + \rho^T h(x),$$

so that sequentially minimizing over x and maximizing over π (in both orders) produces the result in (9.4). This occurs through a *Lagrangian dual problem* to (9.3) as

$$\max_{\pi \geq 0, \rho} \inf_x f(x) + \pi^T g(x) + \rho^T h(x), \quad (9.5)$$

which is always a lower bound on the objective in (9.3) (*weak duality*), and, under the regularity conditions, yields equal optimal values in (9.3) and (9.4) (*strong duality*). In many cases, the Lagrangian can also be interpreted with the *conjugate* function of f, defined as

$$f^*(\pi) = \sup_x \{\pi^T x - f(x)\},$$

which is also a convex function if f is convex.

Our algorithms often apply to the Lagrangian to obtain *convergence*, i.e., a sequence of solutions, $x^\nu \to x^*$. In some cases, we also approximate the function so that $f^\nu \to f$ in some way. If this convergence is *pointwise*, then $f^\nu(x) \to f(x)$ for each x individually. If the convergence is *uniform* on a set X, then, for any $\epsilon > 0$, there exists $N(\epsilon)$ such that for all $\nu \geq N(\epsilon)$ and all $x \in X$, $|f^\nu(x) - f(x)| < \epsilon$.

Part II

Basic Properties

3
Basic Properties and Theory

This chapter considers the basic properties and theory of stochastic programming. As throughout this book, the emphasis is on results that have direct application in the solution of stochastic programs. Proofs are included for those results we consider most central to the overall development.

The main properties we consider are formulations of deterministic equivalent programs to a stochastic program, the forms of the feasible region and objective function, and conditions for optimality and solution stability. Our focus is on stochastic programs with recourse, and, in particular, for stochastic linear programs. The first section describes two-stage versions of these problems in detail. It assumes some knowledge of convex sets and functions.

Sections 2 to 5 add extensions to the results in Section 1 by allowing additional forms of constraints, objectives, and decision variables. Section 2 considers problems with probabilistic or chance constraints that occur with some fixed probability. Section 3 considers problems with integer variables, while Section 4 extends results to include nonlinear functions. Section 5 concludes the chapter with multiple-stage problems.

3.1   Two-Stage Stochastic Linear Programs with Fixed Recourse

a. Formulation

As in Chapter 2, we first form the basic two-stage stochastic linear program with fixed recourse. It is repeated here for clarity.

$$\min z = c^T x + E_{\boldsymbol{\xi}}[\min q(\omega)^T y(\omega)] \tag{1.1}$$
$$\text{s.t. } Ax = b,$$
$$T(\omega)x + Wy(\omega) = h(\omega),$$
$$x \geq 0, \ y(\omega) \geq 0,$$

where c is a known vector in $\Re^{n_1}$, b a known vector in $\Re^{m_1}$, A and W are known matrices of size $m_1 \times n_1$ and $m_2 \times n_2$, respectively, and W is called the *recourse matrix*, which we assume here is fixed. This allows us to characterize the feasibility region in a convenient manner for computation. If W is not fixed, we may have difficulties, as shown next.

For each $\omega, T(\omega)$ is $m_2 \times n_1, q(\omega) \in \Re^{n_2}$ and $h(\omega) \in \Re^{m_2}$. Piecing together the stochastic components of the problem, we obtain a vector $\xi^T(\omega) = (q(\omega)^T, h(\omega)^T, T_1.(\omega), \ldots, T_{m_2}.(\omega))$ with $N = n_2 + m_2 + (m_2 \times n_1)$ components, where $T_i.(\omega)$ is the ith row of the *technology matrix* $T(\omega)$. As before, $E_{\boldsymbol{\xi}}$ represents the mathematical expectation with respect to $\boldsymbol{\xi}$. Let also $\Xi \subseteq \Re^N$ be the support of $\boldsymbol{\xi}$, i.e., the smallest closed subset in $\Re^N$ such that $P\{\boldsymbol{\xi} \in \Xi\} = 1$. As said in Section 2.4, the constraints are assumed to hold almost surely.

Problem (1.1) is equivalent to the so-called deterministic equivalent program (D.E.P.):

$$\min z = c^T x + \mathcal{Q}(x) \tag{1.2}$$
$$\text{s.t. } Ax = b,$$
$$x \geq 0,$$

where

$$\mathcal{Q}(x) = E_{\boldsymbol{\xi}} Q(x, \xi(\omega)) \tag{1.3}$$

and

$$Q(x, \xi(\omega)) = \min_{y}\{q(\omega)^T y | Wy = h(\omega) - T(\omega)x, y \geq 0\}. \tag{1.4}$$

Examples of formulations (1.1) and (1.2-4) have been given in Chapter 1. In the farmer's problem, x represents the surfaces devoted to each crop, $\boldsymbol{\xi}$ represents the yields so that only the technology matrix $T(\omega)$ is stochastic (because prices q and requirements h are fixed), and y represents the sales and purchases of the various crops. Formulations (1.1) and (1.2)–(1.4) apply for both discrete and continuous random variables. Examples with continuous random yields have also been given for the farmer's problem.

This representation clearly illustrates the sequence of events in the recourse problem. First-stage decisions x are taken in the presence of uncertainty about future realizations of $\boldsymbol{\xi}$. In the second stage, the actual value of $\boldsymbol{\xi}$ becomes known and some corrective actions or recourse decisions y can be taken. First-stage decisions are, however, chosen by taking their future effects into account. These future effects are measured by the value function or recourse function, $\mathcal{Q}(x)$, which computes the expected value of taking decision x.

When T is nonstochastic, the original formulation (1.2)–(1.4) can be replaced by

$$\min z = c^T x + \Psi(\chi) \tag{1.5}$$
$$\text{s.t. } Ax = b,$$
$$Tx - \chi = 0,$$
$$x \geq 0,$$

where $\Psi(\chi) = E_{\boldsymbol{\xi}}\psi(\chi, \xi(\omega))$ and $\psi(\chi, \xi(\omega)) = \min\{q(\omega)^T y \,|\, Wy = h(\omega) - \chi, y \geq 0\}$. This formulation stresses the fact that choosing x corresponds to generating an m_2-dimensional *tender* $\chi = Tx$ to be bid against the outcomes $h(\omega)$ of the random events.

The difficulty inherent in stochastic programming clearly lies in the computational burden of computing $\mathcal{Q}(x)$ for all x in (1.2)–(1.4), or $\Psi(\chi)$ for all χ in (1.5). It is no surprise therefore that the properties of the deterministic equivalent program in general and of the functions $\mathcal{Q}(x)$ or $\Psi(\chi)$ have been extensively studied. The next sections present some of the known properties. Section 5.1 presents a general algorithm for solving (1.2)–(1.4) when $\boldsymbol{\xi}$ is a discrete random variable.

b. Feasibility sets

Although in the rest of the book we restrict ourselves to the fixed recourse case defined in the previous section, here we study the situation where the recourse matrix W can be random. This is because the main issues about definitions of second-stage feasibility sets depend on whether W is fixed.

For fixed x and ξ, the value $Q(x, \boldsymbol{\xi})$ of the second-stage program is given by

$$Q(x, \boldsymbol{\xi}) = \min_y \{q(\omega)^T y \,|\, W(\omega)y = h(\omega) - T(\omega)x, y \geq 0\}. \tag{1.6}$$

When the mathematical program (1.6) is unbounded below or infeasible, the value of the second-stage program is defined to be $-\infty$ or $+\infty$, respectively.

The expected second-stage value function is, as given in (1.3),

$$\mathcal{Q}(x) = E_{\boldsymbol{\xi}} Q(x, \boldsymbol{\xi}).$$

Let us first consider the situation when $\boldsymbol{\xi}$ is a finite discrete random variable, namely, $\boldsymbol{\xi} \in \Xi$ with Ξ a finite or countable set.

The second-stage value function is then the weighted sum of the $Q(x, \boldsymbol{\xi})$ values for the various possible realizations of $\boldsymbol{\xi}$. To make the definition complete, we make the additional convention $+\infty + (-\infty) = +\infty$. This corresponds to a conservative attitude, rejecting any first-stage decision that could lead to an undefined recourse action even if there is some realization of the random vector inducing an infinitely low-cost function. Let $K_1 = \{x | Ax = b, x \geq 0\}$ be the set determined by the fixed constraints, namely, those that do not depend on the particular realization of the random vector, and let $K_2 = \{x | \mathcal{Q}(x) < \infty\}$ be the second-stage feasibility set. We may now redefine the deterministic equivalent program as follows

$$\min \ z(x) = c^T x + \mathcal{Q}(x)$$
$$\text{s.t. } x \in K_1 \cap K_2.$$

From a practical point of view, it is not absolutely necessary to have a complete description of the region of finiteness of $\mathcal{Q}(x)$. On the other hand, it is desirable to be able to check if a particular first-stage decision x leads to a finite second-stage value without having to compute that value. The definition of K_2 is not useful in that respect. We therefore consider an alternative definition. Let

$$K_2(\xi) = \{x | Q(x, \xi) < +\infty\}$$

be the elementary feasibility sets and

$$K_2^P = \{x | \text{ for all } \xi \in \Xi,$$
$$y \geq 0 \text{ exists s.t. } Wy = h - T \cdot x\}$$
$$= \cap_{\xi \in \Xi} K_2(\xi).$$

The set K_2^P is said to define the *possibility interpretation* of second-stage feasibility sets. A decision x belongs to the set K_2^P if, for all "possible" values of the random vector $\boldsymbol{\xi}$, a feasible second-stage decision y can be taken. Note that the decision y can be different from one value $\boldsymbol{\xi}$ to another, although this is not stressed in the notation.

Theorem 1.
a. For each ξ, the elementary feasibility set is a closed convex polyhedron, hence the set K_2^P is closed and convex.
b. When Ξ is finite, then K_2^P is also polyhedral and coincides with K_2.

Proof. For each ξ, $K_2(\xi)$ is defined by a set of linear constraints, which suffices to prove a. A particular x belongs to K_2 if $\mathcal{Q}(x)$ is bounded above. Because $\mathcal{Q}(x)$ is a positively weighted sum of finitely many $Q(x, \xi)$ values and, due to our convention, $+\infty + (-\infty)$, $\mathcal{Q}(x)$ is bounded above only if each

$Q(x, \xi)$ is bounded above, which implies x belongs to $K_2(\xi)$ for all ξ, which in turn implies x belongs to K_2^P. Similarly, if x belongs to K_2^P, $Q(x, \xi)$ is bounded above for all values, which implies $Q(x)$ is bounded above and x belongs to K_2. $\square$

The case where $\boldsymbol{\xi}$ is a continuous random variable may lead to some difficulties. To illustrate this, consider an example where the second stage is defined by

$$Q(x, \xi) = \min_y \{y | \xi y = 1 - x, y \geq 0\}$$

where $\boldsymbol{\xi}$ has a triangular distribution on $[0, 1]$, namely, $P(\boldsymbol{\xi} \leq u) = u^2$. Note that here W reduces to a 1×1 matrix and is the only random element. For all ξ in $(0, 1]$, the optimal y is $\frac{1-x}{\xi}$, so that

$$K_2(\xi) = \{x | x \leq 1\}$$

and

$$Q(x, \xi) = \frac{1 - x}{\xi}, \text{ for } x \leq 1.$$

When $\xi = 0$, no y exists such that $0 \cdot y = 1 - x$, unless $x = 1$, so that

$$K_2(0) = \{x | x = 1\}.$$

Now, for $x \neq 1$, $Q(x, 0)$ should normally be $+\infty$. However, because the probability that $\boldsymbol{\xi} = 0$ is zero, the convention is to take $Q(x, 0) = 0$. This corresponds to defining $0 \cdot \infty = 0$. A more detailed justification can be found in Walkup and Wets [1967].

Hence,

$$K_2^P = \{x | x = 1\} \cap \{x | x \leq 1\} = \{x | x = 1\}$$

while

$$Q(x) = \int_0^1 \frac{1 - x}{\xi} \cdot 2\xi d\xi = 2(1 - x) \quad \text{for all } x \leq 1,$$

so that $K_2 = \{x | x \leq 1\}$ and K_2^P is strictly contained in K_2. The difference between the two sets relates to the fact that a point is not in K_2^P as soon as it is infeasible for some ξ value, regardless of the distribution of $\boldsymbol{\xi}$, while K_2 does not consider infeasibilities occurring with zero probability.

Fortunately, this kind of difficulty rarely occurs for programs with a fixed W matrix. It never occurs when the random vector satisfies some conditions.

Another difficulty that could arise and would cause the sets K_2^P and K_2 to be different, would be to have $Q(x, \xi)$ bounded above with probability one and yet to have $Q(x)$, the expectation of $Q(x, \xi)$, unbounded.

Proposition 2. *If $\boldsymbol{\xi}$ has finite second moments, then*

$$P(\omega | Q(x, \boldsymbol{\xi}) < \infty) = 1 \quad \text{implies } Q(x) < \infty.$$

To illustrate why this might be true, consider particular x and ξ values. The second-stage program is the linear program

$$Q(x,\xi) = \min\{q(\omega)^T y | Wy = h(\omega) - T(\omega)x, y \geq 0\}.$$

Solving this linear program for given x and ξ amounts to finding some square submatrix B of W, called a *basis* (see Section 2.9), such that $y_B = B^{-1}(h(\omega) - T(\omega)x), y_N = 0$, and $q_B(\omega)^T B^{-1}W \leq q(\omega)^T$, where y_B is the subvector associated with the columns of B and y_N includes the remaining components of y.

It follows that

$$Q(x,\xi) = q_B(\omega)^T B^{-1}(h(\omega) - T(\omega)x).$$

Now, assume $Q(x,\xi)$ is bounded above with probability one and imagine for a while that the same basis B would be optimal for all x and all ξ. Then, ξ having finite second moments is a sufficient condition for $\mathcal{Q}(x)$ to be bounded because it implies $E_\xi(q_B^T B^{-1}h)$ and $E_\xi(q_B^T B^{-1}T \cdot x)$ are both bounded above. In general the optimal basis B is different for different x and ξ values so that a more general proof taking care of different submatrices of W is needed. This is done in detail in Walkup and Wets [1967].

In the next two theorems, we use the definition of pos $W = \{t | Wy = t, y \geq 0\}$ as the *positive hull* of W. We observe that pos W is a closed set.

Theorem 3. *For a stochastic program with fixed recourse where ξ has finite second moments, the sets K_2 and K_2^P coincide.*

Proof: (Note: This proof uses some concepts from measure theory.) First consider $x \in K_2^P$. This implies $Q(x,\xi) < \infty$ with probability one, so that, by the proposition, $\mathcal{Q}(x)$ is bounded above and $x \in K_2$.

Now, consider $x \in K_2$. It follows that $\{\xi | Q(x,\xi) < \infty\}$ is a set of measure one. Observe that $Q(x,\xi) < \infty$ is equivalent to $h(\omega) - T(\omega)x \in$ pos W and that $h(\omega) - T(\omega)x$ is a linear function of ξ, and $\{\xi \in \sum | Q(x,\xi) < \infty\}$ is a closed subset of $\sum$ of measure one, for any set $\sum$ of measure one. In particular, $\{\xi \in \Xi | Q(x,\xi) < \infty\}$ is a closed subset of Ξ having measure one. By definition of Ξ, this set can only be Ξ itself, so that $\{\xi | Q(x,\xi) < \infty\} \subseteq \Xi$ and therefore $x \in K_2^P$. $\square$

Note however that W being fixed and ξ having finite moments are just sufficient conditions for K_2 and K_2^P to coincide. Other, more general, sufficient conditions can be found in Walkup and Wets [1967].

Note also that a third definition of the second-stage feasibility set could be given as $\{x | Q(x,\xi) < \infty$ with probability one$\}$. For problems with fixed recourse where ξ has finite second moments, this set also coincides with K_2 and K_2^P. In the sequel, we simply speak of K_2, the second-stage feasibility set.

Theorem 4. *When W is fixed and ξ has finite second moments:*

(a) K_2 is closed and convex.

(b) If T is fixed, K_2 is polyhedral.

(c) Let Ξ_T be the support of the distribution of **T**. If $h(\boldsymbol{\xi})$ and $T(\boldsymbol{\xi})$ are independent and Ξ_T is polyhedral, then K_2 is polyhedral.

Proof: The proof of (a) is elementary under the possibility representation of K_2. If T is fixed, $x \in K_2$ if and only if $h(\xi) - Tx \in \text{pos } W$ for all $\xi \in \Xi_h$, where Ξ_h is the support of the distribution of $h(\xi)$.

Consider some x and ξ s.t. $h(\xi) - Tx \notin \text{pos } W$. Then there must exist some hyperplane, say $\{x|\sigma^T x = 0\}$ that separates $h(\xi) - Tx$ from pos W. This hyperplane must satisfy $\sigma^T t \leq 0$ for $t \in \text{pos } W$ and $\sigma^T(h(\xi) - Tx) > 0$. Because W is fixed, there can only be finitely many different such hyperplanes, so that $h(\xi) - Tx \in \text{pos } W$ is equivalent to $W^*(h(\xi) - Tx) \leq 0$ for some matrix W^*. This matrix, called the *polar matrix* of W, is obtained by choosing some minimal set of separating hyperplanes. The set is minimal if removing any hyperplane would no longer guarantee the equivalence between $h(\xi) - Tx \in \text{pos } W$ and $W^*(h(\xi) - Tx) \leq 0$ for all x and ξ in Ξ_h. It follows that $x \in K_2$ if and only if $W^*(h(\xi) - Tx) \leq 0$ for all ξ in Ξ. This can still be an infinite system of linear inequalities due to $h(\xi)$. We may, however, replace this system by

$$(W^*T)_{i.}x \geq u_i^* = \sup_{h(\xi) \in \Xi_h} W_{i.}^* h(\xi), \quad i = 1, \ldots, l, \qquad (1.7)$$

where $W_{i.}^*$ is the ith row of W^* and l is the finite number of rows of W^*. If for some i, u_i^* is unbounded, then the problem is infeasible and the result in (b) is trivially satisfied. If, for all i, $u_i^* < \infty$, then the system (1.7) constitutes a finite system of linear inequalities defining the polyhedron $K_2 = \{x|W^*Tx \geq u^*\}$ where u^* is the vector whose ith component is u_i^*. This proves (b). When T is stochastic, a relation similar to (1.7) holds, which, unless Ξ_T is finite, defines an infinite system of inequalities. Whenever Ξ_T is polyhedral, (c) can be proved by working on the extremal elements of Ξ_T. This is done in Wets [1974, Corollary 4.13]. $\square$

c. Second-stage value function

We first start by properties of $Q(x, \xi)$, assuming it is not $-\infty$.

Theorem 5. *For a stochastic program with fixed recourse, $Q(x, \xi)$ is*

(a) a piecewise linear convex function in (h, T);

(b) a piecewise linear concave function in q;

(c) a piecewise linear convex function in x for all x in $K = K_1 \cap K_2$.

Proof. To prove convexity in (a) and (c), we just need to prove that $f(b) = \min\{q^T y | W y = b\}$ is a convex function in b. We consider two different vectors, say b_1 and b_2, and some convex combination $b_\lambda = \lambda b_1 + (1 - \lambda)b_2$, $\lambda \in (0, 1)$.

Let y_1^* and y_2^* be some optimal solution of $\min\{q^T y | W y = b\}$ for $b = b_1$ and $b = b_2$, respectively. Then, $\lambda y_1^* + (1 - \lambda)y_2^*$ is a feasible solution of $\min\{q^T y | W y = b_\lambda\}$. Now, let y_λ^* be an optimal solution of this last problem. We thus have

$$f(b_\lambda) = q^T y_\lambda^* \leq q^T (\lambda y_1^* + (1 - \lambda)y_2^*)$$
$$= \lambda q^T y_1^* + (1 - \lambda)q^T y_2^* = \lambda f(b_1) + (1 - \lambda)f(b_2),$$

which proves the required proposition. A similar proof can be given to show concavity in q. Piecewise linearity follows from the existence of finitely many different optimal bases for the second-stage program. A detailed proof is given in Walkup and Wets [1969]. This fact will be illustrated and used in the *L*-shaped algorithm of Section 5.1. $\square$

Another property is evident from parametric solutions of linear programs when q and T are fixed. Notice that

$$Q(x, q, \lambda(h') + Tx, T) = \lambda Q(x, q, h' + Tx, T) \qquad (1.8)$$

for any $\lambda \geq 0$ because a dual optimal solution for $h = h' + Tx$ is also dual feasible for $h = \lambda(h') + Tx$ and complementary with y^* optimal for $h = h' + Tx$. Because λy^* is also feasible for $h = \lambda(h') + Tx$, λy^* is optimal for $h = \lambda(h') + Tx$, demonstrating (1.8). This says that $Q(x, q, h' + Tx, T)$ is a *positively homogeneous* function of h'. From the convexity of $Q(x, q, h' + Tx, T)$ in $h = h' + Tx$, this function is also sublinear (see Theorem 4.7 of Rockafellar [1969]) in h'. This property is central to some bounding procedures described in Chapter 9.

Complete descriptions of $Q(x, \xi)$ are also often useful. Finding the distribution induced on $Q(x, \xi)$ is often the goal of these descriptions. This information can then be used to find $\mathcal{Q}$ or to address other risk criteria that may not be given by the expectation functional (e.g., the probability of losing some percentage of one's wealth). The description of the distribution of $Q(x, \xi)$ is called the *distribution problem*. Its solution is quite difficult although some methods exist (see Wets [1980b] and Bereanu [1980]). Approximations are generally required as in Dempster and Papagaki-Papoulias [1980]; because these results are not central to our solution development, we will not go into further detail. We concentrate on properties of the expectation functional $\mathcal{Q}(x)$ as follows.

Theorem 6. *For a stochastic program with fixed recourse where ξ has finite second moments,*

(a) $\mathcal{Q}(x)$ is a Lipschitzian convex function and is finite on K_2.

(b) When $\boldsymbol{\xi}$ is finite, $Q(x)$ is piecewise linear.

(c) If $F(\boldsymbol{\xi})$ is an absolutely continuous distribution, $Q(x)$ is differentiable on K_2.

Proof. Convexity and finiteness in (a) are immediate. Part (b) follows directly. A proof of the Lipschitz condition can be found in Wets [1972] or Kall [1976], who also give conditions for $Q(x)$ to be differentiable. $\square$

Although many of the proofs of these results become intricate in general, the outcomes are relatively easy to apply.

When the random variables are appropriately described by a finite distribution, the constraint set K_2 is best defined by the possibility interpretation and is easily seen to be polyhedral. The second-stage recourse function $Q(x)$ is piecewise linear and convex on K_2. The decomposition techniques of Chapter 5 then apply. This is a category of programs for which computational methods can be made efficient, as we shall see.

When the random variables cannot be described by a finite distribution, they can usually be associated with some probability density. Many common probability densities are absolutely continuous and have finite second moment, so the constraints set definitions K_2 and K_2^P coincide and the second-stage value function $Q(x)$ is differentiable and convex. Classical nonlinear programming techniques could then be applied. A typical example was given in the farmer's problem in Chapter 1. There, a convex differentiable function $Q(x)$ was constructed analytically. It is easily understood that analytical expressions can reasonably be found only for small second-stage problems or problems with a very specific structure such as separability.

In general, one can only compute $Q(x)$ by numerical integration of $Q(x, \xi)$, for a given value of x. Most nonlinear techniques would also require the gradients of $Q(x)$, which in turn require numerical integration. An introduction to numerical integration appears in Chapter 9. From there, we come to the conclusion that numerical integration, as of today, produces an effective computational method only when the random vector is of small dimensionality. As a consequence, the practical solution of stochastic programs having continuous random variables is, in general, a difficult problem. One line of approach is to approximate the random variable by a discrete one and let the discretization be finer and finer, hoping that the solutions of the successive problems with discrete random variables will converge to the optimal solution of the problem with a continuous random variable. This is also discussed in Chapter 9. It is sufficient at this point to observe that approximation is a second reason for constructing efficient methods for stochastic programs with finite random variables.

d. Special cases: relatively complete, complete, and simple recourse

The previous sections presented properties for general problems. In particular instances, the feasible regions and objective values have special properties that are particularly useful in computation. One advantage can be obtained if every solution x that satisfies the first-period constraints, $Ax = b$, also has a feasible completion in the second stage. In other words, $K_1 \subset K_2$. In this case, we say that the stochastic program has *relatively complete recourse*. If, for the example with stochastic W in Section 1.b, we had the first-period constraints $x \leq 1$, then this problem would have relatively complete recourse.

Although relatively complete recourse is very useful in practice and in many of the theoretical results that follow, it may be difficult to identify because it requires some knowledge of the sets K_1 and K_2. A special type of relatively complete recourse may, however, often be identified from the structure of the W. This form, called *complete recourse*, holds when there exists $y \geq 0$ such that $Wy = t$ for all $t \in \Re^{m_2}$.

Complete recourse is also represented by pos $W = \Re^{m_2}$ (the positive cone spanned by the columns of W includes $\Re^{m_2}$), and says that W contains a positive linear basis of $\Re^{m_2}$. Complete recourse is often added to a model to ensure that no outcome can produce infeasible results. With most practical problems, this should be the case. In some instances, complete recourse may not be apparent. An algorithm in Wets and Witzgall [1967] can be used in this situation to determine whether W contains a positive linear basis.

A special type of complete recourse offers additional computational advantages to stochastic programming solutions. This case is the generalization of the news vendor problem introduced in Section 1.1. It is called *simple recourse*. For a simple recourse problem, $W = [I, -I]$, $\mathbf{y}$ is divided correspondingly as $(\mathbf{y}^+, \mathbf{y}^-)$, and $\mathbf{q} = (\mathbf{q}^+, \mathbf{q}^-)$. Note that, in this case, the optimal values of $y_i^+(\omega), y_i^-(\omega)$ are determined purely by the sign of $h_i(\omega) - T_i.(\omega)x$ provided that $\mathbf{q}_i^+ + \mathbf{q}_i^- \geq 0$ with probability one. This finiteness result is in the following theorem.

Theorem 7. *Suppose the two-stage stochastic program in (1.1) is feasible and has simple recourse and that $\boldsymbol{\xi}$ has finite second moments. Then $Q(x)$ is finite if and only if $\mathbf{q}_i^+ + \mathbf{q}_i^- \geq 0$ with probability one.*

Proof: If $q_i^+(\omega) + q_i^-(\omega) < 0$ for $\omega \in \Omega_1$ where $P(\Omega_1) > 0$, then, for any feasible x in (1.1), for all $\omega \in \Omega_1$ where $h_i(\omega) - T_i.(\omega)x > 0$, let $y_i^+(\omega) = h_i(\omega) - T_i.(\omega)x + u$, $y_i^-(\omega) = u$. By letting $u \to \infty$, $Q(x, \omega) \to -\infty$. A similar argument applies if $h_i(\omega) - T_i.(\omega)x \leq 0$, so $Q(x)$ is not finite.

If $\mathbf{q}_i^+ + \mathbf{q}_i^- \geq 0$ with probability one, then $Q(x, \omega) = \sum_{i=1}^{m_2}(q_i^+(\omega)(h_i(\omega) - T_i.(\omega)x)^+ + q_i^-(\omega)(-h_i(\omega) + T_i.(\omega)x)^+)$, which is finite for all ω. Using Proposition 2, we obtain the result. $\square$

We, therefore, assume that $\mathbf{q}_i^+ + \mathbf{q}_i^- \geq 0$ with probability one and can write $\mathcal{Q}(x)$ as $\sum_{i=1}^{m_2} \mathcal{Q}_i(x)$, where $\mathcal{Q}_i(x) = E_\omega[Q_i(x, \xi(\omega))]$, and $Q_i(x, \xi(\omega)) = q_i^+(\omega)(h_i(\omega) - T_i.(\omega)x)^+ + q_i^-(\omega)(-h(\omega) + T_i.(\omega)x))^+$. When q and T are fixed, this characterization of Q allows its expression as a separable function in the remaining random components $\mathbf{h}_i$. Often, in this case, $T_i.x$ is substituted with χ_i and Ψ is substituted for $\mathcal{Q}$ so that $\mathcal{Q}(x) = \Psi(\chi)$. We then obtain $\Psi(\chi) = \sum_{i=1}^{m_2} \Psi_i(\chi_i)$ where $\Psi_i(\chi) = E_{\mathbf{h}_i}[\psi_i(\chi_i, \mathbf{h}_i)]$ and $\psi_i(\chi_i, h_i) = q_i^+(h_i - \chi_i)^+ + q_i^-(-h_i + \chi_i)^+$. We, however, continue to use $\mathcal{Q}(x)$ to maintain consistency with our previous results.

We can define the objective function even further. In this case, let $\mathbf{h}_i$ have an associated distribution function F_i, mean value $\bar{h}_i$, and let $q_i = q_i^+ + q_i^-$. We can then write $\mathcal{Q}_i(x)$ as

$$\mathcal{Q}_i(x) = q_i^+\bar{h}_i - (q_i^+ - q_i F_i(T_i.x))T_i.x - q_i \int_{h_i \leq T_i.x} h_i dF_i(h_i). \quad (1.9)$$

Of particular importance in optimization is the subdifferential of this function, which has the following simple form:

$$\partial \mathcal{Q}_i(x) = \{\pi(T_i.)^T | -q_i^+ + q_i F_i(T_i.x) \leq \pi \leq -q_i^+ + q_i F_i^+(T_i.x)\}, \quad (1.10)$$

where $F_i^+(h) = \lim_{t \downarrow h} F_i(t)$. These results can be used to obtain specific optimality conditions. These general conditions are the subject of the next part of this section.

e. Optimality conditions

In this subsection, we consider optimality conditions for stochastic programs. Our goal in describing these conditions is to show the special conditions that can apply to stochastic programs and to show how stochastic programs may differ from other mathematical programs. In particular, we give the additional assumptions that guarantee necessary and sufficient conditions for two-stage stochastic linear programs. The following sections contain generalizations.

The deterministic equivalent problem in (1.2) provides the framework for optimality conditions, but several questions arise.

1. When is a solution to (1.2) attainable?

2. What form do the optimality conditions take and how can they be simplified?

3. How stable is an optimal solution to (1.2) to changes in the parameters and distributions?

4. What types of dual problems can be formulated to accompany (1.2) and do they obtain bounds on optimal values?

This subsection briefly describes answers to these questions. Further details are contained in Kall [1976], Wets [1974, 1990], and Dempster [1980]. Our aim is to give only the basic results that may be useful in formulating, solving, and analyzing practical stochastic programs.

From the previous section, supposing that ξ has finite second moments, we know that Q is Lipschitzian. We can then apply a direct subgradient result. A question is, however, whether the solution of (1.2) can indeed be obtained, i.e., whether the optimal objective value is finite and attained by some value of x.

To see that this question is indeed relevant, consider the following example. Find

$$\inf\{E_\xi[y^+(\xi)]|y^+(\xi), y^-(\xi) \geq 0, x + y^+(\xi) - y^-(\xi) = \xi, \text{ a.s.}\}, \quad (1.11)$$

where ξ is, for example, negative exponentially distributed on $[0, \infty)$. For any finite value of x, (1.11) has a positive value, but the infimum is zero.

The following theorem gives some sufficient conditions to guarantee that a solution to (1.2) exists. In the following, we use rc to denote the *recession cone*, $\{v|u + \lambda v \in S, \text{ for all } \lambda \geq 0 \text{ and } u \in S\}$ when applied to a set, S, and the recession value, $\sup_{x \in \text{dom} f}(f(x + v) - f(x))$ when applied to a proper convex function, f.

Theorem 8. *Suppose that the random elements ξ have finite second moments and one of the following:*

(a) *the feasible region K is bounded; or*

(b) *the recourse function Q is eventually linear in all recession directions of K, i.e., $Q(x + \lambda v) = Q(x + \bar{\lambda}v) + (\lambda - \bar{\lambda})rcQ(v)$ for some $\bar{\lambda} \geq 0$ (dependent on x), all $\lambda \geq \bar{\lambda}$, and some constant recession value, $rcQ(v)$, for all v such that $x + \lambda v \in K$ for all $x \in K$ and $\lambda \geq 0$.*

Then, if problem (1.2) has a finite optimal value, it is attained for some $x \in \Re^n$.

Proof: The proof given (a) follows immediately by noting that the objective is convex and finite on K, which is by the assumption compact. The only possibility for not attaining an optimum is, therefore, when the optimal value is only attained asymptotically. By (b), along any recession direction v, we must have $rcQ(v) \geq 0$ for a finite value of $Q(x + \lambda v)$. Hence, the optimal value must be attained. $\square$

As shown in Wets [1974], if T is fixed and Ξ is compact, the condition in (b) is obtained. In the exercises, we will show that (b) may not hold if either of these conditions is relaxed.

We now assume that an optimal solution can be attained as we would expect in most practical situations. For optimization, we would like to describe the characteristics of such points. The general deterministic equivalent form gives us the following result in terms of Karush-Kuhn-Tucker conditions.

Theorem 9. *Suppose (1.2) has a finite optimal value. A solution* $x^* \in K_1$, *is optimal in (1.2) if and only if there exists some* $\lambda^* \in \Re^{m_1}$, $\mu^* \in \Re^{n_1}_+$, $\mu^{*T} x^* = 0$, *such that,*

$$-c + A^T \lambda^* + \mu^* \in \partial \mathcal{Q}(x^*). \tag{1.12}$$

Proof. From the optimization of a convex function over a convex region (see, for example, Bazaraa and Shetty [1979, Theorem 3.4.3]), we have that $c^T x + \mathcal{Q}(x)$ has a subgradient η at x^* such that $\eta^T (x - x^*) \geq 0$ for all $x \in K_1$ if and only if x^* minimizes $c^T x + \mathcal{Q}(x)$ over K_1. We can write the set, $\{\eta | \eta^T (x - x^*) \geq 0$ for all $x \in K_1\}$, as $\{\eta | \eta = A^T \lambda + \mu$, for some $\mu \geq 0, \mu^T x^* = 0\}$. Hence, the general optimality condition states that a nonempty intersection of $\{\eta | \eta = A^T \lambda + \mu$, for some $\mu \geq 0, \mu^T x^* = 0\}$ and $\partial(c^T x^* + \mathcal{Q}(x^*)) = c + \partial \mathcal{Q}(x^*)$ is necessary and sufficient for the optimality of x^*. $\square$

This result can be combined with our previous results on simple recourse functions to obtain specific conditions for that problem as in the following.

Corollary 10. *Suppose (1.1) has simple recourse and a finite optimal value. Then* $x^* \in K_1$ *is optimal in (1.2) corresponding to this problem if and only if there exists some* $\lambda^* \in \Re^{m_1}$, $\mu^* \in \Re^{n_1}_+$, $\mu^{*T} x^* = 0$, π_i^* *such that* $-(q_i^+ - q_i F_i(T_i.x^*)) \leq \pi_i^* \leq -(q_i^+ - q_i F_i^+(T_i.x^*))$ *and*

$$-c + A^T \lambda^* + \mu^* - (\pi^*)^T T = 0. \tag{1.13}$$

Proof. This is a direct application of (1.10) and Theorem 9. $\square$

Inclusion (1.12) suggests that a subgradient method or other nondifferentiable optimization procedure may be used to solve (1.2). While this is true, we note that finite realizations of the random vector lead to equivalent linear programs (although of large scale), while absolutely continuous distributions lead to a differentiable recourse function $\mathcal{Q}$.

Obviously if $\mathcal{Q}$ is differentiable, we can replace $\partial \mathcal{Q}(x^*)$ with $\nabla \mathcal{Q}(x^*)$ to obtain:

$$c + \nabla \mathcal{Q}(x^*) = A^T \lambda^* + \mu^* \tag{1.14}$$

in place of (1.12). Possible algorithms based on convex minimization subject to linear constraints are then admissible as in MINOS (Murtagh and Saunders [1983]).

The main practical possibilities for solutions of (1.2) then appear as examples of either large-scale linear programming or smooth nonlinear optimization. The main difficulty is, however, in characterizing $\partial \mathcal{Q}$ because even evaluating this function is difficult. This evaluation is, however, decomposable into subgradients of the recourse for each realization of ξ.

Theorem 11. *If $x \in K$, then*

$$\partial \mathcal{Q}(x) = E_\omega \partial Q(x, \xi(\omega)) + N(K_2, x), \qquad (1.15)$$

where $N(K_2, x) = \{v | v^T y \leq 0, \forall\, y \text{ such that } x + y \in K_2\}$, the normal cone to K_2 at x.

Proof: From the theory of subdifferentials of random convex functions with finite expectations (see, for example, Wets [1990, Proposition 2.11]),

$$\partial \mathcal{Q}(x) = E_\omega \partial Q(x, \xi(\omega)) + rc[\partial \mathcal{Q}(x)], \qquad (1.16)$$

where again rc denotes the recession cone, $\{v | u + \lambda v \in \partial \mathcal{Q}(x), \text{ for all } \lambda \geq 0 \text{ and } u \in \partial \mathcal{Q}(x)\}$. This set is equivalently $\{v | y^T(u + \lambda v) + \mathcal{Q}(x) \leq \mathcal{Q}(x + y) \text{ for all } \lambda \geq 0 \text{ and } y\}$. Hence, $v \in rc[\partial \mathcal{Q}(x)]$ if and only if $y^T v \leq 0$ for all y such that $\mathcal{Q}(x + y) < \infty$. Because $K_2 = \{x | \mathcal{Q}(x) < \infty\}$, the result follows. $\square$

This theorem indeed provides the basis for the results on the differentiability of $\mathcal{Q}$. In the exercises, we illustrate more of the characteristics of optimal solutions. Also note that if the problem has relatively complete recourse, then, for any y such that $x + y \in K_1$, we must also have $x + y \in K_2$. Hence, $N(K_2, x) \subset N(K_1, x) = \{v | v = A^T \lambda + \mu, \mu^T x = 0, \mu \geq 0\}$. This yields the following corollary to Theorems 9 and 11.

Corollary 12. *If (1.2) has relatively complete recourse, a solution x^* is optimal in (1.2) if and only if there exists some $\lambda^* \in \Re^{m_1}$, $\mu^* \in \Re_+^{n_1}$, $\mu^{*T} x^* = 0$, such that*

$$-c + A^T \lambda^* + \mu^* \in E_\omega \partial Q(x, \xi(\omega)). \qquad (1.17)$$

f. Stability and nonanticipativity

Another practical concern is whether the optimal solution set is also stable, i.e., whether it changes continuously in some sense when parameters of the problem change continuously. Although this may be of concern when considering changing problem conditions, we do not develop this theory in detail. The main results are that stability is achieved (i.e., some optimal solution of an original problem is close to some optimal solution of a perturbed problem) if problem (1.2) has complete recourse and the set of recourse problem dual solutions, $\{\pi | \pi^T W \leq q(\omega)^T\}$, is nonempty with probability one. For further details, we refer to Robinson and Wets [1987] and Römisch and Schultz [1991b].

Another approach to optimality conditions is to consider problem (1.2), in which $y(\omega)$ again becomes an explicit part of the problem and the nonanticipativity constraints also become explicit. The advantage in this representation is that we may obtain information on the value of future information. It also leads naturally to algorithms based on relaxing nonanticipativity.

We discuss the main results in this characterization briefly. The following development assumes some knowledge of measure theory and can be skipped by those unfamiliar with these concepts.

In general, for this approach, we wish to have a different x, y pair for every realization of the random outcomes. We then wish to restrict the x decisions to be the same for almost all outcomes. This says that the decision, $(x(\omega), y(\omega))$, is a function (with suitable properties) on Ω. We restrict this to some space, X, of measurable functions on Ω, for example, the p-integrable functions, $\mathcal{L}_p(\Omega, \mathcal{B}, \mu; \Re^n)$, for some $1 \leq p \leq \infty$. (For background on these concepts, see, for example, Royden [1968].) The general version of (1.2) (with certain restrictions) then becomes:

$$\inf_{(x(\omega), y(\omega)) \in X} \int_{\Omega} (c^T x(\omega) + q(\omega)^T y(\omega)) \mu(d\omega)$$

$$\text{s. t. } Ax(\omega) = b, a.s.,$$
$$E_{\Omega}(x(\omega)) - x(\omega) = 0, a.s.,$$
$$T(\omega)x(\omega) + Wy(\omega) = h(\omega), a.s.,$$
$$x(\omega), y(\omega) \geq 0, a.s. \qquad (1.18)$$

Problem (1.18) is equivalent to (1.2) if, for example, X is the space of essentially bounded functions on Ω and K is bounded for (1.2). The two formulations are not necessarily the same, however, as in the problem given in Exercise 10.

The condition that the x decision is taken before realizing the random outcomes is reflected in the second set of constraints in (1.18). These constraints are called *nonanticipativity constraints*. They imply that almost all $x(\omega)$ values are the same.

The only difference in optimality conditions from (1.12) is that we include explicit multipliers for the nonanticipativity constraints. For continuous distributions, these multipliers may, however, have a difficult representation unless (1.18) has relatively complete recourse. The difficulty is that we cannot guarantee boundedness of the multipliers and may not be able to obtain an integrable function to represent them. This difficulty is caused when future constraints restrict the set of feasible solutions at the first stage.

For finite distributions, (1.18) is, however, an implementable problem structure that is used in several algorithms discussed here. In this case, with K possible realizations of ξ with probabilities $p^k, k = 1, \ldots, K$, the problem becomes:

$$\inf_{(x^k,y^k),k=1,\ldots,K} \sum_{k=1}^{K} p^k (c^T x^k + (q^k)^T y^k)$$

$$\text{s. t. } Ax^k = b, k = 1,\ldots,K,$$

$$\left(\sum_{j \neq k} p^j x^j \right) + (p^k - 1)x^k = 0, k = 1,\ldots,K,$$

$$T^k x^k + W y^k = h^k, k = 1,\ldots,K,$$

$$x^k, y^k \geq 0, k = 1,\ldots,K. \qquad (1.19)$$

Notice that (1.19) almost completely decomposes into K separate problems for the K realizations. The only links are in the second set of constraints that impose nonanticipativity. An aim of computation is to take advantage of this structure.

Consider the optimality conditions for (1.19). We wish to illustrate the difficulties that may occur when continuous distributions are allowed. A solution (x^{k*}, y^{k*}), $k = 1,\ldots,K$, is optimal for (1.19) if and only if there exist $(\lambda^{k*}, \rho^{k*}, \pi^{k*})$ such that

$$p^k(c_j - \lambda^{k*T} a_{\cdot j} - \sum_{l \neq k} p^l \rho_j^{l*} - (-1 + p^k)\rho_j^{k*} - \pi^{*T} T_{\cdot j}^k) \geq 0, k = 1,\ldots,K,$$

$$j = 1,\ldots,n_1, \qquad (1.20)$$

$$(c_j - \lambda^{k*T} a_{\cdot j} - \sum_{l \neq k} p^k \rho_j^{k*} p^k - (-1 + p^k)\rho_j^{k*} - \pi^{*T} T_{\cdot j}^k)x_j^{k*} = 0, k = 1,\ldots,K,$$

$$j = 1,\ldots,n_1, \qquad (1.21)$$

$$p^k(q_j^k - \pi^{*T} W_{\cdot j}) \geq 0, k = 1,\ldots,K, j = 1,\ldots,n_2, \qquad (1.22)$$

$$p^k(q_j^k - \pi^{*T} W_{\cdot j})y_j^{k*} = 0, k = 1,\ldots,K, j = 1,\ldots,n_2, \qquad (1.23)$$

where we have effectively multiplied the constraints in (1.19) by p_k to obtain the form in (1.20)–(1.23). We may also add the condition,

$$\sum_{k=1,\ldots,K} p^k \rho^{k*} = 0, \qquad (1.24)$$

without changing the feasibility of (1.20)–(1.23). This is true because, if $\sum_{k=1,\ldots,K} p^k \rho^{k*} = \kappa$ for some $\kappa \neq 0$ is part of a feasible solution to (1.20)–(1.23), then so is $\rho^{k'} = \rho^{k*} - \kappa$. A problem arises if more realizations are included in the formulation (i.e., K increases) and $\rho^{k'}$ becomes unbounded.

For example, consider the following example (see also Rockafellar and Wets [1976a]). We wish to find $\min_x \{x | x \geq 0, x - y = \boldsymbol{\xi}, a.s., \mathbf{y} \geq 0\}$, where $\boldsymbol{\xi}$ is uniformly distributed on k/K for $k = 0,\ldots,K-1$. In this case,

3.1 Two-Stage Stochastic Linear Programs with Fixed Recourse   99

the optimal solution is $x^* = \frac{K-1}{K}$ and $y^{k*} = \frac{K-1-k}{K}$ for $k = 0, \ldots, K$. The multipliers satisfying (1.20)–(1.24) are $\rho^{k*} = 1$, $\pi^{k*} = 0$ for $k = 0, \ldots, K - 2$, and $\rho^{K-1*} = -(K - 1)$ and $\pi^{K-1*} = K + 2$. Note that as K increases, ρ^* approaches a distribution with a singular value at one. The difficulty is that ρ^{K-1*} is unbounded so that bounded convergence cannot apply. If relatively complete recourse is assumed, however, then all elements of ρ^* are bounded (see Exercise 11). No singular values are necessary.

In this example, the continuous distribution would tend toward a singular multiplier for some value of ω (i.e., a multiplier with mass one at a single point). If this is the case, we must have that the solution to the dual of the recourse problem is unbounded, or the recourse problem is infeasible for x^* feasible in the first stage. This possibility is eliminated by imposing the relatively complete recourse assumption.

With relatively complete recourse, we can state the following optimality conditions for a solution $(x^*(\omega), y^*(\omega))$ to (1.19). The theorem appears in other ways in Hiriart-Urruty [1978], Rockafellar and Wets [1976a, 1976b], Birge and Qi [1993], and elsewhere. We only note that regularity conditions (other than relatively complete recourse) follow from the linearity of the constraints.

Theorem 13. *Assuming that (1.18) with* $X = \mathcal{L}_\infty(\Omega, \mathcal{B}, \mu; \Re^{n_1+n_2})$ *is feasible, has a bounded optimal value, and satisfies relatively complete recourse, a solution* $(x^*(\omega), y^*(\omega))$ *is optimal in (1.18) if and only if there exist integrable functions on* Ω, $(\lambda^*(\omega), \rho^*(\omega), \pi^*(\omega))$, *such that*

$$c_j - \lambda^*(\omega)A_{\cdot j} - \rho^*(\omega) - \pi^{*T}(\omega)T_{\cdot j}(\omega) \geq 0, a.s., j = 1, \ldots, n_1, \quad (1.25)$$

$$(c_j - \lambda^*(\omega)A_{\cdot j} - \rho^*(\omega) - \pi^{*T}(\omega)T_{\cdot j}(\omega))x_j^*(\omega) = 0, a.s., j = 1, \ldots, n_1, \quad (1.26)$$

$$q_j(\omega) - \pi^{*T}(\omega)W_{\cdot j} \geq 0, a.s., j = 1, \ldots, n_2, \quad (1.27)$$

$$(q_j(\omega) - \pi^{*T}(\omega)W_{\cdot j})y_j^*(\omega) = 0, a.s., j = 1, \ldots, n_2, \quad (1.28)$$

and

$$E_\omega[\rho^*(\omega)] = 0. \quad (1.29)$$

Proof. We first show the sufficiency of these conditions directly. If (1.25)–(1.29) are satisfied, then for any $(x(\omega), y(\omega))$ (with expected value (x, y)) such that $(x^*(\omega) + x(\omega), y^*(\omega) + y(\omega))$ is feasible in (1.18), then integrating over ω, summing over j in (1.26), and using (1.27), we obtain that $c^T x - E_\omega[\pi^{*T}(\omega)T(\omega)]x \geq 0$. We also have that $q(\omega)^T y(\omega) \geq \pi^{*T}(\omega)Wy(\omega) = -\pi^{*T}(\omega)T(\omega)x$. Hence, $c^T x + E_\omega[q(\omega)^T y(\omega)] \geq 0$, giving the optimality of $(x^*(\omega), y^*(\omega))$.

For necessity, we use the equivalence of (1.18) and (1.2), and Corollary 12. In this case, let λ^* from (1.12) replace $\lambda^*(\omega)$ in (1.25). Let $\pi^*(\omega)$ be the optimal dual value in the recourse problem in (1.4). Thus, $E_\omega[\partial Q(x^*, \xi(\omega))] =$

$E_\omega[-\pi^{*T}(\omega)T(\omega)]$. Now, if we let $\rho^*(\omega) = E_\omega[-\pi^{*T}(\omega)T] - \pi^{*T}(\omega)T(\omega)$, we obtain all the conditions in (1.25)–(1.29). $\square$

The results in this section give conditions that can be useful in algorithms and in checking the optimality of stochastic programming solutions. Dual problems can also be formulated based on these conditions either to obtain bounds on optimal solutions by finding corresponding feasible dual solutions or to give an alternative solution procedure that can be used directly or in some combined primal-dual approach (see, for example, Bazaraa and Shetty [1979]). The dual problem directly obtained from (1.25)–(1.29) is to find $(\lambda(\omega), \rho(\omega), \pi(\omega))$ on the dual space to X to maximize

$$E_\omega[\lambda^T(\omega)b + \pi^T(\omega)h(\omega)] \text{ subject to} \qquad (1.30)$$

$$c_j - \lambda(\omega)A_{.j} - \rho(\omega) - \pi^T(\omega)T_{.j}(\omega) \geq 0, a.s., j = 1, \ldots, n_1, \qquad (1.31)$$

$$q_j(\omega) - \pi^T(\omega)W_{.j} \geq 0, a.s., j = 1, \ldots, n_2, \qquad (1.32)$$

and

$$E_\omega[\rho(\omega)] = 0. \qquad (1.33)$$

This fits the general duality framework used by Klein Haneveld [1985] where further details on the properties of these dual problems may be found. Rockafellar and Wets [1976a, 1976b] also discuss this alternative viewpoint with an analysis based on perturbations of both primal and dual forms. Discussion of alternative dual spaces appears in Eisner and Olsen [1975]. In general, Problem (1.18) attains its minimum with a bounded region, and the supremum in (1.30)–(1.33) gives the same value. Relatively complete recourse, or a similar requirement, is necessary to obtain that the dual optimum is also attained. With unbounded regions or without relatively complete recourse, as we have seen, we may have that an optimal solution is not attained for either (1.19) or (1.30)–(1.33). In this case, it is possible that the corresponding dual problem does not have the same optimal value and the two problems exhibit a duality gap. The exercises explore this possibility further.

Exercises

1. Let a second-stage program be defined as

$$
\min 2y_1 + y_2
$$
$$
\text{s.t. } y_1 + 2y_2 \geq \xi_1 - x_1,
$$
$$
y_1 + y_2 \geq \xi_2 - x_1 - x_2,
$$
$$
0 \leq y_1 \leq 1, 0 \leq y_2 \leq 1.
$$

(a) Find $K_2(\xi)$ for all ξ. (Hint: Use the bounds on y_1 and y_2 to bound the left-hand side.)

(b) Let $\boldsymbol{\xi}_1$ and $\boldsymbol{\xi}_2$ be two independent continuous random variables. Assume they both have uniform density over $[2, 4]$.

   i. What is K_2^P?
   ii. What is K_2?
   iii. Let u_i^* be defined as in (1.7). What are u_1^* and u_2^* in this example?

2. Let the second stage of a stochastic program be

$$
\min 2y_1 + y_2
$$
$$
\text{s.t. } y_1 - y_2 \leq 2 - \boldsymbol{\xi}x_1,
$$
$$
y_2 \leq x_2,
$$
$$
0 \leq y_1, y_2.
$$

Find $K_2(\boldsymbol{\xi})$ and K_2 for:

(a) $\boldsymbol{\xi} \sim U[0, 1]$.

(b) $\boldsymbol{\xi} \sim Poisson(\lambda), \lambda > 0$.

What properties do you expect for K_2?

3. Consider the following second-stage program:

$$
Q(x, \xi) = \min\{y | y \geq \xi, y \geq x\} .
$$

For simplicity, assume $x \geq 0$.
Let $\boldsymbol{\xi}$ have density

$$
f(\xi) = \frac{2}{\xi^3}, \ \xi \geq 1.
$$

Show that $K_2^P \neq K_2$. Compare this with the statement of Theorem 3.

4. Let a second-stage program be defined as

$$\min 2y_1 + y_2$$
$$\text{s. t.} \;\; y_1 + y_2 \geq 1 - x_1,$$
$$y_1 \geq \boldsymbol{\xi} - x_1 - x_2,$$
$$y_1, y_2 \geq 0.$$

(a) Show that this program has complete recourse if $\boldsymbol{\xi}$ has finite expectation.

(b) Assuming $0 \leq x_1 \leq 1, 0 \leq x_2 \leq 1$, show that the following are optimal second-stage solutions:

If $\xi \geq x_1 + x_2 \Rightarrow y_1^* = \xi - x_1 - x_2 : y_2^* = (1 - \xi + x_2)^+$ where $(a)^+ = \max(a; 0)$.

If $\xi \leq x_1 + x_2 \Rightarrow y_1^* = 0; y_2^* = 1 - x_1$.

It follows that

$$Q(x, \xi) = \begin{cases} 1 - x_1 & \text{for } 0 \leq \xi < x_1 + x_2, \\ \xi + 1 - 2x_1 - x_2 & \text{for } x_1 + x_2 \leq \xi \leq 1 + x_2, \\ 2(\xi - x_1 - x_2) & \text{for } 1 + x_2 \leq \xi. \end{cases}$$

Check that $Q(x, \xi)$ has properties (a) and (c) in Theorem 5.

(c) Assume $\boldsymbol{\xi} \sim U[0, 2]$. After a tedious integration that probably only the authors of this book will go through, one obtains $\mathcal{Q}(x) = \frac{1}{4}(x_1^2 + 2x_2^2 + 2x_1 x_2 - 8x_1 - 6x_2 + 9)$. Check that the relevant properties of Theorem 6 are satisfied.

5. Let a second-stage program be defined as

$$\min \xi y_1 + y_2$$
$$\text{s. t.} \;\; y_1 + y_2 \geq 1 - x_1,$$
$$y_1 \geq 1 - x_1 - x_2,$$
$$y_1, y_2 \geq 0.$$

Assume $0 \leq x_1, x_2 \leq 1$. Obtain $Q(x, \xi)$ and observe that it is concave in ξ.

6. Prove the positive homogeneity property in (1.8).

7. Derive the simple recourse results in (1.9) and (1.10).

8. Show that the news vendor problem is a special case of a simple recourse problem.

9. Consider the following example:

$$\min \; -x + E_{(t(\omega),h(\omega))}[y^+(\omega) + y^-(\omega)]$$
$$\text{s. t. } t(\omega)x + y^+(\omega) - y^-(\omega) = h(\omega), \; a.s.,$$
$$x, y^+(\omega), y^-(\omega) \geq 0, \; a.s.,$$

where $\mathbf{h}, \mathbf{t}$ are uniformly distributed on the unit circle, $h^2 + t^2 \leq 1$. Find $\mathcal{Q}(x)$ and show that it is not eventually linear for $x \to \infty$ (Wets [1974]).

10. Suppose you wish to solve (1.11) in the form of (1.18) over $(x(\omega), y(\omega)) \in \mathcal{L}_\infty(\Omega, \mathcal{B}, \mu : \Re^{n_1+n_2})$. What is the optimal value? How does this differ using (1.2)?

11. This exercise uses approximation results to give an alternative proof of Theorem 13. As shown in Chapter 9, if a discrete distribution approaches a continuous distribution (in distribution) and problem (1.2) has a bounded optimal solution and the bounded second moment property, then a limiting optimal solution for the discrete distributions is an optimal solution using the continuous distribution. This also implies that recourse solutions, y^*, converge and that the optimality conditions in (1.25)–(1.29) are obtained as long as the ρ^{k*} in the discrete approximations are uniformly bounded. Show that relatively complete recourse implies uniform boundedness of some ρ^{k*} for any discrete approximation approaching a continuous distribution in (1.18). (Hint: Construct a system of equations that must be violated for some iteration ν of the discretization and for any bound M on the largest value of ρ^{k*} if the ρ^{k*} are not uniformly bounded. Then show that the complementary system implies no relatively complete recourse.)

3.2   Probabilistic or Chance Constraints

As mentioned in Chapter 2, in some models, constraints need not hold *almost surely* as we have assumed to this point. They can instead hold with some probability or reliability level. These *probabilistic*, or *chance*, constraints take the form:

$$P\{A^i(\omega)x \geq h^i(\omega)\} \geq \alpha^i, \qquad (2.1)$$

where $0 < \alpha^i < 1$ and $i = 1, \ldots, I$ is an index of the constraints that must hold jointly. We can, of course, model these constraints in a general expectational form $E_\omega(f^i(\omega, x(\omega)) \geq \alpha^i$ where f^i is an indicator of $\{\omega | A^i(\omega)x \geq h^i(\omega)\}$ but we would then have to deal with a discontinuous function.

In chance-constrained programming (see, e.g., Charnes and Cooper [1963]), the objective is often an expectational functional as we used earlier (the *E-model*), or it may be the variance of some result (the *V-model*) or the probability of some occurrence (such as satisfying the constraints) (the *P-model*). Another variation includes an objective that is a quantile of a random function (see, e.g., Kibzun and Kurbakovskiy [1991] and Kibzun and Kan [1996]).

The main results with probabilistic constraints refer to forms of deterministic equivalents for constraints of the form in (2.1). Provided the deterministic equivalents of these constraints and objectives have the desired convexity properties, these functions can be added to the recourse problems given earlier (or used as objectives). In this way, all our previous results apply to chance-constrained programming with suitable function characteristics.

The main goal in problems with probabilistic constraints is, therefore, to determine deterministic equivalents and their properties. To maintain consistency with the recourse problem results, we let

$$K_1^i(\alpha^i) = \{x | P(A^i(\omega)x \geq h^i(\omega)) \geq \alpha^i\}, \qquad (2.2)$$

where $0 < \alpha^i \leq 1$ and $\cap_i K_1^i(1) = K_1$ as in Section 1. Unfortunately, $K_1^i(\alpha^i)$ need not be convex or even connected. Suppose, for example that $\Omega = \{\omega_1, \omega_2\}$, $P[\omega_1] = P[\omega_2] = \frac{1}{2}$,

$$A^i(\omega_1) = A^i(\omega_2) = \begin{pmatrix} 1 \\ -1 \end{pmatrix}$$

$$h^i(\omega_1) = \begin{pmatrix} 0 \\ -1 \end{pmatrix}$$

$$h^i(\omega_2) = \begin{pmatrix} 2 \\ -3 \end{pmatrix}. \qquad (2.3)$$

For $0 < \alpha^i \leq \frac{1}{2}$, $K_1^i(\alpha^i) = [0,1] \cup [2,3]$.

When each i corresponds to a distinct linear constraint and A^i is a fixed row vector, then obtaining a deterministic equivalent of (2.2) is fairly straightforward. In this case, $P(A^i x \geq h^i(\omega)) = F^i(A^i x)$, where F^i is the distribution function of $\mathbf{h}^i$. Hence, $K_1^i(\alpha^i) = \{x | F^i(A^i x) \geq \alpha^i\}$, which immediately yields a deterministic equivalent form. In general, however, the constraints must hold jointly so that the set I is a singleton. This situation corresponds to requiring an α-confidence interval that x is feasible. We assume this in the remainder of this section and drop the superscript i indicating the set of joint constraints.

The results to determine the deterministic equivalent often involve manipulations of probability distributions that use measure theory. The remainder of this section is intended for readers familiar with this area. One of the main results in probabilistic constraints is that, in the joint constraint case, a large class of probability measures on $h(\omega)$ (for A fixed) leads to convex and closed $K_1(\alpha)$. A probability measure P is in this class of *quasi-concave* measures if for any convex measurable sets U and V and any $0 \leq \lambda \leq 1$,

$$P((1-\lambda)U + \lambda V) \geq \min\{P(U), P(V)\}. \tag{2.4}$$

The use of this and a special form, called *logarithmically concave measures*, began with Prékopa [1971, 1973]. General discussions also appear in Prékopa [1980, 1995], Kallberg and Ziemba [1983] concerning related utility functions, and the surveys of Wets [1983b, 1990] which include the following theorem.

Theorem 14. *Suppose A is fixed and $\mathbf{h}$ has an associated quasi-concave probability measure P. Then $K_1(\alpha)$ is a closed convex set for $0 \leq \alpha \leq 1$.*

Proof: Let $\mathcal{H}(x) = \{h|Ax \geq h\}$. Suppose $x(\lambda) = \lambda x^1 + (1-\lambda)x^2$ where $x^1, x^2 \in K_1(\alpha)$. Suppose $h^1 \in \mathcal{H}(x^1)$ and $h^2 \in \mathcal{H}(x^2)$. Then $\lambda h^1 + (1-\lambda)h^2 \leq Ax(\lambda)$, so $\mathcal{H}(x(\lambda)) \supset \lambda \mathcal{H}(x^1) + (1-\lambda)\mathcal{H}(x^2)$. Hence, $P(\{Ax(\lambda) \geq h\}) = P(\mathcal{H}(x(\lambda)) \geq P(\lambda \mathcal{H}(x^1) + (1-\lambda)\mathcal{H}(x^2)) \geq \alpha$. Thus, $K_1(\alpha)$ is convex.

For closure, suppose that $x^\nu \to \bar{x}$, where $x^\nu \in K_1(\alpha)$. Consider $\mathcal{H}(x^\nu)$. If $h \leq Ax^{\nu_i}$ for some subsequence $\{\nu_i\}$ of $\{\nu\}$, then $h \leq A\bar{x}$. Hence $\limsup_\nu \mathcal{H}(x^\nu) \subseteq \mathcal{H}(\bar{x})$, so $P(\mathcal{H}(\bar{x})) \geq P(\limsup_\nu \mathcal{H}(x^\nu)) \geq \limsup_\nu P(\mathcal{H}(x^\nu)) \geq \alpha$. $\square$

The relevance of this result stems from the large class of probability measures which fit these conditions. Some extent of this class is given in the following result of Borell [1975], which we state without proof.

Theorem 15. *If f is the density of a continuous probability distribution in $\Re^m$ and $f^{-(\frac{1}{m})}$ is convex on $\Re^m$, then the probability measure*

$$P(B) = \int_B f(x)dx,$$

defined for all Borel sets B in $\Re^m$ is quasi-concave. $\square$

In particular, this result states that any density of the form $f(x) = e^{-l(x)}$ for some convex function l yields a quasi-concave probability measure. These measures include the multivariate normal, beta, and Dirichlet distributions and are logarithmically concave (because, for $0 \leq \lambda \leq 1$, $P((1-\lambda)U + \lambda V) \geq P(U)^\lambda P(V)^{1-\lambda}$ for all Borel sets U and V) as studied by Prékopa. These distributions lead to computable deterministic equivalents as, for example, in the following theorem.

Theorem 16. *Suppose A is fixed and the components $\mathbf{h}_i, i = 1, \ldots, m_1$, of $\mathbf{h}$ are stochastically independent random variables with logarithmically concave probability measures, P_i, and distribution functions, F_i, then $K_1(\alpha) = \{x| \sum_{i=1}^{m_1} ln(F_i(A_i.x)) \geq ln\alpha\}$ and is convex.*

Proof: From the independence assumption, $P[Ax \geq \mathbf{h}] = \Pi_{i=1}^{m_1} P_i[A_i.x \geq \mathbf{h}_i] = \Pi_{i=1}^{m_1} F_i(A_i.x)$. So, $K_1(\alpha) = \{x|\Pi_{i=1}^{m_1} F_i(A_i.x) \geq \alpha\}$. Taking logarithms (which is a monotonically increasing function), we obtain $K_1(\alpha) = \{x| \sum_{i=1}^{m_1} ln(F_i(A_i.x)) \geq ln\alpha\}$. Because

$$F_i(A_i.(\lambda x^1 + (1-\lambda)x^2)) = P_i(\mathbf{h}_i \leq A_i.(\lambda x^1 + (1-\lambda)x^2))$$
$$\geq P_i(\lambda\{\mathbf{h}_i \leq A_i.x^1\} + (1-\lambda)\{\mathbf{h}_i \leq A_i.x^2)\})$$
$$\geq (P_i(\{\mathbf{h}_i \leq A_i.x^1\})^\lambda (P_i(\{\mathbf{h}_i \leq A_i.x^2\})^{1-\lambda}$$
$$= (F_i(A_i.x^1)^\lambda)(F_i(A_i.x^2)^{1-\lambda}),$$

the logarithm of $F_i(A_i.x)$ is a concave function, and $K_1(\alpha)$ is convex. $\square$

Logarithmically concave distribution functions include the increasing failure rate functions (see Miller and Wagner [1965] and Parikh [1968]) that are common in reliability studies. Other types of quasi-concave measures include the multivariate t and F distributions. Because these distributions include those most commonly used in multivariate analysis, it appears that, with continuous distributions and fixed A, the convexity of the solution set is generally assured.

When A is also random, the convexity of the solution set is, however, not as clear. The following theorem from Prékopa [1974], given without proof, shows this result for normal distributions with fixed covariance structure across columns of A and h.

Theorem 17. *If* $\mathbf{A}_1., \ldots, \mathbf{A}_{n_1}., \mathbf{h}$ *have a joint normal distribution with a common covariance structure, a matrix C, such that* $E[(\mathbf{A}_i. - E(\mathbf{A}_i.))(\mathbf{A}_j. - E(\mathbf{A}_j.))^T] = r_{ij}C$ *for i, j in* $1, \ldots, n_1$, *and*

$$E[(\mathbf{A}_i. - E(\mathbf{A}_i.))(\mathbf{h} - E(\mathbf{h}))] = s_i C$$

for $i = 1, \ldots, n_1$, where r_{ij} and s_i are constants for all i and j, then $K_1(\alpha)$ is convex for $\alpha \geq \frac{1}{2}$. $\square$

Stronger results than Theorem 17 are difficult to obtain. In general, one must rely on approximations to the deterministic equivalent that maintain convexity although the original solution set may not be convex. We will consider some of these approximations in Chapter 9.

Some other specific examples where A may be random include single constraints (see Exercise 5). In the case of $h \equiv 0$ and normally distributed A, the deterministic equivalent is again readily obtainable as in the following from Parikh [1968].

Theorem 18. *Suppose that $m_1 = 1$, $h_1 = 0$, and $\mathbf{A}_1.$ has mean $\bar{A}_1.$ and covariance matrix C_1, then $K_1(\alpha) = \{x|\bar{A}_1.x - \Phi^{-1}(\alpha)\sqrt{x^T C_1 x} \geq 0\}$, where Φ is the standard normal distribution function.*

Proof: Observe that $\mathbf{A}_1.x$ is normally distributed with mean, $\bar{A}_1.x$, and variance, $x^T C_1 x$. If $x^T C_1 x = 0$, then the result is immediate. If not, then

$\frac{\mathbf{A}_1.x - \bar{A}_1.x}{\sqrt{x^T C_1 x}}$ is a standard normal random variable with cumulative Φ, and

$$P(\mathbf{A}_1.x \geq 0) = P(\frac{\mathbf{A}_1.x - \bar{A}_1.x}{\sqrt{x^T C_1 x}} \geq \frac{-\bar{A}_1.x}{\sqrt{x^T C_1 x}})$$
$$= P(\frac{\mathbf{A}_1.x - \bar{A}_1.x}{\sqrt{x^T C_1 x}} \leq \frac{\bar{A}_1.x}{\sqrt{x^T C_1 x}})$$
$$= \Phi(\frac{\bar{A}_1.x}{\sqrt{x^T C_1 x}}).$$

Substitution in the definition of $K_1(\alpha)$ yields the result. $\square$

Finally in this chapter, we would like to show some of the similarities between models with probabilistic constraints and problems with recourse. As stated in Chapter 2, models with probabilistic constraints and models with recourse can often lead to the same optimal solutions. Some other aspects of the modeling process may favor one over the other (see, e.g., Hogan, Morris, and Thompson [1981, 1984], Charnes and Cooper [1983]), but, these differences generally just represent decision makers' different attitudes toward risk.

We use an example from Parikh [1968] to relate simple recourse and chance-constrained problems. Consider the following problem with probabilistic constraints:

$$\min c^T x$$
$$\text{s. t. } Ax = b,$$
$$P_i[T_i.x \geq \mathbf{h}_i] \geq \alpha_i, i = 1, \ldots, m_2,$$
$$x \geq 0, \tag{2.5}$$

where P_i is the probability measure of $\mathbf{h}_i$ and F_i is the distribution function for $\mathbf{h}_i$. For the deterministic equivalent to (2.5), we just let $F_i(h_i^*) = \alpha_i$, to obtain:

$$\min c^T x$$
$$\text{s. t. } Ax = b,$$
$$T_i.x \geq h_i^*, i = 1, \ldots, m_2,$$
$$x \geq 0. \tag{2.6}$$

Suppose we solve (2.6) and obtain an optimal x^* and optimal dual solution $\{\lambda^*, \pi^*\}$, where $c^T x^* = b^T \lambda^* + h^{*T} \pi^*$. If $\pi_i^* = 0$, let $q_i^+ = 0$ and, if $\pi_i^* > 0$, let $q_i^+ = \frac{\pi_i^*}{1 - \alpha_i}$. An equivalent stochastic program with simple recourse to (2.5) is then:

$$\min c^T x + E_{\mathbf{h}}[q^+ \mathbf{y}^+]$$
$$\text{s. t. } Ax = b,$$
$$T_i.x + \mathbf{y}_i^+ - \mathbf{y}_i^- = \mathbf{h}_i, i = 1, \ldots, m_2,$$

$$x, y^+, y^- \geq 0. \tag{2.7}$$

For problems (2.5) and (2.7) to be equivalent, we mean that any x^* optimal in (2.5) corresponds to some (x^*, y^{*+}) optimal in (2.7) for a suitable definition of q^+ and that any (x^*, y^{*+}) optimal in (2.7) corresponds to x^* optimal in (2.5) for a suitable definition of α_i. We show the first part of this equivalence in the following theorem.

Theorem 19. *For the q_i^+ defined as a function of some optimal π^* for the dual to (2.5), if x^* is optimal in (2.5), there exists $\mathbf{y}^{*+} \geq 0$ a.s. such that $(x^*, \mathbf{y}^{*+})$ is optimal in (2.7).*

Proof. First, let x^* be optimal in (2.5). It must also be optimal in (2.6) with dual variables, $\{\lambda^*, \pi^*\}$. We must have $\pi^* \geq 0$,

$$c^T - \lambda^{*T}A - \pi^{*T}T \geq 0,$$
$$Tx^* - h^* \geq 0,$$
$$(c^T - \lambda^{*T}A - \pi^{*T}T)x^* = 0, \text{ and}$$
$$\pi^{*T}(Tx^* - h^*) = 0. \tag{2.8}$$

Now, for x^* to be optimal in (2.7), consider the optimality conditions (1.13) from Corollary 10. These conditions state that if there exists λ^* such that

$$c^T - \lambda^{*T}A - \sum_{i=1}^{m_2} T_{i\cdot}(q_i^+ - q_i F_i(T_{i\cdot}x^*)) \geq 0,$$
$$(c^T - \lambda^{*T}A - \sum_{i=1}^{m_2} T_{i\cdot}(q_i^+ - q_i F_i(T_{i\cdot}x^*)))x^* = 0. \tag{2.9}$$

Substituting for $\pi_i^* = q_i^+(1 - \alpha_i)$ in (2.8) and noting from the complementarity condition that $\alpha_i = F_i(h_i^*) = F_i(T_{i\cdot}x^*)$ if $\pi_i^* > 0$, we obtain

$$c^T - \lambda^{*T}A - \pi^{*T}T = c^T - \lambda^{*T}A - \sum_{i=1}^{m_2} T_{i\cdot}(q_i^+(1 - F_i(T_{i\cdot}x^*)))$$
$$= c^T - \lambda^{*T}A - \sum_{i=1}^{m_2} T_{i\cdot}(q_i^+ - q_i F_i(T_{i\cdot}x^*)) \tag{2.10}$$

from the definitions and noting that $\pi_i^* > 0$ if and only if $q_i^+ > 0$. From (2.10), we can verify the conditions in (2.9) and obtain the optimality of x^* in (2.7). $\square$

If we assume x^* is optimal in (2.7), we can reverse the argument to show that x^* is also optimal in (2.5) for some value of α_i. This result (from Symonds [1968]) is Exercise 7. Further equivalences are discussed in Gartska [1980]. We note that all of these equivalences are somewhat weak because they require a priori knowledge of the optimal solution to one of the problems (see also the discussion in Gartska and Wets [1974]).

Exercises

1. Suppose a single probabilistic constraint with fixed A and that $\mathbf{h}$ has an exponential distribution with mean λ. What is the resulting deterministic equivalent constraint for $K_1(\alpha)$?

2. For the example in (2.3), what happens for $\frac{1}{2} < \alpha^i \leq 1$?

3. Can you construct an example with continuous random variables where $K_1(\alpha)$ is not connected? (Hint: Try a multimodal distribution such as a random choice of one of two bivariate normal random variables.)

4. Extend Theorem 14 to allow any set of convex constraints, $g_i(x, \xi(\omega)) \leq 0$, $i = 1, \ldots, m$.

5. Suppose a single linear constraint in $K_1(\alpha)$ where the components of A and h have a joint normal distribution. Show that $K_1(\alpha)$ is also convex in this case for $\alpha \geq \frac{1}{2}$. (Hint: The random variable, $\mathbf{A}_1.x - \mathbf{h}_1$, is also normally distributed.)

6. Show that $\sqrt{x^T C_1 x}$ is a convex function of x.

7. Prove the converse of Theorem 19 by finding an appropriate α_i so that x^* optimal in (2.7) is also optimal in (2.5).

3.3   Stochastic Integer Programs

a. Recourse problems

The general formulation of a two-stage integer program resembles that of the general linear case presented in Section 1. It simply requires that some variables, in either the first stage or the second stage, are integer. As we have seen in the examples in Chapter 1, in many practical situations the restrictions are, in fact, that the variables must be binary, i.e., they can only take the value zero or one. Formally, we may write

$$\min_{x \in X} z = c^T x + E_{\boldsymbol{\xi}} \min\{q(\omega)^T y | Wy = h(\omega) - T(\omega)x, y \in Y\}$$

$$\text{s.t. } Ax = b,$$

where the definitions of $c, b, \boldsymbol{\xi}, A, W, T$, and h are as before. However, X and/or Y contains some integrality or binary restrictions on x and/or y. With this definition, we may again define a deterministic equivalent program of the form

$$\min_{x \in X} z = c^T x + \mathcal{Q}(x)$$

$$\text{s.t. } Ax = b$$

with $\mathcal{Q}(x)$ the expected value of the second stage defined as in Section 1.2.
In this section, we are interested in the properties of $\mathcal{Q}(x)$ and $K_2 = \{x|\mathcal{Q}(x) < \infty\}$. Clearly, if the only integrality restrictions are in X, the properties of $\mathcal{Q}(x)$ and K_2 are the same as in the continuous case. The main interesting cases are those in which some integrality restrictions are present in the second stage. The properties of $Q(x,\xi)$ for given ξ are those of the value function of an integer program in terms of its right-hand side. This problem has received much attention in the field of integer programming (see, e.g., Blair and Jeroslow [1982] or Nemhauser and Wolsey [1988]). In addition to being *subadditive*, the value function of an integer program can be obtained by starting from a linear function and finitely often repeating the operations of sums, maxima, and non-negative multiples of functions already obtained and rounding up to the nearest integer. Functions so obtained are known as *Gomory functions* (see again Blair and Jeroslow [1982] or Nemhauser and Wolsey [1988]). Clearly, the maximum and rounding up operations imply undesirable properties for $Q(x,\xi)$, $\mathcal{Q}(x)$, and K_2, as we now illustrate.

Proposition 20. *The expected recourse function $Q(x)$ of an integer program is in general nonconvex and discontinuous.*

Example 1

We illustrate the proposition in the following simple example where the first stage contains a single decision variable $x \geq 0$ and the second-stage recourse function is defined as:

$$Q(x,\xi) = \min\{2y_1 + y_2 | y_1 \geq x - \xi, y_2 \geq \xi - x, y \geq 0, \text{ integer}\}. \quad (3.1)$$

Assume ξ can take on the values one and two with equal probability $1/2$. Let $\lceil a \rceil$ denote the smallest integer greater than or equal to a (the rounding up operation) and $\lfloor a \rfloor$ the truncation or rounding down operation ($\lfloor a \rfloor = -\lceil -a \rceil$). Consider $\xi = 1$. For $x \leq 1$, the optimal second-stage solution is $y_1 = 0$, $y_2 = \lceil 1 - x \rceil$. For $x \geq 1$, it is $y_1 = \lceil x - 1 \rceil$, $y_2 = 0$. Hence, $Q(x,1) = \max\{2(\lceil x - 1 \rceil), \lceil 1 - x \rceil\}$, a typical Gomory function. It is discontinuous at $x = 1$. Nonconvexity can be illustrated by $Q(0.5,1) > 0.5Q(0,1)| + 0.5Q(1,1)$. Similarly, $Q(x,2) = \max\{2(\lceil x - 2 \rceil), \lceil 2 - x \rceil\}$. The three functions, $Q(x,1)$, $Q(x,2)$, and $\mathcal{Q}(x)$ are represented in Figure 1.

The recourse function, $\mathcal{Q}(x)$, is clearly discontinuous in all positive integers. Nonconvexity can be illustrated by $\mathcal{Q}(1.5) = 1.5 > 0.5\mathcal{Q}(1) + 0.5\mathcal{Q}(2) = .75$. Thus $\mathcal{Q}(x)$ has none of the properties that one may wish for to design an algorithmic procedure. Note, however, that a convexity-related property exists in the case of simple integer recourse (Proposition 8.4) and that it applies to this example.

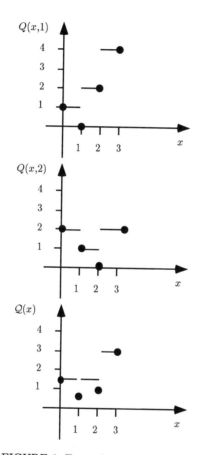

FIGURE 1. Example of discontinuity.

Continuity of the recourse function can be regained when the random variable is absolutely continuous (Stougie [1987]).

Proposition 21. *The expected recourse function $Q(x)$ of an integer program with an absolutely continuous random variable is continuous.*

Note, however, that despite Proposition 21, the recourse function $Q(x)$ remains, in general, nonconvex.

Example 2

Consider Example 1 but with the (continuous) random variable defined by its cumulative distribution,

$$F(t) = P(\boldsymbol{\xi} \le t) = 2 - 2/t, 1 \le t \le 2.$$

Consider $1 < x < 2$. For $1 \le \xi < x$, we have $0 < x - \xi < 1$, hence $y_1 = 1, y_2 = 0$, while for $x < \xi \le 2$, we have $0 < \xi - x \le 1$, hence $y_1 = 0, y_2 = 1$.

It follows that

$$Q(x) = \int_1^x 2dF(t) + \int_x^2 1dF(t) = 2F(x) + 1 - F(x)$$
$$= F(x) + 1 = 3 - 2/x,$$

which is easily seen to be nonconvex.

Properties are just as poor in terms of feasibility sets. As in the continuous case, we may define the second-stage feasibility set for a fixed value of $\boldsymbol{\xi}$ as $K_2(\xi(\omega)) = \{x \mid \text{there exists } y \text{ s.t. } Wy = h(\omega) - T(\omega)x, y \in Y\}$ where $\xi(\omega)$ is formed by the stochastic components of $h(\omega)$ and $T(\omega)$.

Proposition 22. *The second-stage feasibility set $K_2(\xi)$ is in general nonconvex.*

Proof. Because $K_2(\xi) = \{x \mid Q(x,\xi) < \infty\}$, nonconvexity of $K_2(\xi)$ immediately follows from nonconvexity of $Q(x,\xi)$. $\Box$

A simple example suffices to illustrate this possibility.

Example 3

Let the second stage of a stochastic program be defined as

$$-y_1 + y_2 \le \boldsymbol{\xi} - x_1, \qquad (3.2)$$
$$y_1 + y_2 \le 2 - x_2, \qquad (3.3)$$
$$y_1, y_2 \ge 0 \text{ and integer.} \qquad (3.4)$$

Assume $\boldsymbol{\xi}$ takes on the values 1 and 2 with equal probability 1/2. We then construct $K_2(1)$.

By (3.3), $x_2 \leq 2$ is a necessary condition for second-stage feasibility. For $1 < x_2 \leq 2$, the only feasible integer satisfying (3.3) is $y_1 = y_2 = 0$. This point is also feasible for (3.2) if $\xi - x_1 \geq 0$, i.e., if $x_1 \leq 1$.

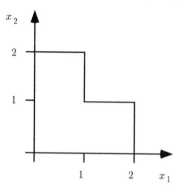

FIGURE 2. Feasibility set for Example 3.

For $0 < x_2 \leq 1$, the integer points y satisfying (3.3) are $(0,0), (0,1), (1,0)$. The one yielding the smallest left-hand side (and thus the most likely to yield points in $K_2(1)$) is $(1,0)$. It requires $\xi - x_1 \geq -1$, i.e., $x_1 \leq 2$. Hence $K_2(1)$ is as in Figure 2 and is clearly nonconvex. It may be represented as $K_2(1) = \{x| \min\{x_1 - 1, x_2 - 1\} \leq 0, 0 \leq x_1 \leq 2, 0 \leq x_2 \leq 2\}$ and is again a typical Gomory function due to the minimum operation.

We may then define the second-stage feasibility set K_2 as the intersection of $K_2(\xi)$ over all possible ξ values. This definition poses no difficulty when ξ has a discrete distribution. In Example 3, $K_2 = K_2(1)$ and is thus also nonconvex.

Computationally, it might be very useful to have the constraint matrix of the extensive form *totally unimodular*. (Recall that a matrix is totally unimodular if the determinants of all square submatrices are 0, 1, or -1.) This would imply that any solution of the associated stochastic continuous program would be integer when right-hand sides of all constraints are also integer. A widely used sufficient condition for total unimodularity is as follows: all coefficients are 0, 1, or -1; every variable has at most two nonzero coefficients and constraints can be separated in two groups such that, if a variable has two nonzero coefficients and if they are of the same sign, the two associated rows belong to different sets and if they are of opposite signs they belong to the same set.

To help understand the sufficiency condition, consider the following matrix

$$\begin{pmatrix} 1 & 0 & 1 & -1 \\ 0 & 1 & 1 & 0 \\ -1 & 1 & 0 & 1 \end{pmatrix}$$

as an example. For this matrix, one set consists of Rows 1 and 3, and the second set contains just Row 2. The constraint matrix of the extensive form of a nontrivial stochastic program cannot satisfy this sufficient condition. For simplicity, consider the case of a fixed T matrix. Assume that any variable that has a nonzero coefficient in T also has a nonzero coefficient in A. Then, if $|\Xi| \geq 2$, the constraint matrix of the extensive form contains a submatrix

$$
\begin{pmatrix} A \\ T \\ T \end{pmatrix}
$$

that has at least three nonzero coefficients. Thus, only very special cases (a random T matrix with every column having a nonzero element in only one realization, for example) could lead to totally unimodular matrices.

Last but not least, it should be clear that just finding $\mathcal{Q}(x)$ for a given x becomes an extremely difficult task for a general integer second stage. This is especially true because there is no hope to use sensitivity analysis or some sort of bunching procedure (see Section 5.4) to find $Q(x, \xi)$ for neighboring values of ξ. Cases where $\mathcal{Q}(x)$ can be computed or even approximated in a reasonable amount of time should thus be considered exceptions. One such exception is provided in the next section.

b. Simple integer recourse

Let ξ be a random vector with support Ξ in $\Re^m$, expectation μ, and cumulative distribution F with $F(t) = P\{\xi \leq t\}, t \in R^m$. A two-stage stochastic program with simple integer recourse is as follows:

$$
\begin{aligned}
SIR \qquad \min z= &\ c^T x + E_{\xi}\{\min(q^+)^T y^+ + (q^-)^T y^- \,| \\
&\ y^+ \geq \xi - Tx, y^- \geq Tx - \xi, \\
&\ y^+ \in Z_+^m, y^- \in Z_+^m\} \\
\text{s.t. } &\ Ax= b, x \in X,
\end{aligned} \tag{3.5}
$$

where X typically defines either non-negative continuous or non-negative integer decision variables and where we use $\xi = \mathbf{h}$ because both T and q are known and fixed. As in the continuous case, we may replace the second-stage value function $\mathcal{Q}(x)$ by a separable sum over the various coordinates. Let $\chi = Tx$ be a tender to be bid against future outcomes. Then $\mathcal{Q}(x)$ is separable in the components χ_i.

$$
\mathcal{Q}(x) = \sum_{i=1}^m \psi_i(\chi_i), \tag{3.6}
$$

with

$$
\psi_i(\chi_i) = E_{\xi_i} \psi_i(\chi_i, \xi_i) \tag{3.7}
$$

and

$$
\psi_i(\chi_i, \xi_i)= \min\{q_i^+ y_i^+ + q_i^- y_i^- \,| y_i^+ \geq \xi_i - \chi_i,
$$

$$y_i^- \geq \chi_i - \xi_i, \; y_i^+, y_i^- \in Z_+\}. \tag{3.8}$$

As in the continuous case, any error made in bidding χ_i versus ξ_i must be compensated for in the second stage, but this compensation must now be an integer.

Now define the expected shortage as

$$u_i(\chi_i) = E\lceil \xi_i - \chi_i \rceil^+$$

and the expected surplus as

$$v_i(\chi_i) = E\lceil \chi_i - \xi_i \rceil^+,$$

where $\lceil x \rceil^+ = \max\{\lceil x \rceil, 0\}$. It follows that $\psi_i(\chi_i)$ is simply

$$\psi_i(\chi_i) = q_i^+ u_i(\chi_i) + q_i^- v_i(\chi_i).$$

As is reasonable from the definition of SIR, we assume $q_i^+ \geq 0, q_i^- \geq 0$.

Studying SIR is thus simply studying the expected shortage and surplus. Unless necessary, we drop the indices in the sequel. Let ξ be some random variable and $x \in \Re$. The expected shortage is

$$u(x) = E\lceil \xi - x \rceil^+ \tag{3.9}$$

and the expected surplus is

$$v(x) = E\lceil x - \xi \rceil^+. \tag{3.10}$$

For easy reference, we also define their continuous counterparts. Let the continuous expected shortage be

$$\hat{u}(x) = E(\xi - x)^+ \tag{3.11}$$

and the continuous expected surplus be

$$\hat{v}(x) = E(x - \xi)^+. \tag{3.12}$$

First observe that Example 1 (and 2) is a case of a stochastic program with simple recourse, from which we know that $u(x) + v(x)$ is in general non-convex and discontinuous unless ξ has an absolutely continuous probability distribution function. We thus limit our ambitions to study finiteness and computational tractability for $u(.)$ and $v(.)$. The following results appear in Louveaux and van der Vlerk [1993].

Proposition 23. *The expected shortage function is a non-negative non-decreasing extended real-valued function. It is finite for all $x \in \Re$ if and only if $\mu^+ = E \max\{\xi, 0\}$ is finite.*

Proof. We only give the proof for finiteness because the other results are immediate. First, observe that for all t in $\Re$,

$$(t-x)^+ \leq \lceil t-x \rceil^+ \leq (t-x+1)^+ \leq (t-x)^+ + 1.$$

Taking expectation yields

$$\hat{u}(x) \leq u(x) \leq \hat{u}(x-1) \leq \hat{u}(x) + 1. \tag{3.13}$$

The result follows as $\hat{u}(x)$ is finite if and only if μ^+ is finite. $\square$

We now provide a computational formula for $u(x)$.

Theorem 24. *Let $\boldsymbol{\xi}$ be a random variable with cumulative distribution function F. Then*

$$u(x) = \sum_{k=0}^{\infty} (1 - F(x+k)). \tag{3.14}$$

Proof. Following the previous definitions, we have:

$$
\begin{aligned}
\sum_{k=0}^{\infty}(1 - F(x+k)) &= \sum_{k=0}^{\infty} P\{\boldsymbol{\xi} - x > k\} \\
&= \sum_{k=0}^{\infty} \sum_{j=k+1}^{\infty} P\{\lceil \boldsymbol{\xi} - x \rceil^+ = j\} \\
&= \sum_{j=1}^{\infty} \sum_{k=0}^{j-1} P\{\lceil \boldsymbol{\xi} - x \rceil^+ = j\} \\
&= \sum_{j=1}^{\infty} j P\{\lceil \boldsymbol{\xi} - x \rceil^+ = j\} = E\lceil \boldsymbol{\xi} - x \rceil^+ = u(x),
\end{aligned}
$$

which completes the proof. $\square$

Similar results hold for $v(x)$.

Theorem 25. *Let $\boldsymbol{\xi}$ be a random variable with $\hat{F}(t) = P\{\boldsymbol{\xi} < t\}$ and $\mu^- = E\boldsymbol{\xi}^-$. Then v is a non-negative nondecreasing extended real-valued function, which is finite for all $x \in \Re$ if and only if μ^- is finite. Moreover,*

$$v(x) = \sum_{k=0}^{\infty} \hat{F}(x-k). \tag{3.15}$$

Theorems 24 and 25 provide workable formulas for a number of cases.

Case a. Clearly, if $\boldsymbol{\xi}$ has a finite range, then (3.14) and (3.15) reduce to a finite computation.

Example 4

Let $\boldsymbol{\xi}$ have a uniform density on $[0, a]$. Consider $0 \leq x \leq a$. Then

$$
u(x) = \sum_{k=0}^{\infty} (1 - F(x+k)) = \sum_{k=0}^{\lceil a-x \rceil^+ - 1} (1 - F(x+k))
$$

$$
= \sum_{k=0}^{\lceil a-x \rceil^+ - 1} \left(1 - \frac{x+k}{a}\right)
$$

$$
= \lceil a - x \rceil^+ \left(1 - \frac{x}{a}\right) - \frac{\lceil a - x \rceil^+ (\lceil a - x \rceil^+ - 1)}{2a}.
$$

Observe that $\lceil a - x \rceil^+$ is piecewise constant. Hence, $u(x)$ is piecewise linear and convex.

Similarly, one computes

$$
v(x) = \frac{x(\lfloor x \rfloor + 1)}{a} - \frac{\lfloor x \rfloor (\lfloor x \rfloor + 1)}{2a}.
$$

Again, $v(x)$ is piecewise linear and convex. It follows that a simple integer recourse program with uniform densities is a piecewise linear convex program whose second-stage recourse function is easily computable.

Case b. For some continuous random variables, we may obtain analytical expressions for $u(x)$ and $v(x)$.

Example 5

Let $\boldsymbol{\xi}$ follow an exponential distribution with parameter $\lambda > 0$. Then, for $x \geq 0$,

$$
u(x) = \sum_{k=0}^{\infty} (1 - F(x+k)) = \sum_{k=0}^{\infty} e^{-\lambda(x+k)} = \frac{e^{-\lambda x}}{1 - e^{-\lambda}},
$$

while

$$
v(x) = \sum_{k=0}^{\infty} F(x - k) = \lfloor x \rfloor + 1 - e^{-\lambda(x - \lfloor x \rfloor)} \cdot \sum_{k=0}^{\lfloor x \rfloor} e^{-\lambda k}
$$

$$
= \lfloor x \rfloor + 1 - \left(\frac{e^{-\lambda(x - \lfloor x \rfloor)} - e^{-\lambda(x+1)}}{1 - e^{-\lambda}}\right).
$$

Observe that $v(x)$ is nonconvex (as it would be $u(x)$ for $x \leq 0$).

Case c. Finite computation can also be obtained when $\Xi \in Z$. From Theorems 24 and 25, we derive the following corollary.

Corollary 26. *For all $n \in Z_+$, we have*

$$u(x + n) = u(x) - \sum_{k=0}^{n-1}(1 - F(x + k)) \qquad (3.16)$$

and

$$v(x + n) = v(x) + \sum_{k=1}^{n} \hat{F}(x + k)). \qquad (3.17)$$

Corollary 27. *Let ξ be a discrete random variable with support $\Xi \in Z$. Then*

$$u(x) = \begin{cases} \mu^+ - \lfloor x \rfloor - \sum_{k=\lfloor x \rfloor}^{-1} F(k) & \text{if } x < 0, \\ \mu^+ - \lfloor x \rfloor + \sum_{k=0}^{\lfloor x \rfloor - 1} F(k) & \text{if } x \geq 0. \end{cases}$$

Proof. Because $\Xi \in Z$, $F(t) = F(\lfloor t \rfloor)$, for all $t \in \Re$. Hence, $u(x) = u(\lfloor x \rfloor)$ for all $x \in R$. Now, $u(0) = \mu^+$. Then apply (3.16) to obtain the result. $\square$

Corollary 28. *Let ξ be a discrete random variable with support $\Xi \in Z$. Then*

$$v(x) = \begin{cases} \mu^- - \sum_{k=\lceil x \rceil}^{-1} F(k) & \text{if } x < 0, \\ \mu^- + \sum_{k=0}^{\lceil x \rceil - 1} F(k) & \text{if } x \geq 0. \end{cases}$$

Thus, here the finite computation comes from the finiteness of $\lceil x \rceil$.

Case d. Finally, we may have a random variable that does not fall in any of the given categories. We may then resort to approximations.

Theorem 29. *Let ξ be a random variable with cumulative distribution function, F. Then*

$$\hat{u}(x) \leq u(x) \leq \hat{u}(x) + 1 - F(x). \qquad (3.18)$$

Proof. The first inequality was given in (3.13). Because $1 - F(t)$ is nonincreasing, we have for any $x \in \Re$ and any $k \in \{1, 2, \cdots\}$ that

$$1 - F(x + k) \leq 1 - F(t), t \in [x + k - 1, x + k).$$

Hence,

$$\sum_{k=1}^{\infty}(1 - F(x + k)) \leq \int_{x}^{\infty}(1 - F(t))dt.$$

Adding $1 - F(x)$ to both sides gives the desired result. $\square$

Theorem 30. *Let ξ be a random variable with cumulative distribution function F. Let n be some integer, $n \geq 1$. Define*

$$u_n(x) = \sum_{k=0}^{n-1}(1 - F(x+k)) + \hat{u}(x+n). \tag{3.19}$$

Then

$$u_n(x) \leq u(x) \leq u_n(x) + 1 - F(x+n). \tag{3.20}$$

Proof: The proof follows directly from Theorem 29 and formula (3.16). $\square$

To approximate $u(x)$ within an accuracy ε, we have to compute the first n terms in $u(x)$, where n is chosen so that $F(x+n) \geq 1 - \varepsilon$ and $\hat{u}(x+n)$, which involves computing one integral.

Example 6

Let ξ follow a normal distribution with mean μ and variance σ^2, i.e., $N(\mu, \sigma^2)$, with cumulative distribution function F and probability density function f. Integrating by parts, one obtains:

$$
\begin{aligned}
u_n(x) &= \sum_{k=0}^{n-1}(1 - F(x+k)) + \int_{x+n}^{\infty}(1 - F(t))dt \\
&= \sum_{k=0}^{n-1}(1 - F(x+k)) - (x+n)(1 - F(x+n)) + \int_{x+n}^{\infty} tf(t)dt.
\end{aligned}
$$

Using $tf(t) = \mu f(t) - \sigma^2 f'(t)$, it follows that

$$u_n(x) = \sum_{k=0}^{n-1}(1 - F(x+k)) + (\mu - x - n)(1 - F(x+n)) + \sigma^2 f(x+n).$$

Similar results apply for $v(x)$.

Theorem 31. *Let ξ be a random variable with cumulative distribution function $\hat{F}(t) = P\{\xi < t\}$. Then*

$$\hat{v}(x) \leq v(x) \leq \hat{v}(x) + \hat{F}(x). \tag{3.21}$$

Let n be some integer, $n \geq 1$. Define

$$v_n(x) = \sum_{k=0}^{n-1}\hat{F}(x-k) + \hat{v}(x-n). \tag{3.22}$$

Then

$$v_n(x) \leq v(x) \leq v_n(x) + \hat{F}(x-n). \tag{3.23}$$

Example 6 (continued)

Let ξ follow an $N(\mu, \sigma^2)$ distribution, with cumulative distribution function F and probability density function f. Then

$$v_n(x) = \sum_{k=0}^{n-1} F(x-k) + (x-n-\mu)F(x-n) + \sigma^2 f(x-n).$$

As a conclusion, expected shortage, expected surplus, and thus simple integer recourse functions can be computed in finitely many steps either in an exact manner or within a prespecified tolerance ε. Deeper studies of continuity and differentiability properties of the recourse function can be found in Stougie [1987], Louveaux and van der Vlerk [1993], and Schultz [1993].

c. Probabilistic constraints

Probabilistic constraints involving integer decision variables may generally be treated in exactly the same manner as if they involved continuous decision variables. One need only take the intersection of their deterministic equivalents with the integrality requirements. The question is then how to obtain a polyhedral representation of this intersection. This problem sometimes has quite nice solutions. Here we provide an example of probabilistic constraints in routing problems.

Let $V = \{v_1, v_2, \cdots, v_n\}$ be a set of vertices, typically representing customers. Let v_0 represent the depot and let $V_0 = V \cup \{v_0\}$. A route is an ordered sequence $L = \{i_0 = 0, i_1, i_2, \cdots, i_k, i_{k+1} = 0\}$, with $k \leq n$, starting and ending at the depot and visiting each customer at most once. Clearly, if $k < n$, more than one vehicle is needed to visit all customers. Assume a vehicle of given capacity C follows each route, collecting customers' demands d_i. If demands d_i are random, it may turn out that at some point of a given route, the vehicle cannot load a customer's demand. This is clearly an undesirable feature, which is usually referred to as a *failure of the route*. A probabilistic constraint for the capacitated routing requires that only routes with a small probability of failure are considered feasible:

$$P(\text{failure on any route}) \leq \alpha. \tag{3.24}$$

We now show, as in Laporte, Louveaux, and Mercure [1989], that any route that violates (3.24) can be eliminated by a linear inequality. For any route L, let $S = \{i_1, i_2, \cdots, i_k\}$ be the index set of visited customers. Violation of (3.24) occurs if

$$P\left(\sum_{i \in S} \mathbf{d}_i > C\right) > \alpha. \tag{3.25}$$

Let $V_\alpha(S)$ denote the smallest number of vehicles required to serve S so that the probability of failure in S does not exceed α, i.e., $V_\alpha(S)$ is the smallest integer such that

$$P\left(\sum_{i \in S} \mathbf{d}_i > C \cdot V_\alpha(S)\right) \leq \alpha. \tag{3.26}$$

Now, let $\bar{S}$ denote the complement of S versus V_0, i.e., $\bar{S} = V_0 \backslash S$. Then the following *subtour elimination constraint* imposes, in a linear fashion, that at least $V_\alpha(S)$ vehicles are needed to cover demand in S:

$$\sum_{i \in S, j \in \bar{S} \text{ or } i \in \bar{S}, j \in S} x_{ij} \geq 2V_\alpha(S), \tag{3.27}$$

where, as usual, $x_{ij} = 1$ when arc ij is traveled in the solution and $x_{ij} = 0$ otherwise. It follows that routes that violate (3.24) can be eliminated when needed by the linear constraint (3.27). Observe that this result is obtained without any assumption on the random variables. Also observe that (3.27) is not the deterministic equivalent of (3.24). This should be clear from the fact that an analytical expression for (3.24) is difficult to write. Finally, observe that in practice, as for many random variables, the probability distribution of $\sum_{i \in S} \mathbf{d}_i$ is easily obtained. The computation of $V_\alpha(S)$ in (3.26) poses no difficulty. Additional results appear in the survey on stochastic vehicle routing by Gendreau, Laporte, and Séguin [1996].

Exercises

1. Consider the following second-stage integer program:

   $$Q(x, \xi) = \max\{4y_1 + y_2 | y_1 + y_2 \leq \xi x, y_1 \leq 2, y_2 \leq 1, y \text{ integer}\}.$$

   (a) Obtain y_1^*, y_2^*, and $Q(x, \xi)$ as Gomory functions.
   (b) Consider $\xi = 1$. Observe that $Q(x, 1)$ is piecewise constant on four pieces ($x < 1$, $1 \leq x < 2$, $2 \leq x < 3$, $3 \leq x$).
   (c) Now assume ξ is uniformly distributed over $[0, 2]$. Obtain $\mathcal{Q}(x)$ on four pieces ($x < 0.5$, $0.5 \leq x < 1$, $1 \leq x < 1.5$, $1.5 \leq x$). Check the nonconcavity of $\mathcal{Q}(x)$. Observe that $\mathcal{Q}(x)$ is concave on each piece separately, but that $\mathcal{Q}(x)$ is not (compare, e.g., $\mathcal{Q}(1)$ to $1/2\mathcal{Q}(3/4) + 1/2\mathcal{Q}(5/4)$.

2. Consider ξ uniformly distributed over $[0, 1]$ and $0 \leq x \leq 1$. Show that $u(x) + v(x) = 1$.

3. Consider ξ uniformly distributed over $[0, 2]$.

   (a) Compute $u(x)$ directly from definition (3.9) and check with the result in Example 4. Observe that $u(x)$ is piecewise linear, convex, and continuous.

(b) Compute $\hat{u}(x)$.

(c) Show that $u(x) - \hat{u}(x)$ is decreasing in x.

4. Consider ξ that is Poisson distributed with parameter three. Compute $u(3)$.

5. (a) Let ξ be normally distributed with mean zero and variance one. What is the accuracy level of $u_3(0)$ versus $u(0)$.

(b) Let ξ be normally distributed with mean μ and variance σ^2. Show that $u(\mu)$ is independent of μ. Is the accuracy of $u_n(\mu)$, n given, increasing or decreasing with σ^2?

3.4 Two-Stage Stochastic Nonlinear Programs with Recourse

In this section, we generalize the results from the previous sections to problems with nonlinear functions. The results extend directly so the treatment here will be brief. The basic types of results we would like to obtain concern the structure of the feasible region, the optimal value function, and optimality conditions. As a note of caution, some of the results in this chapter refer to concepts from measure theory.

We begin with a definition of the two-stage stochastic nonlinear program with (additive) recourse. The additive form of the recourse is used to obtain separation of first- and second-period problems as in (1.2). This problem has the form:

$$\inf z = f^1(x) + \mathcal{Q}(x)$$
$$\text{s. t. } g_i^1(x) \le 0, i = 1,\ldots,\bar{m}_1,$$
$$g_i^1(x) = 0, i = \bar{m}_1 + 1,\ldots,m_1, \quad (4.1)$$

where $\mathcal{Q}(x) = E_\omega[Q(x,\omega)]$ and

$$Q(x,\omega) = \inf f^2(y(\omega),\omega)$$
$$\text{s. t. } t_i^2(x,\omega) + g_i^2(y(\omega),\omega) \le 0, \ i = 1,\ldots,\bar{m}_2,$$
$$t_i^2(x,\omega) + g_i^2(y(\omega),\omega) = 0, \ i = \bar{m}_2 + 1,\ldots,m_2, \quad (4.2)$$

where all functions $f^2(\cdot,\omega), t_i^2(\cdot,\omega)$, and $g_i^2(\cdot,\omega)$ are continuous for any fixed ω and measurable in ω for any fixed first argument. Given this assumption, $Q(x,\omega)$ is measurable (Exercise 1) and hence $\mathcal{Q}(x)$ is well-defined.

We make the following definitions consistent with Section 1.

$$K_1 \equiv \{x|g_i^1(x) \le 0, i = 1,\ldots,\bar{m}_1; g_i^1(x) = 0, i = \bar{m}_1 + 1,\ldots,m_1\},$$

$$K_2(\omega) = \{x|\exists y(\omega)|t_i^2(x,\omega) + g_i^2(y(\omega),\omega) \le 0, i = 1,\ldots,\bar{m}_2;$$

$$t_i^2(x,\omega) + g_i^2(y(\omega),\omega) = 0, i = \bar{m}_2 + 1, \ldots, m_2\},$$

and

$$K_2 = \{x | \mathcal{Q}(x) < \infty\}.$$

We have not forced fixed recourse in Problem 4.1 because the second-period constraint functions may depend on ω and on $y(\omega)$. For linear programs, we assumed fixed recourse so we could describe the feasible region in terms of intersections of feasible regions for each random outcome. We could also follow this approach here but the conditions for this result depend directly on the form of the objective and constraint functions. We explore these possibilities in Exercise 1 but we continue here with the more general case.

We also only allow the first-period decision x to act separately in the constraints of the recourse problem. We make this restriction so we can develop optimality conditions that are separable between the first- and second-period variables. Other formulations might allow for nonseparable constraints and dependence of the objective on x. In this way, we could model distributional changes through a transformation determined by x.

We make some other assumptions, however, to allow results along the lines of the previous section. These conditions ensure regularity for the application of necessary and sufficient optimality conditions.

1. *Convexity.* The function f^1 is convex on $\Re^{n_1}$, g_i^1 is convex on $\Re^{n_1}$ for $i = 1, \ldots, \bar{m}_1$, g_i^1 is affine on $\Re^{n_1}$ for $i = \bar{m}_1 + 1, \ldots, m_1$, $f^2(\cdot, \omega)$ is convex on $\Re^{n_2}$ for all $\omega \in \Omega$, $g_i^2(\cdot, \omega)$ is convex on $\Re^{n_2}$ for all $i = 1, \ldots, \bar{m}_2$ and for all $\omega \in \Omega$, $g_i^2(\cdot, \omega)$ is affine on $\Re^{n_2}$ for $i = \bar{m}_2 + 1, \ldots, m_2$ and for all $\omega \in \Omega$, and $t_i^2(\cdot, \omega)$ is convex on $\Re^{n_1}$ for all $i = 1, \ldots, \bar{m}_2$ and for all $\omega \in \Omega$, and $t_i^2(\cdot, \omega)$ is affine on $\Re^{n_1}$ for $i = \bar{m}_2 + 1, \ldots, m_2$.

2. *Slater condition.* If $\mathcal{Q}(x) < \infty$, for almost all $\omega \in \Omega$, there exists some $y(\omega)$ such that $t_i^2(x,\omega) + g_i^2(\omega, y(\omega)) < 0$ for $i = 1, \ldots, \bar{m}_2$ and $t_i^2(x,\omega) + g_i^2(\omega, y(\omega)) = 0$ for $i = \bar{m}_2 + 1, \ldots, m_2$.

The main purpose of these assumptions is to ensure that the resulting deterministic equivalent nonlinear program is also convex. The following theorem gives conditions for convexity of the recourse function. It follows directly from the definitions.

Theorem 32. *Under Assumptions 1 and 2, the recourse function $Q(x, \omega)$ is a convex function of x for all $\omega \in \Omega$.*

Proof. Let y_1 solve the optimization problem in (4.2) for x_1 and let y_2 solve the corresponding problem for x_2. Consider $x = \lambda x_1 + (1 - \lambda)x_2$. In this case, $t_i^2(\lambda x_1 + (1-\lambda)x_2, \omega) + g_i^2(\lambda y_1 + (1-\lambda)y_2, \omega) \le \lambda t_i^2(x_1, \omega) + (1 - \lambda)t_i^2(x_2, \omega) + \lambda g_i^2(y_1, \omega) + (1 - \lambda)g_i^2(y_2, \omega) \le 0$ for each $i = 1, \ldots, \bar{m}_2$. We also have that $t_i^2(\lambda x_1 + (1-\lambda)x_2, \omega) + g_i^2(\lambda y_1 + (1-\lambda)y_2, \omega) = \lambda t_i^2(x_1, \omega) +$

$(1-\lambda)t_i^2(x,\omega)+\lambda g_i^2(y_1,\omega)+(1-\lambda)g_i^2(y_2,\omega) = 0$ for each $i = \bar{m}_2+1,\ldots,m_2$. So, $Q(\lambda x_1 + (1-\lambda)x_2,\omega) \le f^2(\lambda y_1 + (1-\lambda)y_2,\omega) \le \lambda f^2(x_1,\omega) + (1-\lambda)f^2(x_2,\omega) = \lambda Q(x_1,\omega) + (1-\lambda)Q(x_2,\omega)$, giving the result. $\square$

We can also obtain continuity of the recourse function if we assume the recourse feasible region is bounded.

Theorem 33. *If the recourse feasible region is bounded for any $x \in \Re^{n_1}$, then the function $Q(x,\omega)$ is lower semicontinuous in x for all $\omega \in \Omega$ (i.e., $Q(x,\omega)$ is a closed convex function).*

Proof. Proving lower semicontinuity is equivalent (see, e.g., Rockafellar [1969]) to showing that

$$\liminf_{x\to\bar{x}} Q(x,\omega) \ge Q(\bar{x},\omega)$$

for any $\bar{x} \in \Re^{n_1}$, $x \to \bar{x}$, and $\omega \in \Omega$. Suppose a sequence $x^\nu \to \bar{x}$. We can assume that $Q(x^\nu,\omega) < \infty$ for all ν because there is either a subsequence of $\{x^\nu\}$ that is finite valued in Q or the result holds trivially.

We therefore have $t_i^2(x^\nu,\omega) + g_i^2(y^\nu(\omega),\omega) \le 0$ for $i = 1,\ldots,\bar{m}_2$ and $t_i^2(x^\nu,\omega) + g_i^2(y^\nu(\omega),\omega) = 0$ for $i = \bar{m}_2 + 1,\ldots,m_2$ and for some $y^\nu(\omega)$. Hence, by continuity of each of these functions and the boundedness assumption, the $\{y^\nu(\omega)\}$ sequence must have some limit point, e.g., $\bar{y}(\omega)$. Thus, $t_i^2(\bar{x},\omega) + g_i^2(\bar{y}(\omega),\omega) \le 0$ for $i = 1,\ldots,\bar{m}_2$ and $t_i^2(\bar{x},\omega) + g_i^2(\bar{y}(\omega),\omega) = 0$ for $i = \bar{m}_2 + 1,\ldots,m_2$. So, $\bar{x}$ is feasible and $Q(\bar{x},\omega) \le f^2(\bar{x},\omega) = \lim_\nu f^2(x^\nu,\omega) = \lim_\nu Q(x^\nu,\omega)$. $\square$

Because integration is a linear operation on the convex function Q, we obtain the following corollaries.

Corollary 34. *The expected recourse function $\mathcal{Q}(x)$ is a convex function in x.*

Corollary 35. *The feasibility set $K_2 = \{x|\mathcal{Q}(x) < \infty\}$ is closed and convex.*

Corollary 36. *Under the conditions in Theorem 33, $\mathcal{Q}$ is a lower semicontinuous function on x.*

This corollary then leads directly to the following attainability result.

Theorem 37. *Suppose the conditions in Theorem 33, K_1 is bounded, f^1 continuous, g_i^1 and g_i^2 continuous for each i, and $K_1 \cap K_2 \ne \emptyset$. Then (4.1) has a finite optimal solution and the infimum is attained.*

Proof. From Corollary 34, $\mathcal{Q}$ is continuous on its effective domain. The continuity of g_i^1 also implies that K_1 is closed so the optimization is for a continuous, convex function over the nonempty, compact region $K_1 \cap K_2$. $\square$

Other results may follow for specific cases from Fenchel's duality theorem (see Rockafellar [1969]). In some cases, it may be difficult to decompose the feasibility set K_2 into $\cap_\omega K_2(\omega)$. It is possible if f^2 is always dominated by some integrable function in ω for any $y(\omega)$ feasible in the recourse problem for all x. This might be verifiable if, for example, the feasible recourse region is bounded for all $x \in K_1$. Another possibility is for special functions such as the quadratic function in Exercise 2.

We can now proceed to state optimality conditions for (4.1) as in Theorem 9. As a reminder from Section 2.9, in the following, we use ri to indicate *relative interior*.

Theorem 38. *If there exists x such that $x \in ri(dom(f^1(x)))$ and $x \in ri(dom(\mathcal{Q}(x)))$ and $g_i^1(x) < 0$ for all $i = 1, \ldots, \bar{m}_1$ and $g_i^1(x) = 0$ for all $i = \bar{m}_1 + 1, \ldots, m_1$, then x^* is optimal in (4.1) if and only if $x^* \in K_1$ and there exists $\lambda_i^* \geq 0$, $i = 1, \ldots, \bar{m}_1$, λ_i^*, $i = \bar{m}_1 + 1, \ldots, m_1$, such that $\lambda_i^* g_i^1(x^*) = 0$, $i = 1, \ldots, \bar{m}_1$, and*

$$0 \in \partial f^1(x^*) + \partial \mathcal{Q}(x^*) + \sum_{i=1}^{m_1} \lambda_i^* \partial g_i^1(x^*). \tag{4.3}$$

Proof. This result is a direct extension of the general optimality conditions in nonlinear programming (see, e.g., Rockafellar [1969, Theorem 28.3]). $\square$

For most practical purposes, we need to obtain some decomposition of $\partial \mathcal{Q}(x)$ into subgradients of the $Q(x, \omega)$. The same argument as in Theorem 11 applies here so that

$$\partial \mathcal{Q}(x) = E_\omega[\partial Q(x, \omega)] + N(K_2, x) \tag{4.4}$$

for all $x \in K$. Moreover, if we have relatively complete recourse, we can remove the normal cone term in (4.4).

We can also develop optimality conditions that apply to the problem with explicit constraints on nonanticipativity as in Section 1. In this case, Problem 4.1 becomes

$$\inf_{(x(\omega),y(\omega))\in X} \int_\Omega (f^1(x(\omega)) + f^2(y(\omega), \omega))\mu(d\omega)$$

$$\text{s. t. } g_i^1(x(\omega)) \leq 0, a.s.,$$
$$i = 1, \ldots, \bar{m}_1,$$
$$g_i^1(x(\omega)) = 0, a.s.,$$
$$i = \bar{m}_1 + 1, \ldots, m_1,$$
$$E_\Omega(x(\omega)) - x(\omega) = 0, a.s.,$$
$$t_i^2(x(\omega), \omega) + g_i^2(y(\omega), \omega) \leq 0, a.s.,$$

$$i = 1, \ldots, \bar{m}_2,$$
$$t_i^2(x(\omega), \omega) + g_i^2(y(\omega), \omega) = 0, a.s.,$$
$$i = \bar{m}_2 + 1, \ldots, m_2,$$
$$x(\omega), y(\omega) \geq 0, a.s. \qquad (4.5)$$

The optimality results appear in the following theorem.

Theorem 39. *Assume that (4.5) with $X = \mathcal{L}_\infty(\Omega, \mathcal{B}, \mu; \Re^{n_1+n_2})$ is feasible, has a bounded optimal value, satisfies relatively complete recourse, and that a feasible solution $(x^*(\omega), y^*(\omega))$ is at a point satisfying the linear independence condition that any vector in $\partial f^2(y^*(\omega), \omega)$ cannot be written as a combination of some strict subset of representative vectors from $\partial g_i^2(y^*(\omega), \omega)$ for i such that $t_i^2(x^*(\omega), \omega) + g_i^2(y^*(\omega), \omega) = 0$. Then $(x^*(\omega), y^*(\omega))$ is optimal in (4.5) if and only if there exist integrable functions on Ω, $(\lambda^*(\omega), \rho^*(\omega), \pi^*(\omega))$, such that, for almost all ω,*

$$\rho^*(\omega) \in \partial f^1(x^*(\omega)) + \sum_{i=1}^{m_1} \lambda_i^*(\omega)\partial g_i^1(x^*(\omega)) + \sum_{i=1}^{m_2} \pi_i^*(\omega)\partial t_i^2(x^*(\omega), \omega), \quad (4.6)$$

$$\lambda_i^*(\omega) \geq 0, \lambda_i^*(\omega)g_i^1(x^*(\omega)) = 0, i = 1, \ldots, \bar{m}_1, \qquad (4.7)$$

$$0 \in \partial f^2(y^*(\omega), \omega) + \sum_{i=1}^{m_2} \pi_i^*(\omega)\partial g_i^2(y^*(\omega), \omega), \qquad (4.8)$$

$$\pi_i^*(\omega) \geq 0, \pi_i^*(\omega)g_i^2(y^*(\omega), \omega) = 0, i = 1, \ldots, \bar{m}_2, \qquad (4.9)$$

and

$$E_\omega[\rho^*(\omega)] = 0. \qquad (4.10)$$

Proof. The proof of Theorem 39 is similar to the proof of Theorem 13. We first observe that the conditions are sufficient (Exercise 4) and then develop the necessary conditions using (4.5). In this case, we just need to find $\partial Q(x, \omega)$. By the regularity assumption (see again, e.g., Bazaraa and Shetty [1979, Theorem 6.2.4]), $Q(x, \omega) = \sup_{\{\pi(\omega):\pi_i(\omega)\geq 0, i=1,\ldots,\bar{m}_2\}} \inf_{y(\omega)}\{f^2(\omega, y(\omega)) + \sum_{i=1}^{m_2} \pi_i(\omega)(g_i^2(\omega, y(\omega)) + t_i^2(x(\omega), \omega))\}$. For the infimum problem, we have through the regularity condition that if $y^*(\omega)$ is a minimizer given $x^*(\omega)$ and $\pi^*(\omega)$ with $\pi_i^*(\omega) \geq 0, i = 1, \ldots, \bar{m}_2$, then $0 \in \partial f^2(y^*(\omega), \omega) + \sum_{i=1}^{m_2} \pi_i^*(\omega)\partial g_i^2(y^*(\omega), \omega)$.

Because any vector $\eta_i \in \partial t_i^2(x^*(\omega), \omega)$ satisfies $\eta_i^T(x-x^*) \le t_i^2(x(\omega), \omega) - t_i^2(x^*(\omega), \omega)$ by assumption,

$$\sum_{i=1}^{m_2} \pi_i^*(\omega)\eta_i^T(x - x^*)$$

$$\le f^2(y^*(\omega), \omega) + \sum_{i=1}^{m_2} \pi_i^*(\omega)(g_i^2(y^*(\omega), \omega) + t_i^2(x(\omega), \omega))$$

$$- f^2(y^*(\omega), \omega) - \sum_{i=1}^{m_2} \pi_i^*(\omega)(g_i^2(y^*(\omega), \omega) + t_i^2(x^*(\omega), \omega))$$

$$\le Q(x, \omega) - Q(x^*, \omega).$$

Next, suppose there exists some η such that $\eta \in \partial Q(x^*, \omega)$ but $\eta \ne \sum_i \pi_i^* \eta_i$ for any $\eta_i \in \partial t_i^2(x^*(\omega), \omega)$ then, by the Farkas lemma, there exists some z such that $z^T\eta > 0$ and $\pi_i^* z^T \eta_i \le 0$ for all i and π_i^* such that $0 \in \partial f^2(y^*(\omega), \omega) + \sum_{i=1}^{m_2} \pi_i^*(\omega)\partial g_i^2(y^*(\omega), \omega)$. However, $z^T\eta > 0$ implies that $Q(x^* + z, \omega) > Q(x^*, \omega)$ while $\pi_i^* z^T \eta_i \le 0$, which then implies that $-t_i^2(x + z, \omega) + t_i^2(x, \omega) \le 0$ and, for any constraint i with some positive π_i^*, the directional derivative of t_i^2 at x^* in the direction z is $t_i^{2\prime}(x^*; z) = \sup_{\eta_i \in \partial t_i^2(x^*)}\{z^T\eta_i\} \le 0$. Hence for all $t_i^2(x^*, \omega) + g_i^2(y^*(\omega), \omega) = 0$, $t_i^2(x^* + z, \omega) \le t_i^2(x^*, \omega)$ for some nonzero z. Thus, the $y^*(\omega)$ is still feasible in the recourse problem with $x(\omega) = x^* + z$ and $Q(x^* + z, \omega) \le Q(x^*, \omega)$.

So, we must have that $\{\sum_{i=1}^{m_2} \pi_i^*(\omega)\eta_i^T(x - x^*)|\eta_i \in \partial t_i^2(x^*(\omega), \omega)\}$ and $0 \in \partial f^2(y^*(\omega), \omega) + \sum_{i=1}^{m_2} \pi_i^*(\omega)\partial g_i^2(y^*(\omega), \omega) = \partial Q(x^*, \omega)$. Now, we can define $\rho^*(\omega)$ as $-E_\omega[\sum_{i=1}^{m_2} \pi_i^*(\omega)\eta_i] + \sum_{i=1}^{m_2} \pi_i^*(\omega)\eta_i$ to obtain the result as in Theorem 12. $\square$

Again the ρ functions represent the value of information in each of the scenarios under ω. These results can also be generalized to allow for non-separability between the first and second stage but for our computational descriptions, this is generally not necessary.

Exercises

1. Show that the assumptions made when defining (4.1) and (4.2) imply that $Q(x, \omega)$ is a measurable function of ω for all x. (Hint: Find $\{\omega | Q(x, \omega) \le \alpha\}$ for any α using a countable covering of $\Re^{n_2}$.)

2. Suppose f^2 is a convex, quadratic function on $\Re^{n_2}$ for each $\omega \in \Omega$ and the constraints g_i^2 and h_j^2 are affine on $\Re^{n_2}$ for all $i = 1, \ldots, m_2$ and $j = 1, \ldots, m_2 - \bar{m}_2$. What conditions on $\xi(\omega)$ can guarantee that $K_2 = \cap_\omega K_2(\omega)$?

3. Construct an example in which the recourse function $Q(x, \omega)$ is not lower semicontinuous. (Hint: Try to make the only feasible recourse

action tend to ∞ while the first-period action tends to some finite value.)

4. Show that conditions in (4.6)–(4.10) are sufficient to obtain optimality in (4.5).

3.5 Multistage Stochastic Programs with Recourse

The previous sections in this chapter concerned stochastic programs with two stages. Most practical decision problems, however, involve a sequence of decisions that react to outcomes that evolve over time. In this section, we will consider the stochastic programming approach to these multistage problems.

We present the same basic results as in previous chapters. We describe the basic structure of feasible solutions, objective values, and conditions for optimality. We begin again with the linear, fixed recourse, finite horizon framework because this model has been the most widely implemented. We then continue with more general approaches.

We start with implicit nonanticipativity constraints as in the previous sections. The multistage stochastic linear program with fixed recourse then takes the following form.

$$\min z = c^1 x^1 + E_{\xi^2}[\min c^2(\omega)x^2(\omega^2) + \cdots + E_{\xi^H}[\min c^H(\omega)x^H(\omega^H)]\cdots]$$
$$\text{s. t. } W^1 x^1 = h^1,$$
$$T^1(\omega)x^1 + W^2 x^2(\omega^2) = h^2(\omega),$$
$$\cdots:$$
$$T^{H-1}(\omega)x^{H-1}(\omega^{H-1}) + W^H x^H(\omega^H) = h^H(\omega),$$
$$x^1 \geq 0; x^t(\omega^t) \geq 0, \ t = 2, \ldots, H; \tag{5.1}$$

where c^1 is a known vector in $\Re^{n_1}$, h^1 is a known vector in $\Re^{m_1}$, $\xi^t(\omega)^T = (c^t(\omega)^T, h^t(\omega)^T, T_{1\cdot}^{t-1}(\omega), \ldots, T_{m_t\cdot}^{t-1})$ is a random N_t-vector defined on (Ω, Σ^t, P) (where $\Sigma^t \subset \Sigma^{t+1}$) for all $t = 2, \ldots, H$, and each W^t is a known $m_t \times n_t$ matrix. The decisions x depend on the history up to time t, which we indicate by ω^t. We also suppose that Ξ^t is the support of ξ^t.

We first describe the deterministic equivalent form of this problem in terms of a dynamic program. If the stages are 1 to H, we can define states as $x^t(\omega^t)$. Noting that the only interaction between periods is through this realization, we can define a dynamic programming type of recursion. For terminal conditions, we have

$$Q^H(x^{H-1}, \xi^H(\omega)) = \min c^H(\omega)x^H(\omega)$$
$$\text{s. t. } W^H x^H(\omega) = h^H(\omega) - T^{H-1}(\omega)x^{H-1},$$
$$x^H(\omega) \geq 0. \tag{5.2}$$

Letting $Q^{t+1}(x^t) = E_{\xi^{t+1}}[Q^{t+1}(x^t, \xi^{t+1}(\omega))]$ for all t, we obtain the recursion for $t = 2, \ldots, H - 1$,

$$Q^t(x^{t-1}, \xi^t(\omega)) = \min c^t(\omega)x^t(\omega) + Q^{t+1}(x^t)$$
$$\text{s. t. } W^t x^t(\omega) = h^t(\omega) - T^{t-1}(\omega)x^{t-1},$$
$$x^t(\omega) \geq 0, \qquad (5.3)$$

where we use x^t to indicate the state of the system. Other state information in terms of the realizations of the random parameters up to time t should be included if the distribution of ξ^t is not independent of the past outcomes. The value we seek is:

$$\min z = c^1 x^1 + Q(x^1)$$
$$\text{s. t. } W^1 x^1 = h^1,$$
$$x^1 \geq 0, \qquad (5.4)$$

which has the same form as the two-stage deterministic equivalent program. Examples of this formulation appeared in Chapter 1 in terms of the capacity expansion and finance problems. The recourse represented reactions to actual demand for power in the first case and yield realizations in the second case.

We would again like to obtain properties of the problems in (5.2)–(5.4) that allow uses of mathematical programming procedures such as decomposition. We concentrate first on the form of the feasible regions for problems of the form (5.3). Let these be

$$K^t = \{x^t | Q^{t+1}(x^t) < \infty\}.$$

We have the following result which helps in the development of several algorithms for multistage stochastic programs.

Theorem 40. *The sets K^t and functions $Q^{t+1}(x^t)$ are convex for $t = 1, \ldots, H - 1$ and, if Ξ^t is finite for $t = 1, \ldots, H$, then K^t and $Q^{t+1}(x^t)$ are polyhedral.*

Proof. Proceed by induction. Because $Q^H(x^{H-1}, \xi^H(\omega))$ is convex for all $\xi^H(\omega)$, so is $Q^H(x^{H-1})$. We can then carry this back to each $t < T - 1$. The same applies for the polyhedrality property because finite numbers of realizations lead to each $Q^{t+1}(x^t)$'s being the sum of a finite number of polyhedral functions, which is then polyhedral. $\square$

We note that we may also describe the feasibility sets K^t in terms of intersections of feasibility sets for each outcome if we have finite second moments for ξ^t in each period. This result is also true when we have a finite number of possible realizations of the future outcomes. In this case, the set of possible future sequences of outcomes are called *scenarios*.

The description of scenarios is often made on a tree such as that in Figure 3. Here, there are eight scenarios that are evident in the last stage

($H = 4$). In previous stages ($t < 4$), we have a more limited number of possible realizations, which we call the *stage t scenarios*. Each of these period t scenarios is said to have a single *ancestor* scenario in stage ($t - 1$) and perhaps several *descendant* scenarios in stage ($t + 1$). We note that different scenarios at stage t may correspond to the same ξ^t realizations and are only distinguished by differences in their ancestors.

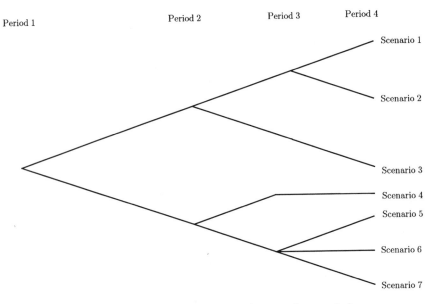

FIGURE 3. A tree of seven scenarios over four periods.

The deterministic equivalent program to (5.1) with a finite number of scenarios is still a linear program. It has the structural form indicated in Figure 4, where we use an additional superscript to index distinct values of W^t and T^t. This is often called *arborescent* form and can be exploited in large-scale optimization approaches as in Kallio and Porteus [1977]. A difficulty is still, however, that these problems become extremely large as the number of stages increases, even if only a few realizations are allowed in each stage.

In some problems, however, we can avoid much of this difficulty if the interactions between consecutive stages are sufficiently weak. This is the case in the capacity expansion problem described in Chapter 1. Here, capacity carried over from one stage to the next is not affected by the demand in that stage. Decisions about the amount of capacity to install can be made at the beginning and then the future only involves reactions to these outcomes. Problems with this form are called *block separable*.

Formally, we have the following definition for block separability (see Louveaux [1986]).

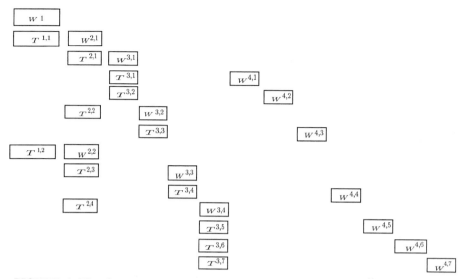

FIGURE 4. The deterministic equivalent matrix for a problem with seven scenarios in four periods.

Definition 41. A multistage stochastic linear program (5.1) has *block separable recourse* if for all periods $t = 1, \ldots, H$ and all ω, the decision vectors, $x^t(\omega)$, can be written as $x^t(\omega) = (w^t(\omega), y^t(\omega))$ where w^t represents aggregate level decisions and y^t represents detailed level decisions. The constraints also follow these partitions:

1. The stage t objective contribution is $c^t x^t(\omega) = r^t w^t(\omega) + q^t y^t(\omega)$.

2. The constraint matrix W^t is block diagonal:

$$W^t = \begin{pmatrix} A^t & 0 \\ 0 & B^t \end{pmatrix}. \tag{5.5}$$

3. The other components of the constraints are random but we assume that for each realization of ω, $T^t(\omega)$ and $h^t(\omega)$ can be written:

$$T^t(\omega) = \begin{pmatrix} R^t(\omega) & 0 \\ S^t(\omega) & 0 \end{pmatrix} \text{ and } h^t(\omega) = \begin{pmatrix} b^t(\omega) \\ d^t(\omega) \end{pmatrix}, \tag{5.6}$$

where the zero components of T^t correspond to the detailed level variables.

Notice that (3) in the definition implies that detailed level variables have no direct effect on future constraints. This is the fundamental advantage of block separability.

With block separable recourse, we may rewrite $Q^t(x^{t-1}, \xi^t(\omega))$ as the sum of two quantities, $Q_w^t(w^{t-1}, \xi^t(\omega)) + Q_y^t(w^{t-1}, \xi^t(\omega))$, where we need not include the y^{t-1} terms in x^{t-1},

$$Q_w^t(w^{t-1}, \xi^t(\omega)) = \min r^t(\omega)w^t(\omega) + Q^{t+1}(x^t)$$
$$\text{s. t. } A^t w^t(\omega) = b^t(\omega) - R^{t-1}(\omega)w^{t-1},$$
$$w^t(\omega) \geq 0, \tag{5.7}$$

and

$$Q_y^t(w^{t-1}, \xi^t(\omega)) = \min q^t(\omega)y^t(\omega)$$
$$\text{s. t. } B^t y^t(\omega) = d^t(\omega) - S^{t-1}(\omega)w^{t-1},$$
$$y^t(\omega) \geq 0. \tag{5.8}$$

The great advantage of block separability is that we need not consider nesting among the detailed level decisions. In this way, the w variables can all be pulled together into a first stage of aggregate level decisions. The second stage is then composed of the detailed level decisions. Note that if the b^t and R^t are known, then the block separable problem is equivalent to a similarly sized two-stage stochastic linear program.

Separability is indeed a very useful property for stochastic programs. Computational methods should try to exploit it whenever it is inherent in the problem because it may reduce work by orders of magnitude. We will also see in Chapter 11 that separability can be added to a problem (with some error that can be bounded). This approach opens many possible applications with large numbers of random variables.

Another modeling approach that may have some computational advantage appears in Grinold [1976]. This approach extends from analyses of stochastic programs as examples of Markov decision process. He assumes that ω^t belongs to some finite set $1, \ldots, k_t$, that the probabilities are determined by $p_{ij} = P\{\omega^{t+1} = j | \omega^t = i\}$ for all t, and that $T^t = T^t(\omega^t, \omega^{t+1})$. In this framework, he can obtain an approximation that again obtains a form of separability of future decisions from previous outcomes. We discuss more approximation approaches in Chapter 11.

We now consider generalizations into nonlinear functions and infinite horizons. The general results of the previous section can be extended here directly. We concentrate on some areas where differences may occur and, for notational convenience, concentrate just on the description of problems in the form with explicit nonanticipativity constraints. More detailed descriptions of these problems appear in the papers by Rockafellar and Wets [1976a,1976b], Dempster [1988], Flåm [1985], and Birge and Dempster [1992].

For our basic model, we assume that within each period t, there is a convex random extended-value function of x^t and x^{t+1} that includes the original objective and any constraints linking the decisions in periods t and $t + 1$. The constraints are, therefore, implicitly contained in the objective function.

The only explicit constraints are to maintain nonanticipativity. The remaining development again supposes some familiarity with concepts from measure theory. Proofs are omitted because we are most interested in the results for computational considerations. In this model, the random parameter is a data process, $\omega := \{\omega^t : t = 0, \ldots\}$. The decisions are a process $\mathbf{x} := \{x^t : t = 0, \ldots\}$ such that x is a measurable function $x : \omega \to x(\omega)$. The space of the decision processes is again the space of *essentially bounded functions*, $L_\infty^n := L_\infty(\Omega \times N, \Sigma \times \mathcal{P}(N), \mu \times \#; \Re^n)$, where $\mathcal{P}$ is the power set and $\#$ is the counting measure. In distinguishing information from period to period, we associate a filtration with the data process. $F := \{\Sigma^t\}_{t=1}^\infty$, where $\Sigma^t := \sigma(\bar{\omega}^t)$ is the σ-field of the *history process* $\bar{\omega}^t := \{\omega^0, \ldots, \omega^t\}$, and the Σ^t satisfy $\{0, \Omega\} \subset \Sigma_0 \subset \cdots \subset \Sigma$.

Nonanticipativity of the decision process at time t implies that decisions must only depend on the data up to time t, i.e., $\mathbf{x}^t$ must be Σ^t–measurable. An alternative characterization of this nonanticipative property is that $\mathbf{x}_t = E\{\mathbf{x}^t | \Sigma^t\}$ a.s., $t = 0, \ldots$, where $E\{\cdot | \Sigma^t\}$ is conditional expectation with respect to the σ-field Σ^t. Using the projection operator $\Pi^t : z \to \Pi^t z := E\{z | \Sigma^t\}, t = 0, \ldots$, this is equivalent to

$$(I - \Pi_t)x^t = 0, \ t = 0, \ldots. \tag{5.9}$$

We then let $\mathcal{N}$ denote the closed linear subspace of nonanticipative processes in L_∞^n.

The general multistage stochastic programming model is to find

$$\inf_{\mathbf{x} \in \mathcal{N}} E \sum_{t=0}^H f^t(\omega, x^t(\omega), x_{t+1}(\omega)), \tag{5.10}$$

where "E" is expectation with respect to Σ. Using our random variable boldface notation, expression (5.10) then becomes

$$\inf_{\mathbf{x} \in \mathcal{N}} E \sum_{t=0}^H \mathbf{f}^t(\mathbf{x}_t, \mathbf{x}^{t+1}), \tag{5.11}$$

with objective $z(\mathbf{x}) := E \sum_{t=0}^H \mathbf{f}_t(\mathbf{x}^t, \mathbf{x}^{t+1})$.

We develop optimality conditions that allow $H \to \infty$. The conditions are basically the same as in previous sections (in terms of some assumption about relatively complete recourse and some regularity condition), but we need some additional assumptions to control multipliers at $H = \infty$. We assume in (5.11) that the objective components $\mathbf{f}^t$ are proper convex normal integrands (in this case, proper, lower semicontinuous convex functions for each t, see Rockafellar [1976b]) .

Without additional restrictions, the objective in (5.11) may be infinite. We can avoid this difficulty by defining a policy $\mathbf{x}^* := \{\mathbf{x}^0, \mathbf{x}^{1*}, \ldots\}$ as *optimal (weakly)* as in McKenzie [1976] if it is not *overtaken* by any other policy, i.e., if there does not exist $\mathbf{x}'$ such that

$$\limsup_{\tau \to \infty} E \sum_{t=0}^{\tau}(\mathbf{f}^t(\mathbf{x}'^{,t},\mathbf{x}'^{,t+1}) - \mathbf{f}^t(\mathbf{x}^{*t},\mathbf{x}^{*t+1})) \leq -\epsilon, \qquad (5.12)$$

where $\epsilon > 0$. This type of optimality has some advantages over other infinite horizon optimality conditions (such as *average optimality*). If we also make some assumptions about the growth of values so that we can always subtract some value from a weakly optimal solution to obtain a solution with zero value, then we can assume that (5.11) has finite optimal value.

The following theorem states the basic necessary conditions for an optimal solution to (5.11). We do not give the proof that appears in Birge and Dempster [1992]. Note that relatively complete recourse is replaced here with *nonanticipative feasibility* as in Flåm [1986]. Dempster [1981] obtains similar results using relatively complete recourse. In general, the two conditions are similar but not identical (see Exercises 4 and 5).

Theorem 42. *Suppose $\mathbf{x}^*$ is optimal in (5.11) with finite infimum and the following.*

(a) *(nonanticipative feasibility) For any $\mathbf{x} \in$ dom z (i.e., such that $E\sum_{t=0}^{\infty}\mathbf{f}^t(\mathbf{x}^t,\mathbf{x}^{t+1}) < \infty$), the projection of $\mathbf{x}$ into $\mathcal{N}$, $\Pi\mathbf{x}$, is such that $E\sum_{t=0}^{\infty}\mathbf{f}^t(\Pi\mathbf{x}^t,\Pi\mathbf{x}^{t+1}) < \infty$.*

(b) *(strict feasibility) For some $\mathbf{x} \in \mathcal{N}$, such that $E\sum_{t=0}^{\infty}\mathbf{f}^t(\mathbf{x}^t,\mathbf{x}^{t+1}) < \infty$, there exists $\delta > 0$ such that for all $\|\mathbf{y} - \mathbf{x}\| < \delta, \mathbf{y} \in L_\infty^n$, $E\sum_{t=0}^{\infty}\mathbf{f}^t(\mathbf{y}^t,\mathbf{y}^{t+1}) < \infty$.*

(c) *(finite horizon continuation approximation) For any $\mathbf{x} \in$ dom z, there exists $\mathbf{x}'$ such that for all H_k in some sequence $\{H_1,H_2,...\}$, $(\mathbf{x}^{H_k},\mathbf{x}'^{,H_k+1},\mathbf{x}'^{,H_k+2},...)$ is also feasible, and the transition cost to $\mathbf{x}'$ is such that $|E[\mathbf{f}^{H_k-1}(\mathbf{x}^{H_k-1},\mathbf{x}^{H_k}) + \mathbf{f}^{H_k}(\mathbf{x}^{H_k},\mathbf{x}'^{,H_k+1})]| \to 0$ as $k \to \infty$ and $|E[\mathbf{f}^{H_k-1}(\mathbf{x}^{H_k-1},\mathbf{x}^{H_k}) + \mathbf{f}^{H_k}(\mathbf{x}^{H_k},\mathbf{x}'^{,H_k+1})]| \geq |E[\mathbf{f}^{H_k-1}(\mathbf{x}^{H_k-1},\mathbf{x}^{H_k})+\mathbf{f}^{H_k}(\mathbf{x}^{H_k},\mathbf{x}^{H_k+1})]|$ for $k = 1,...$.*

Then $\mathbf{x}^*$ is optimal with given initial conditions $\mathbf{x}^0$ if and only if there exist $\mathbf{p}^t \in L_1^n(\Sigma), t = 0,...$ such that

i. *$\mathbf{p}^t$ is nonanticipative, i.e. $\mathbf{p}_t = E\{\mathbf{p}^t|\Sigma_t\}$ a.s. for $t = 0,....$*

ii. *$E^0(\mathbf{f}^0(\mathbf{x}^0,\mathbf{x}^1) - \mathbf{p}^0\mathbf{x}^0 + \mathbf{p}^1\mathbf{x}^1)$ is a.s. minimized by $\mathbf{x}^{*1} = \mathbf{x}^1$ over $\mathbf{x}^1 = E\{\mathbf{x}^1 \mid \Sigma^1\}$, and, for $t > 0$, $E(\mathbf{f}^t(\mathbf{x}^t,\mathbf{x}^{t+1})-\mathbf{p}^t\mathbf{x}^t+ \mathbf{p}^{t+1}\mathbf{x}^{t+1})$ is a.s. minimized by $(\mathbf{x}^{*t},\mathbf{x}^{*t+1}) = (\mathbf{x}^t,\mathbf{x}^{t+1})$ over $\mathbf{x}^t = E\{\mathbf{x}^t \mid \Sigma^t\}$ and $\mathbf{x}^{t+1} = E\{\mathbf{x}^{t+1} \mid \Sigma^{t+1}\}$, and*

iii. *$E\,\mathbf{p}^{t_k}(\mathbf{x}^{t_k} - \mathbf{x}^{*t_k}) \to 0$ as $t_k \to \infty$, for all $\mathbf{x} \in$ dom z.*

Proof: The proof is given in Birge and Dempster [1992] and follows Flåm [1985, 1986]. The result establishes a price system, $\mathbf{p}^t$, on the value of information. $\square$

This basic result can be extended to results with constraints in the same way as necessary conditions in the previous sections. The only requirement is to describe the subdifferentials of f^t in terms of an objective and constraint functions (see Exercise 6). The significance of the result is that we may again decompose the multistage problem into individual period t problems. In this way, optimization may be applied at each period provided suitable multipliers are available. This property is the basis for the Lagrangian and progressive hedging algorithms described in Chapter 6.

Exercises

1. Show why the capacity expansion example in Section 1.3 is block separable.

2. State the optimality conditions of Theorem 42 for the financial planning model in Section 1.2.

3. A policy is average optimal if it minimizes $\lim_{H \to \infty} \frac{\sum_{t=1}^{H} f(\mathbf{x}^t, \mathbf{x}^{t+1})}{H}$. Give an example where an average optimal policy is not weakly optimal. (Example: Suppose you can produce up to two items in any period with a cost of one per item. You must meet demand of one in each period. Can you produce more than one in each period and still be average optimal?)

4. Suppose a multistage problem with constraints,

$$x^t(1, \omega) + x^{t+1}(1, \omega) + x^{t+1}(2, \omega) \geq \xi^{t+1}(\omega),$$

where $\boldsymbol{\xi}^t \in [0, 1]$. If $x^{t+1}(i, \omega) \geq 0$ a.s. with no other restrictions, this problem would have relatively complete recourse. Suppose

$$\int_{\Sigma_{t+1}} \xi^{t+1}(\omega) P(d\omega) < 1.$$

Show that this problem does not satisfy nonanticipative feasibility.

5. Suppose in the previous example that $\xi^{t+1}(\omega) \equiv 1 + t$. Show that this problem satisfies nonanticipative feasibility but not relatively complete recourse.

6. Suppose that constraints are explicitly represented by $\mathbf{g}^t(\mathbf{x}^t, \mathbf{x}^{t+1})$ in (5.7) instead of being incorporated into $\mathbf{f}^t$. Use Theorem 42 to find a subdifferential form of the necessary conditions similar to the result in Theorem 39.

4

The Value of Information and the Stochastic Solution

Stochastic programs have the reputation of being computationally difficult to solve. Many people faced with real-world problems are naturally inclined to solve simpler versions. Frequently used simpler versions are, for example, to solve the deterministic program obtained by replacing all random variables by their expected values or to solve several deterministic programs, each corresponding to one particular scenario, and then to combine these different solutions by some heuristic rule.

A natural question is whether these approaches can sometimes be nearly optimal or whether they are totally inaccurate. The theoretical answer to this is given by two concepts: the expected value of perfect information and the value of the stochastic solution. The object of this chapter is to study these two concepts. Section 1 introduces the expected value of perfect information. Section 2 gives the value of the stochastic solution. Some basic inequalities and the relationships between these quantities are given in Sections 3 and 4, respectively. Section 5 provides some examples of these quantities. Section 6 presents additional bounds.

4.1 The Expected Value of Perfect Information

The *expected value of perfect information* (EVPI) measures the maximum amount a decision maker would be ready to pay in return for complete (and accurate) information about the future. In the farmer's problem of Chapter 1, we saw that the farmer would greatly benefit from perfect information about future weather conditions, so that he could allocate his land optimally to the various crops.

The concept of EVPI was first developed in the context of decision analysis and can be found in a classical reference such as Raiffa and Schlaifer [1961]. In the stochastic programming setting, we may define it as follows. Suppose uncertainty can be modeled through a number of scenarios. Let ξ be the random variable whose realizations correspond to the various scenarios. Define

$$\min z(x,\xi)= c^T x + \min\{q^T y \mid Wy = h - Tx, y \geq 0\} \qquad (1.1)$$
$$\text{s.t.}\ \ Ax= b, x \geq 0,$$

as the optimization problem associated with one particular scenario ξ, where, as before, $\xi(\omega)^T = (q(\omega)^T, h(\omega)^T, T_1.(\omega), \ldots, T_{m_2}.(\omega))$. To make the definition complete, we repeat the notation, $K_1 = \{x|Ax = b, x \geq 0\}$ and $K_2(\xi) = \{x|\exists y \geq 0 \text{ s.t. } Wy = h - Tx\}$. We define $z(x,\xi) = +\infty$ if $x \notin K_1 \cap K_2(\xi)$ and $z(x,\xi) = -\infty$ if (1.1) is unbounded below. We again use the convention $+\infty + (-\infty) = +\infty$.

We may also reasonably assume that for all $\xi \in \Xi$, there exists at least one $x \in \Re^{n_1}$ such that $z(x,\xi) < \infty$. (Otherwise, there would exist one scenario for which no feasible solution exists at all. No reasonable stochastic model could be constructed in such a situation.) This assumption implies that, for all $\xi \in \Xi$, there exists at least one feasible solution, which in turn implies the existence of at least one optimal solution. Let $\bar{x}(\xi)$ denote some optimal solution to (1.1). As in a scenario approach, we might be interested in finding all solutions $\bar{x}(\xi)$ of problem (1.1) for all scenarios and the related optimal objective values $z(\bar{x}(\xi),\xi)$.

This search is known as the *distribution problem* (as we mentioned in Section 3.1c) because it looks for the distribution of $\bar{x}(\xi)$ and of $z(\bar{x}(\xi),\xi)$ in terms of ξ. The distribution problem can be seen as a generalization of sensitivity analysis or parametric analysis in linear programming.

Here, we assume we somehow have the ability to find these decisions $\bar{x}(\xi)$ and their objective values $z(\bar{x}(\xi),\xi)$ so that we are in a position to compute the expected value of the optimal solution, known in the literature as the *wait-and-see* solution (WS, see Madansky [1960]) where

$$WS= E_\xi \left[\min_x z(x,\xi)\right]$$
$$= E_\xi z(\bar{x}(\xi),\xi). \qquad (1.2)$$

We may now compare the wait-and-see solution to the so-called *here-and-now* solution corresponding to the recourse problem (RP) defined earlier in Chapter 3 as (1.1), and we may now write that as

$$RP = \min_x E_\xi z(x,\xi), \qquad (1.3)$$

with an optimal solution, x^*.

The expected value of perfect information is, by definition, the difference between the wait-and-see and the here-and-now solution, namely,

$$EVPI = RP - WS. \qquad (1.4)$$

An example was given in Chapter 1 in the farmer's problem. The wait-and-see solution value was $-\$115,406$ (when converted to a minimization problem), while the recourse solution value was $-\$108,390$. The expected value of perfect information for the farmer was then $\$7016$.

This is how much the farmer would be ready to pay each year to obtain perfect information on next summer's weather. A meteorologist could reasonably ask him to pay part of this amount to support meteorological research.

4.2   The Value of the Stochastic Solution

For practical purposes, many people would believe that finding the wait-and-see solution or equivalently solving the distribution problem is still too much work (or impossible if perfect information is just not available at any price). This is especially difficult because the wait-and-see approach delivers a set of solutions instead of one solution that would be implementable.

A natural temptation is to solve a much simpler problem: the one obtained by replacing all random variables by their expected values. This is called the *expected value problem* or *mean value problem*, which is simply

$$EV = \min_x z(x, \bar{\xi}), \qquad (2.1)$$

where $\bar{\xi} = E(\xi)$ denotes the expectation of ξ. Let us denote by $\bar{x}(\bar{\xi})$ an optimal solution to (2.1), called the *expected value solution*. Anyone aware of some stochastic programming or realizing that uncertainty is a fact of life would feel at least a little insecure about advising to take decision $\bar{x}(\bar{\xi})$. Indeed, unless $\bar{x}(\xi)$ is somehow independent of ξ, there is no reason to believe that $\bar{x}(\bar{\xi})$ is in any way near the solution of the recourse problem (1.3).

The value of the stochastic solution (first introduced in Chapter 1) is the concept that precisely measures how good or, more frequently, how bad a decision $\bar{x}(\bar{\xi})$ is in terms of (1.3). We first define the *expected result of using the EV solution* to be

$$EEV = E_\xi(z(\bar{x}(\bar{\xi}), \xi)). \qquad (2.2)$$

The quantity, EEV, measures how $\bar{x}(\bar{\xi})$ performs, allowing second-stage decisions to be chosen optimally as functions of $\bar{x}(\bar{\xi})$ and ξ. The value of the stochastic solution is then defined as

$$VSS = EEV - RP. \qquad (2.3)$$

Recall, for example, that in Section 1.1 this value was found using $EEV = -\$107,240$ and $RP = -\$108,390$, for $VSS = \$1150$. This quantity is the cost of ignoring uncertainty in choosing a decision.

4.3   Basic Inequalities

The following relations between the defined values have been established by Madansky [1960]. Generalizations to nonlinear functions can be found in Mangasarian and Rosen [1964].

Proposition 1.

$$WS \leq RP \leq EEV. \tag{3.1}$$

Proof: For every realization, ξ, we have the relation

$$z(\bar{x}(\xi),\xi) \leq z(x^*,\xi),$$

where, as said before, x^* denotes an optimal solution to the recourse problem (1.3). Taking the expectation of both sides yields the first inequality. x^* being an optimal solution to the recourse problem (1.3) while $\bar{x}(\bar{\xi})$ is just one solution to (1.3) yields the second inequality. $\square$

Proposition 2. *For stochastic programs with fixed objective coefficients, fixed T, and fixed W,*

$$EV \leq WS. \tag{3.2}$$

Proof: Jensen's inequality (Jensen [1906]) states that for any convex function $f(\xi)$ of ξ, $Ef(\xi) \geq f(E\xi)$. To apply this result, we need to show that $f(\xi) = \min_x z(x,\xi)$ is a convex function of $\xi = (h)$. Convexity follows by noting that $\min_x z(x,\xi) = \max_{\sigma,\pi}\{\sigma^T b + \pi^T h | \sigma^T A + \pi^T T \leq c^T, \pi^T W \leq q\}$. Since the constraints of the dual problem are unchanged for all $\xi = (h)$, the epigraph of $f(\xi)$ is the intersection of the epigraphs of the linear functions $\sigma^T b + \pi^T h$ for all feasible (σ^T, π^T). Hence, $f(\xi)$ is convex because it has a convex epigraph. $\square$

Proposition 2 does not hold for general stochastic programs. Indeed, if we consider q only to be stochastic, by Theorem 3.5 the function $z(x,\xi)$ is a concave function of ξ and Jensen's inequality does not apply. An example of a program where $EV > WS$ is given in Exercise 3.

Other bounds can be obtained. We give two more examples of such bounds here.

Proposition 3. *Let x^* represent an optimal solution to the recourse problem (1.3) and let $\bar{x}(\bar{\xi})$ be a solution to the expected value problem (1.5). Then*

$$RP \geq EEV + (x^* - \bar{x}(\bar{\xi}))^T \eta, \tag{3.3}$$

where $\eta \in \partial E_{\xi} z(\bar{x}(\bar{\xi}), \xi)$, the subdifferential set of $E_{\xi} z(x, \xi)$ at $\bar{x}(\bar{\xi})$.

Proof. By convexity of $E_{\xi} z(x, \xi)$, the subgradient inequality applied at point x_1 implies that for any x_2 the relation $E_{\xi} z(x_2, \xi) \geq E_{\xi} z(x_1, \xi) + (x_2 - x_1)^T \eta$ holds. The proposition follows by application of this relation for $x_1 = \bar{x}(\bar{\xi})$ and $x_2 = x^*$, by noting that $RP = E_{\xi} z(x^*, \xi)$ and $EEV = E_{\xi} z(\bar{x}(\bar{\xi}), \xi)$. $\square$

The last bound is obtained by considering a slightly different version of the recourse problem, defined as follows:

$$\min z_u(x, \xi) = c^T x + \min\{q^T y \mid Wy \geq h(\xi) - Tx, y \geq 0\} \tag{3.4}$$
$$\text{s. t. } Ax = b,$$
$$x \geq 0.$$

Problem (3.4) differs from problem (1.1) because in (3.4) only the right-hand side is stochastic and the second-stage constraints are inequalities. It is not difficult to observe that all definitions and relations also apply to z_u. If we further assume that $h(\xi)$ is bounded above, then an additional inequality results.

Proposition 4. *Consider problem (3.4) and the related definition*

$$RP = \min_x E_{\xi} z_u(x, \xi).$$

Assume further that $h(\xi)$ is bounded above by a fixed quantity $h_{\max}$. Let $x_{\max}$ be an optimal solution to $z_u(x, h_{\max})$. Then

$$RP \leq z_u(x_{\max}, h_{\max}). \tag{3.5}$$

Proof. For any ξ in Ξ and any $x \in K_1$, a feasible solution to $Wy \geq h_{\max} - Tx, y \geq 0$, is also a feasible solution to $Wy \geq h(\xi) - Tx, y \geq 0$. Hence $z_u(x, h_{\max}) \geq z_u(x, h(\xi))$. Thus $z_u(x, h_{\max}) \geq E_{\xi} z_u(x, h(\xi))$, hence $z_u(x, h_{\max}) \geq \min_x E_{\xi} z_u(x, h(\xi)) = RP$. $\square$

4.4   The Relationship between EVPI and VSS

The quantities, EVPI and VSS, are often different, as our examples have shown. This section describes the relationships that exist between the two measures of uncertainty effects.

From the inequalities in the previous section, the following proposition holds.

Proposition 5.
a. *For any stochastic program,*

$$0 \leq EVPI, \tag{4.1}$$
$$0 \leq VSS. \tag{4.2}$$

b. *For stochastic programs with fixed recourse matrix and fixed objective coefficients,*

$$EVPI \leq EEV - EV, \tag{4.3}$$
$$VSS \leq EEV - EV. \tag{4.4}$$

The proposition indicates that the EVPI and the VSS are (both) nonnegative (anyone would be surprised if this was not true) and are both bounded above by the same quantity $EEV - EV$, which is easily computable. It follows that when $EV = EEV$, both the EVPI and VSS vanish. A sufficient condition for this to happen is to have $\bar{x}(\boldsymbol{\xi})$ independent of $\boldsymbol{\xi}$. This means that optimal solutions are insensitive to the value of the random elements. In such situations, finding the optimal solution for one particular ξ (or for $\bar{\xi}$) would yield the same result, and it is unnecessary to solve a recourse problem. Such extreme situations rarely occur.

From these observations, three lines of research have been addressed. The first one studies relationships between EVPI and VSS. It is illustrated in the sequel of this paragraph by showing an example where EVPI is zero and VSS is not and an example of the reverse. The second one studies classes of problems for which one can observe or theorize that the EVPI is low. Examples and counterexamples are given in Section 5. The third one studies refined bounds on EVPI and VSS. Results about refined upper and lower bounds on EVPI and VSS appear in Section 6.

We thus end this section by showing examples taken from Birge [1982] that illustrate cases in which one of the two concepts (EVPI and VSS) is null and the other is positive.

a. $EVPI = O$ and $VSS \neq O$

Consider the following problem

$$z(x, \boldsymbol{\xi}) = x_1 + 4x_2 + \min\{y_1 + 10y_2^+ + 10y_2^- \mid y_1 + y_2^+ - y_2^- = \boldsymbol{\xi} + x_1 - 2x_2, y_1 \leq 2, y \geq 0\}$$
$$\text{s. t. } x_1 + x_2 = 1,$$
$$x \geq 0, \tag{4.5}$$

where the random variable ξ follows a uniform density over $[1,3]$. For a given x and ξ, we may conclude that

$$y^*(x,\xi) = \begin{cases} y_1 = \xi + x_1 - 2x_2, y_2 = 0 & \text{if } 0 \le \xi + x_1 - 2x_2 \le 2, \\ y_1 = 2, y_2^+ = \xi + x_1 - 2x_2 - 2 & \text{if } \quad \xi + x_1 - 2x_2 > 2, \\ y_2^- = 2x_2 - \xi - x_1 & \text{if } \quad \xi + x_1 - 2x_2 < 0, \end{cases}$$

so that

$$z(x,\xi) = \begin{cases} 2x_1 + 2x_2 + \xi & \text{if } 0 \le \xi + x_1 - 2x_2 \le 2, \\ -18 + 11x_1 - 16x_2 + 10\xi & \text{if } \quad \xi + x_1 - 2x_2 > 2, \\ -9x_1 + 24x_2 - 10\xi & \text{if } \quad \xi + x_1 - 2x_2 < 0. \end{cases}$$

Given the first-stage constraint $x_1 + x_2 = 1$, one has $z(x,\xi) = 2 + \xi$ in the first of these three regions. Now, using the first-stage constraint and the definition of the regions, one can easily check that $z(x,\xi) \ge 2 + \xi$ in the other two regions. Hence, any $\hat{x} \in \{(x_1, x_2) | x_1 + x_2 = 1, x \ge 0\}$ is an optimal solution of (4.5) for $-x_1 + 2x_2 \le \xi \le 2 - x_1 + 2x_2$, or equivalently for $2 - 3x_1 \le \xi \le 4 - 3x_1$.

In particular, $\left(\frac{1}{3}, \frac{2}{3}\right)$ is optimal for all ξ, $(0, 1)$ is optimal for all $\xi \in [2, 3]$, and $(1, 0)$ is optimal for $\xi = \{1\}$.

Taking $\bar{x}(\xi) = \left(\frac{1}{3}, \frac{2}{3}\right)$ for all ξ leads to the conclusion that $\bar{x}(\xi)$ is identical for all ξ, hence WS $=$ RP $= 4$, so that EVPI $= 0$. On the other hand, solving $z(x, \bar{\xi} = 2)$ may yield a different solution, for example, $\bar{x}(2) = (0, 1)$, with EV $= 4$.

In that case,

$$EEV = E_{\xi \le 2}(24 - 10\xi) + E_{\xi \ge 2}(2 + \xi) = \frac{27}{4},$$

so that VSS $= 11/4$.

Because linear programs often include multiple optimal solutions, this type of situation is far from exceptional.

b. VSS $= O$ and $EVPI \ne O$

We consider the same function $z(x,\xi)$ with $\xi \in \{0, \frac{3}{2}, 3\}$, with each event occurring with probability $1/3$.

For $\xi = 0, \bar{x}(0) = \{x | x_1 + x_2 = 1, \frac{2}{3} \le x_1 \le 1\}$.
For $\xi = 3/2, \bar{x}(3/2) = \{x | x_1 + x_2 = 1, 1/6 \le x_1 \le 5/6\}$.
For $\xi = 3, \bar{x}(3) = \{x | x_1 + x_2 = 1, 0 \le x_1 \le 1/3\}$.
Let us take $\bar{x}(3/2) = (2/3, 1/3)$. Then $EV = z(\bar{x}, 3/2) = 2 + 3/2 = 7/2$, and $EEV = 2 + \frac{1}{3}\left(0 + \frac{3}{2} + 12\right) = 2 + \frac{13}{2} = 13/2$.

No single decision is optimal for the three cases, so we expect EVPI to be nonzero. In the wait-and-see solution, it is possible for all three cases to take a different optimal solution, such as $\bar{x}(0) = (1, 0), \bar{x}(3/2) = (1/2, 1/2)$, and $\bar{x}(3) = (0,1)$, yielding

$$WS = \frac{1}{3}(1+1) + \frac{1}{3}\left(\frac{5}{2}+1\right) + \frac{1}{3}(4+1)$$
$$= \frac{2}{3} + \frac{7}{6} + \frac{5}{3} = \frac{21}{6} = \frac{7}{2}.$$

The recourse solution is obtained by solving the stochastic program $\min E_{\boldsymbol{\xi}}(z(x,\boldsymbol{\xi}))$, which yields $x^* = (2/3, 1/3)$ with the RP value equal to the EEV value. Hence,

$$EV = WS = 7/2 \leq RP = 13/2 = EEV,$$

which means $EVPI = 3$ while $VSS = 0$.

4.5 Examples

There has always been a strong interest in trying to have a better understanding of when the EVPI and VSS take large values and when they take low values. A definite answer to this question would greatly simplify the practice of stochastic programming. Only those programs with large EVPI or VSS would require the solution of a stochastic program. Interested readers may find detailed examples in the field of energy policy and exhaustible resources. Manne [1974] provides an example where EVPI is low, while H.P. Chao [1981] elaborates general conditions for EVPI to be low on a resource exhaustion model. By introducing other types of uncertainty, Louveaux and Smeers [1997] and Birge [1988a] show related examples where EVPI and/or VSS is large.

In this section, we provide simple examples to show that no general answer is available. It is usually felt that using stochastic programming is more relevant when there is more randomness in the problem. To translate this feeling in a more precise statement, we would, for example, expect that for a given problem, EVPI and VSS would increase when the variances of the random variables increase. In the following example, we show that this may or may not be the case.

Example 1

Let $\boldsymbol{\xi}$ be a single random variable taking the two values ξ_1 and ξ_2, with probability p_1 and p_2, respectively, where $p_2 = 1 - p_1$. Let $\bar{\xi} = E[\boldsymbol{\xi}] = 1/2$. Let x be a single decision variable. Consider the recourse problem:

$$\min 6x + 10E_{\boldsymbol{\xi}}|x - \boldsymbol{\xi}|$$
$$\text{s. t. } x \geq 0.$$

(a) Let $\xi_1 = 1/3, \xi_2 = 2/3, p_1 = p_2 = 1/2$ serve as reference. We compute EVPI =2/3 and VSS=1. We also observe that the variance, Var $(\xi) = 1/36$.

(b) Consider the case $\xi_1 = 0, \xi_2 = 1$ again with equal probability $1/2$ (and unchanged expectation). The variance Var (ξ) is now $1/4$, 9 times higher. We now obtain EVPI $= 2$ and VSS $= 3$, showing an example where both values clearly increase with the variance of ξ.

(c) Consider the case $\xi_1 = 0, \xi_2 = 5/8$ with probability $p_1 = 0.2$ and $p_2 = 0.8$, respectively. Again, $\bar{\xi} = 0.5$. Now, Var $(\xi) = 1/16$, larger than in (a). We obtain $EVPI = 2$, larger than in (a) but $VSS = 0$. Knowing this result in advance would mean that the solution of the deterministic problem with $\bar{\xi} = E\xi$ delivers the optimal solution (although EVPI is three times larger than in (a)).

(d) Consider the case $\xi_1 = 0.4, \xi_2 = 0.8$ with $p_1 = 0.75$ and $p_2 = 0.25$, always with $\bar{\xi} = 0.5$. Now, Var $(\xi) = 0.03$, slightly larger than in (a). We now observe $EVPI = 0.4$ and $VSS = 1.1$, namely the opposite behavior from (c), a decrease in EVPI and an increase in VSS.

(e) It is also felt that a more "difficult" stochastic program would induce higher EVPI and VSS. One such case would be to have integer decision variables instead of continuous ones. Exercise 3 of Section 1.1, shows that, with first-stage integer variables, the farming problem sees that VSS remains almost unchanged while EVPI even decreases. On the other hand, Exercise 4 of that section shows that with second-stage integer variables, both EVPI and VSS strongly increase. It would probably not be difficult to reach different conclusions by suitably changing the data.

We may conclude from these simple examples that a general rule is unlikely to be found. One alternative to such a rule is to consider bounds on the information and solution value quantities that require less than complete solutions. We discuss these bounds in the next section.

4.6   Bounds on EVPI and VSS

Bounds on EVPI and VSS rely on constructing intervals for the expected value of solutions of linear programs representing WS, RP, and EEV. The simplest bounds stem from the inequalities in Proposition 5. The EVPI bound was suggested in Avriel and Williams [1970] while the VSS form appears in Birge [1982]. Many other bounds are possible with different limits on the defining quantities. In the remainder of this section, we consider refined bounds that particularly address the value of the stochastic solution. More general approaches to bound expectations of value functions appear in Chapter 9.

The VSS bounds were developed in Birge [1982]. To find them, we consider a simplified version of the stochastic program, where only the right-hand side is stochastic $(\boldsymbol{\xi} = h(\omega))$ and Ξ is finite. Let $\xi^1, \xi^2, \ldots, \xi^K$ index the possible realizations of $\boldsymbol{\xi}$, and $p^k, k = 1, \ldots, K$ be their probabilities. It is customary to refer to each realization ξ^k of $\boldsymbol{\xi}$ as a *scenario k*.

To refine the bounds on VSS, we consider a *reference scenario*, say ξ^r. Two classical reference scenarios are $\bar{\xi}$, the expected value of $\boldsymbol{\xi}$, or the worst-case scenario (for example, the one with the highest demand level for problems when costs have to be minimized under the restriction that demand must be satisfied). Note that in both situations the reference scenario may not correspond to any of the possible scenarios in Ξ. This is obvious for $\bar{\xi}$. The worst-case scenario is, however, a possible scenario when, for example, $\boldsymbol{\xi}$ is formed by components that are independent random variables. If the random variables are not independent, then a meaningful worst-case scenario may be more difficult to construct. Let $p^r = P(\boldsymbol{\xi} = \xi^r)$ be the reference scenario's probability.

The PAIRS subproblem of ξ^r and ξ^k is defined as

$$\min \ z^P(x, \xi^r, \xi^k) = c^T x + p^r q^T y(\xi^r) + (1 - p^r) q^T y(\xi^k)$$
$$\text{s.t.} \ Ax = b,$$
$$Wy(\xi^r) = \xi^r - Tx,$$
$$Wy(\xi^k) = \xi^k - Tx,$$
$$x, y \geq 0.$$

Let $(\bar{x}^k, \bar{y}^k, y(\xi^k))$ denote an optimal solution to the PAIRS subproblem and z_k the optimal objective value $z^P(\bar{x}^k, \bar{y}^k, y(\xi^k))$. We may see the PAIRS subproblem as a stochastic programming problem with two possible realizations ξ^r and ξ^k, with probability p^r and $1 - p^r$, respectively.

Two particular cases of the pairs subproblem are of interest. First, observe that $z^P(x, \xi^r, \xi^r)$ is well-defined and is in fact $z(x, \xi^r)$, the deterministic problem for which the only scenario is the reference scenario. Next, observe that if the reference scenario is not a possible scenario, $p^r = P(\boldsymbol{\xi} = \xi^r) = 0$, then $z^P(x, \xi^r, \xi^k)$ becomes simply $z(x, \xi^k)$.

We now show the relations between the pairs subproblems and the recourse problem. To do this, we define the *sum of pairs expected values*, denoted by SPEV, to be

$$SPEV = \frac{1}{1 - p^r} \sum_{k=1}^{K} p^k \min z^P(x, \xi^r, \xi^k).$$

Again, observe that this definition still makes sense when scenario r is not possible. In that case, however, it is not really a new concept.

Proposition 6. *When the reference scenario is not in Ξ, then SPEV = WS.*

Proof: As we observed before, when $p^r = 0$, the pairs subproblems $z^P(x, \xi^r, \xi^k)$ coincide with $z(x, \xi^k)$. Hence, $SPEV = \sum\limits_{\substack{k=1 \\ k \neq r}}^{K} p^k \min z(x, \xi^k)$, which by definition (1.2) is WS. $\square$

In general, the SPEV is related to WS and RP as follows.

Proposition 7. $WS \leq SPEV \leq RP$.

Proof: Let us first prove the first inequality. By definition,

$$SPEV = \sum_{\substack{k=1 \\ k \neq r}}^{K} p^k \frac{(c^T \bar{x}^k + p^r q^T \bar{y}^k + (1 - p^r) q^T y(\xi^k))}{1 - p^r},$$

where $(\bar{x}^k, \bar{y}^k, y(\xi^k))$ is a solution to the pairs subproblem of ξ^r and ξ^k. By the constraint definition in the pairs subproblem, the solution $(\bar{x}^k, \bar{y}^k)$ is feasible for the problem $z(x, \xi^r)$ so that

$$c^T \bar{x}^k + q^T \bar{y}^k \geq \min z(x, \xi^r) = z_r^*.$$

Weighting $c^T x^k$ with a p^r and a $(1 - p^r)$ term, we obtain:

$$SPEV = \sum_{\substack{k=1 \\ k \neq r}}^{K} \frac{p^k [p^r (c^T \bar{x}^k + q^T \bar{y}^k) + (1 - p^r)(c^T \bar{x}^k + q^T y(\xi^k))]}{1 - p^r},$$

which, by the property just given, is bounded by

$$SPEV \geq \sum_{k \neq r} \frac{p^k \cdot p^r \cdot z_r^*}{1 - p^r} + \sum_{k \neq r} p^k (c^T \bar{x}^k + q^T y(\xi^k)).$$

Now, we simplify the first term and bound $c^T \bar{x}^k + q^T y(\xi^k)$ by z_k^* in the second term, because $(\bar{x}, y(\xi^k))$ is feasible for $\min z(x, \xi^k) = z_k^*$. Thus,

$$SPEV \geq p^r z_r^* + \sum_{k \neq r} p^k z_{k^*} = WS.$$

For the second inequality, let $x^*, y^*(\xi^k), k = 1, \ldots, K$, be an optimal solution to the recourse problem. For simplicity, we assume here that $r \in \Xi$.

By the constraint definitions, $(x^*, y^*(\xi^r), y^*(\xi^k))$ is feasible for the PAIRS subproblem of ξ^r and ξ^k. This implies

$$c^T \bar{x}^k + p^r q^T \bar{y}^k + (1 - p^r) q^T y(\xi^k) \leq c^T x^* + p^r q^T y^*(\xi^r) + (1 - p^r) q^T y^*(\xi^k).$$

If we take the weighted sums of these inequalities for all $k \neq r$, with p^k as the weight of the kth inequality, the weighted sum of the left-hand side elements is, by definition, equal to $(1 - p^r) \cdot SPEV$ and the weighted sum of the right-hand side elements is

$$\sum_{\substack{k=1 \\ k \neq r}}^{K} p^k (c^T x^* + p^r q^T y^*(\xi^r) + (1 - p^r) q^T y^*(\xi^k))$$

$$= (1 - p^r) \left[c^T x^* + p^r q^T y^*(\xi^r) + \sum_{k \neq r} p^k q^T y^*(\xi^k) \right]$$

$$= (1 - p^r) \left[c^T x^* + \sum_{k=1}^{K} p^k q^T y^*(\xi^k) \right] = (1 - p^r) RP,$$

which proves the desired inequality. $\square$

To obtain upper bounds on RP that relate to the pairs subproblem, we generalize the VSS definition. Let $z(x, \xi^r)$ be the deterministic problem associated with scenario ξ^r (remember ξ^r need not necessarily be a possible scenario) and $\bar{x}^r$ an optimal solution to $\min_x z(x, \xi^r)$. We may then define the expected value of the reference scenario,

$$EVRS = E_\xi z(\bar{x}^r, \xi),$$

and the value of a stochastic solution to be

$$VSS = EVRS - RP.$$

Note that VSS is still nonnegative, because $\bar{x}^r$ is either a feasible solution to the recourse problem and $EVRS \geq RP$ or an infeasible solution so that $EVRS = +\infty$.

Now, as before, let $(\bar{x}^k, \bar{y}^k, y(\xi^k))$ be optimal solutions to the pairs subproblem of ξ^r and $\xi^k, k = 1, \ldots, K$. Define the expectations of pairs expected value to be

$$EPEV = \min_{k=1,\ldots,K \cup \{r\}} E_\xi z(\bar{x}^k, \xi).$$

Proposition 8. $RP \leq EPEV \leq EVRS.$

Proof. The three values are the optimal value of the recourse function $\min_x E_\xi z(x, \xi)$ over smaller and smaller feasibility sets: the first one over all feasible x in $K_1 \cap K_2$, the second one over $x \in K_1 \cap K_2 \cap \{\bar{x}^k, k = 1, \dots, K \cup \{r\}\}$, and the third one over $\bar{x}^r \cap K_1 \cap K_2$. $\square$

Putting these two propositions together, one obtains the following theorem.

Theorem 9. $0 \leq EVRS - EPEV \leq VSS \leq EVRS - SPEV \leq EVRS - WS.$

We apply these concepts in the following example.

Example 2

Consider the problem to find:

$$\min \ 3x_1 + 2x_2 + E_{\boldsymbol{\xi}} \min(-15\mathbf{y}_1 - 12\mathbf{y}_2)$$
$$\text{s.t.} \ 3\mathbf{y}_1 + 2\mathbf{y}_2 \leq x_1,$$
$$2\mathbf{y}_1 + 5\mathbf{y}_2 \leq x_2,$$
$$.8\boldsymbol{\xi}_1 \leq \mathbf{y}_1 \leq \boldsymbol{\xi}_1,$$
$$.8\boldsymbol{\xi}_2 \leq \mathbf{y}_2 \leq \boldsymbol{\xi}_2,$$
$$x, y \geq 0,$$

where $\boldsymbol{\xi}_1 = 4$ or 6 and $\boldsymbol{\xi}_2 = 4$ or 8, independently of each other, with probability $1/2$ each.

This example can be seen as an investment decision in two resources x_1 and x_2, which are needed in the second-stage problem to cover at least 80% of the demand. In this situation, the EEV and WS answers are totally inconclusive.

Table 1 gives the various solutions under the four scenarios, the optimal objective values under these scenarios and the WS value. It also describes the EV value under the expected value scenario $\bar{\xi} = (5,6)^T$. Note that this scenario is not one of those possible. The optimal solution $\bar{x}(\bar{\xi}) = (24.6, 34)^T$ is infeasible for the stochastic problem so that EEV is set to be $+\infty$.

It follows from Table 1 that $EV = WS = 9.2 \leq RP \leq EEV = +\infty$. This relation is of no help: we can only conclude from it that EVPI is somewhere between 0 and $+\infty$, and so is VSS. These statements could have been made without any computation.

It is in such situations that the pairs subproblems are of great interest. Because the problem under consideration is an investment problem with demand satisfaction constraints, the most logical reference scenario corresponds to the largest demand, $\xi^r = (6,8)^T$, and not to the mean demand $\bar{\xi}$.

This will force the first-stage decisions to take demand satisfaction under the maximal demand into consideration, so that decisions taken under the

TABLE 1. Solutions and optimal values under the four scenarios and the expected value scenario.

Scenario	First-Stage Solution	Second-Stage Solution	Optimal Value $z(\bar{x}(\xi), \xi)$
1. (4,4)	(18.4, 24)	(4, 3.2)	4.8
2. (6,4)	(24.4, 28)	(6, 3.2)	0.8
3. (4,8)	(24.8, 40)	(4, 6.4)	17.6
4. (6,8)	(30.8, 44)	(6, 6.4)	13.6
			WS = 9.2
$\bar{\xi} = (5,6)$	(24.6, 34)	(5, 4.8)	EV = 9.2
			EEV = $+\infty$

pairs subproblem are feasible for the recourse problem. Due to independence, ξ^r is one of the possible realizations of ξ, with $p^r = 1/4$.

The PAIRS subproblems of ξ^r and ξ^k are

$$\min 3x_1 + 2x_2 - \frac{1}{4}(15y_1^r + 12y_2^r) - \frac{3}{4}(15y_1 + 12y_2)$$

s.t. $\quad x_1 \geq 27.2, \qquad 3y_1^r + 2y_2^r \leq x_1, \qquad 3y_1 + 2y_2 \leq x_1,$

$\qquad x_2 \geq 41.6, \qquad 2y_1^r + 5y_2^r \leq x_2, \qquad 2y_1 + 5y_2 \leq x_2,$

$\qquad\qquad\qquad\qquad 4.8 \leq y_1^r \leq 6, \qquad .8\xi_1^k \leq y_1 \leq \xi_1^k,$

$\qquad\qquad\qquad\qquad 6.4 \leq y_2^r \leq 8, \qquad .8\xi_2^k \leq y_2 \leq \xi_2^k,$

$$y \geq 0.$$

The bounds on x_1 and x_2 are induced by the feasibility for the reference scenarios.

Table 2 gives the solutions of the pairs subproblems for the three sce-narios (other than the reference scenario), the SPEV, the EVRS and the EPEV values.

This time, the relations one can derive from this table are strongly conclusive:

$$WS = 9.2 \leq SPEV = 30.94 \leq RP \leq EPEV = 30.94 \leq EVRS = 40.6$$

implies RP $= 30.94$ and $(27.2, 41.6)^T$ is an optimal solution.

Exercises

1. Show that Proposition 1 still holds if some of the x and/or y must be integer.

TABLE 2. Pairs subproblems solutions.

Pairs Subproblem	First-Stage Solution	Second-Stage under Reference Sc.	Second-Stage under ξ_k	Objective Value z^P
1. (4,4), r	(27.2, 41.6)	(4.8, 6.4)	(4,4)	46.6
2. (6,4), r	(27.2, 41.6)	(4.8, 6.4)	(6,4)	24.1
3. (4,8), r	(27.2, 41.6)	(4.8, 6.4)	(4, 6.72)	22.12
				SPEV = 30.94
		EPEV = $\min_k E_{\boldsymbol{\xi}} z(\bar{x}(\boldsymbol{\xi}^k), \boldsymbol{\xi}) = E_{\boldsymbol{\xi}} z(27.2, 41.6, \boldsymbol{\xi})$ = 30.94		
		EVRS = $E_{\boldsymbol{\xi}} z((30.8, 44), \boldsymbol{\xi}) = 40.6$		

2. Consider Example 3.1 with a single first-stage decision x and $Q(x, \boldsymbol{\xi}) = \min\{2y_1 + y_2 | y_1 \geq x - \boldsymbol{\xi}, y_2 \geq \boldsymbol{\xi} - x, y \geq 0, \text{integer } \}$ with $\boldsymbol{\xi} = 1$ or 2 with probability of 1/2 each. Show:

(a) If x must be integer, then $EV > WS$ for any value of $c \geq 0$.

(b) If x is continuous, then $EV = WS$ for $0 \leq c \leq 1$ and $EV > WS$ for $c > 1$. Beware that y is always integer, the discussion is on x being integer or not.

3. Consider the following stochastic program

$$\min_{x \geq 0} 2x + E_{\boldsymbol{\xi}}\{\boldsymbol{\xi} \cdot y \mid y \geq 1 - x, \ y \geq 0\},$$

and $\boldsymbol{\xi}$ takes on values 1 and 3 with probability 3/4 and 1/4, respectively. Show that in this case $EV > WS$.

4. Consider the following two-stage program:

$$\min 2x_1 + x_2 + E_{\boldsymbol{\xi}}(-3y_1 - 4y_2 | y_1 + 2y_2 \geq \boldsymbol{\xi}_1, y_1 \leq x_1 \, ,$$

$$y_2 \leq x_2, y_2 \leq \boldsymbol{\xi}_2, y \geq 0)$$

s.t. $x_1 + x_2 \leq 7, x_1, x_2 \geq 0,$

where $\boldsymbol{\xi}$ can take the values $\binom{3}{2}, \binom{5}{3}, \binom{7}{3}$ with probability 1/3 each.

(a) Choose the scenario $\binom{7}{3}$ as the reference scenario. Define the problem $z(x, \boldsymbol{\xi})$ for this reference scenario. Its optimal solution gives the optimal first-stage decision $x_1 = 4, x_2 = 3$. Compute the EVRS value.

(b) State the pairs subproblem for $\binom{3}{2}$ and the reference scenario.

(c) The solution of the pairs subproblem for $\binom{3}{2}$ and the reference scenario has first-stage optimal solutions $x_1 = 5, x_2 = 2$; the solution of the pairs subproblem for $\binom{5}{3}$ and the reference scenario has first-stage optimal solutions $x_1 = 4, x_2 = 3$. Compute the values of the two pairs subproblems. Compute the SPEV value. What relation holds for the recourse problem value?

5. Adapt the proofs in Proposition 7 for the case where $r \notin \Xi$.

Part III

Solution Methods

5
Two-Stage Linear Recourse Problems

Computation in stochastic programs with recourse has focused on two-stage problems with finite numbers of realizations. This problem was introduced in the farming example of Chapter 1. As we saw in the capacity expansion model, this problem can also represent multiple stages of decisions with block separable recourse and it provides a foundation for multistage methods. The two-stage problem is, therefore, our primary model for computation.

The general model is to choose some initial decision that minimizes current costs plus the expected value of future recourse actions. With a finite number of second-stage realizations and all linear functions, we can always form the full deterministic equivalent linear program or extensive form. With many realizations, this form of the problem becomes quite large. Methods that ignore the special structure of stochastic linear programs become quite inefficient (as some of the results in Section 3 show). Taking advantage of structure is especially beneficial in stochastic programs and is the focus of much of the algorithmic work in this area.

The method used most frequently is based on building an outer linearization of the recourse cost function and a solution of the first-stage problem plus this linearization. This cutting plane technique is called the *L-shaped method* in stochastic programming. Section 1 describes the basic L-shaped method in some detail, while Sections 2 to 4 continue this development with a discussion of enhancements of the L-shaped method in terms of feasibility, multicuts, and bunching of realizations.

Several variants and extensions of the L-shaped method have been designed. Variants adding nonlinear regularized terms will be studied in Chapter 6. Bounding techniques will be considered in Chapter 9. The use of sampling will be studied in Chapter 10.

The remainder of this chapter discusses alternative algorithms. In Section 5, we will discuss alternative decomposition procedures. The first method is an inner linearization, or Dantzig-Wolfe decomposition approach, that solves the dual of the L-shaped method problem. The other approach is a primal form of inner linearization based on generalized programming.

Section 6 will consider direct approaches to the extensive form through efficient extreme point and interior point methods. We discuss basis factorization and its relationship to decomposition methods. We also present interior point approaches and the use of a special stochastic programming structure for these algorithms.

Additional problem structures can be of further benefit for solving two-stage stochastic linear programs. These structures are generally based on the form of the recourse function. Section 7 will discuss methods for the generalizations of the news vendor problem called *simple recourse problems* and problems involving networks.

5.1   The L-Shaped Method

Consider the general formulation in (3.1.2) or (3.1.5). The basic idea of the L-shaped method is to approximate the nonlinear term in the objective of these problems. A general principle behind this approach is that, because the nonlinear objective term (the *recourse function*) involves a solution of all second-stage recourse linear programs, we want to avoid numerous function evaluations for it. We therefore use that term to build a master problem in x, but we only evaluate the recourse function exactly as a subproblem.

To make this approach possible, we assume that the random vector ξ has finite support. Let $k = 1, \ldots, K$ index its possible realizations and let p_k be their probabilities. Under this assumption, we may now write the deterministic equivalent program in the extensive form. This form is created by associating one set of second-stage decisions, say, y_k, to each realization ξ, i.e., to each realization of q_k, h_k, and T_k. It is a large-scale linear problem that we can define as the *extensive form (EF)*:

$$(EF) \quad \min \quad c^T x + \sum_{k=1}^{K} p_k q_k^T y_k \tag{1.1}$$

$$\text{s.t.} \quad Ax = b,$$
$$T_k x + W y_k = h_k, \quad k = 1, \ldots, K;$$
$$x \geq 0, \quad y_k \geq 0, \quad k = 1, \ldots, K.$$

An example of an extensive form has been given for the farmer's problem in Chapter 1 (model (1.1.2)).

The block structure of the extensive form appears in Figure 1.

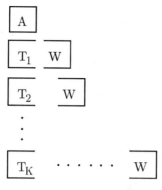

FIGURE 1. Block structure of the two-stage extensive form.

This picture has given rise to the name, *L-shaped method* for the following algorithm. Taking the dual of the extensive form, one obtains a dual block-angular structure, as in Figure 2. Therefore it seems natural to exploit this dual structure by performing a Dantzig-Wolfe [1960] decomposition (inner linearization) of the dual or a Benders [1962] decomposition (outer linearization) of the primal. This method has been extended in stochastic programming to take care of feasibility questions and is known as Van Slyke and Wets's [1969] L-shaped method. It proceeds as follows.

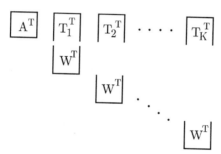

FIGURE 2. Block angular structure of the two-stage dual.

L-Shaped Algorithm

Step 0. Set $r = s = \nu = 0$.

Step 1. Set $\nu = \nu + 1$. Solve the linear program (1.2)–(1.4).

$$\min \ z = c^T x + \theta \tag{1.2}$$
$$\text{s.t. } Ax = b,$$
$$D_\ell x \geq d_\ell, \quad \ell = 1, \ldots, r, \tag{1.3}$$
$$E_\ell x + \theta \geq e_\ell, \quad \ell = 1, \ldots, s, \tag{1.4}$$
$$x \geq 0, \quad \theta \in \Re.$$

Let (x^ν, θ^ν) be an optimal solution. If no constraint (1.4) is present, θ^ν is set equal to $-\infty$ and is not considered in the computation of x^ν.

Step 2. For $k = 1, \ldots, K$ solve the linear program

$$\min \ w' = e^T v^+ + e^T v^- \tag{1.5}$$
$$\text{s.t.} \ Wy + Iv^+ - Iv^- = h_k - T_k x^\nu, \tag{1.6}$$
$$y \geq 0, \quad v^+ \geq 0, \quad v^- \geq 0,$$

where $e^T = (1, \ldots, 1)$, until, for some k, the optimal value $w' > 0$. In this case, let σ^ν be the associated simplex multipliers and define

$$D_{r+1} = (\sigma^\nu)^T T_k \tag{1.7}$$

and

$$d_{r+1} = (\sigma^\nu)^T h_k \tag{1.8}$$

to generate a constraint (called a *feasibility cut*) of type (1.3). Set $r = r+1$, add to the constraint set (1.3), and return to Step 1. If for all k, $w' = 0$, go to Step 3.

Step 3. For $k = 1, \ldots, K$ solve the linear program

$$\min \ w = q_k^T y \tag{1.9}$$
$$\text{s.t.} \ Wy = h_k - T_k x^\nu,$$
$$y \geq 0.$$

Let π_k^ν be the simplex multipliers associated with the optimal solution of Problem k of type (1.9). Define

$$E_{s+1} = \sum_{k=1}^{K} p_k \cdot (\pi_k^\nu)^T T_k \tag{1.10}$$

and

$$e_{s+1} = \sum_{k=1}^{K} p_k \cdot (\pi_k^\nu)^T h_k. \tag{1.11}$$

Let $w^\nu = e_{s+1} - E_{s+1} x^\nu$. If $\theta^\nu \geq w^\nu$, stop; x^ν is an optimal solution. Otherwise, set $s = s+1$, add to the constraint set (1.4), and return to Step 1.

The method consists of solving an approximation of (3.1.2) by using an outer linearization of Q. Two types of constraints are sequentially added: (i) feasibility cuts (1.3) determining $\{x \mid Q(x) < +\infty\}$ and (ii) *optimality cuts* (1.4), which are linear approximations to Q on its domain of finiteness. We first illustrate the optimality cuts.

Example 1

Let

$$Q(x,\xi) = \begin{cases} \xi - x & \text{if } x \leq \xi, \\ x - \xi & \text{if } x \geq \xi, \end{cases}$$

and let ξ take on the values 1, 2, and 4, each with probability 1/3. Assume also $c = 0$ and $0 \leq x \leq 10$.

Figure 3 represents the functions $Q(x,1), Q(x,2), Q(x,4)$ and $\mathcal{Q}(x)$. As indicated in Theorem 3.5, each of these functions is polyhedral. Because the first-stage objective $c^T x$ is zero, $\mathcal{Q}(x)$ is also the function $z(x)$ to be minimized. Assume the starting point is $x^1 = 0$. The sequence of iterations for the L-shaped method is as follows.

Iteration 1:

x^1 is not optimal; send the cut

$$\theta \geq 7/3 - x.$$

Iteration 2:

$x^2 = 10, \theta^2 = -23/3$ is not optimal; send the cut

$$\theta \geq x - 7/3.$$

Iteration 3:

$x^3 = 7/3, \theta^3 = 0$ is not optimal; send the cut

$$\theta \geq \frac{x+1}{3}.$$

Iteration 4:

$x^4 = 1.5, \theta^4 = 2.5/3$ is not optimal; send the cut

$$\theta \geq \frac{5-x}{3}.$$

Iteration 5:

$x^5 = 2, \theta^5 = 1$, which is the optimal solution.

We now constructively prove that constraints of the type (1.4) defined in Step 3 are supporting hyperplanes of $\mathcal{Q}(x)$ and that the algorithm will converge to an optimal solution, provided the constraints (1.3) adequately define feasible points of K_2. (This last provision will be taken care of later in this section.)

First, observe that solving (3.1.2), namely,

$$\min \ c^T x + \mathcal{Q}(x) \tag{1.12}$$

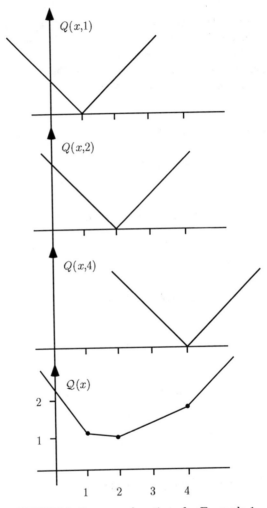

FIGURE 3. Recourse functions for Example 1.

$$\text{s. t. } x \in K_1 \cap K_2,$$

is equivalent to solving

$$\min \ c^T x + \theta \tag{1.13}$$
$$\mathcal{Q}(x) \leq \theta,$$
$$\text{s. t. } x \in K_1 \cap K_2,$$

where, in both problems, $\mathcal{Q}(x)$ is defined as in (3.1.3),

$$\mathcal{Q}(x) = E_\omega Q(x, \xi(\omega))$$

and

$$Q(x, \xi(\omega)) = \min_y \{q(\omega)^T y | Wy = h(\omega) - T(\omega)x, y \geq 0\}$$

as in (3.1.4).

We are thus looking for a finitely convergent algorithm for solving (1.12) or (1.13).

In Step 3 of the algorithm, problem (1.9) is solved repeatedly for each $k = 1, \ldots, K$, yielding optimal simplex multipliers $\pi_k^\nu, k = 1, \ldots, K$. It follows from duality in linear programming that, for each k,

$$Q(x^\nu, \xi_k) = (\pi_k^\nu)^T (h_k - T_k x^\nu).$$

Moreover, by convexity of $Q(x, \xi_k)$, it follows from the subgradient inequality that

$$Q(x, \xi_k) \geq (\pi_k^\nu)^T h_k - (\pi_k^\nu)^T T_k x.$$

We may now take the expectation of these two relations to obtain

$$\mathcal{Q}(x^\nu) = E(\pi^\nu)^T (\mathbf{h} - \mathbf{T} x^\nu) = \sum_{k=1}^K p_k \ (\pi_k^\nu)^T (h_k - T_k x^\nu)$$

and

$$\mathcal{Q}(x) \geq E(\pi^\nu)^T (\mathbf{h} - \mathbf{T} x) = \sum_{k=1}^K p_k (\pi_k^\nu)^T h_k - \left(\sum_{k=1}^K p_k (\pi_k^\nu)^T T_k \right) x,$$

respectively. By $\theta \geq \mathcal{Q}(x)$, it follows that a pair (x, θ) is feasible for (1.13) only if $\theta \geq E(\pi^\nu)^T (\mathbf{h} - \mathbf{T} x)$, which corresponds to (1.4) where E_ℓ and e_ℓ are defined in (1.10) and (1.11).

On the other hand, if (x^ν, θ^ν) is optimal for (1.13), it follows that $\mathcal{Q}(x^\nu) = \theta^\nu$, because θ is unrestricted in (1.13) except for $\theta \geq \mathcal{Q}(x)$. This happens when $\theta^\nu = E(\pi^\nu)^T (\mathbf{h} - \mathbf{T} x^\nu)$, which justifies the termination criterion in Step 3.

This means that at each iteration either $\theta^\nu \geq \mathcal{Q}(x^\nu)$ implying termination or $\theta^\nu < \mathcal{Q}(x^\nu)$. In the latter case, none of the already defined optimality cuts (1.4) adequately imposes $\theta \geq \mathcal{Q}(x)$, so a new set of multipliers

π_k^ν will be defined at x^ν to generate an appropriate constraint (1.4). The finite convergence of the algorithm follows from the fact that there is only a finite number of different combinations of the K multipliers π_k, because each corresponds to one of the finitely many different bases of (1.9).

An alternative proof of convergence could be obtained by showing that Step 3 coincides with an iteration of the subproblems in the Dantzig-Wolfe decomposition of the dual of (1.12) while Step 1 coincides with the master problem. We will consider this approach in Section 5.

We now have to prove that at most a finite number of constraints (1.3) is needed to guarantee $x \in K_2$. Constraints (1.3) are generated in Step 2 of the algorithm.

By definition, $x \in K_2$ is equivalent to

$$x \in \{x| \text{ for } k = 1, \ldots, K, \exists y \geq 0 \text{ s.t. } Wy = h_k - T_k x\}.$$

Referring to a previously introduced notation, this means

$$h_k - T_k x \in \text{pos } W, \text{ for } k = 1, \ldots, K.$$

In Step 2, a subproblem (1.5) is solved that tests whether $h_k - T_k x^\nu$ belongs to pos W for $k = 1, \ldots, K$. If not, this means that for some $k = 1, \ldots, K, h_k - T_k x^\nu \notin \text{pos } W$. Then, there must be a hyperplane separating $h_k - T_k x^\nu$ and pos W. This hyperplane must satisfy $\sigma^T t \leq 0$ for all $t \in \text{pos } W$ and $\sigma^T (h_k - T_k x^\nu) > 0$. In Step 2, this hyperplane is obtained by taking σ for the value σ^ν of the simplex multipliers of the subproblem (1.5) solved in Step 2.

By duality, w' being strictly positive is the same as $(\sigma^\nu)^T (h_k - T_k x^\nu) > 0$. Also, $(\sigma^\nu)^T W \leq 0$ is satisfied because σ^ν is an optimal simplex multiplier and, at the optimum, the reduced costs associated with y must be non-negative.

Therefore, σ^ν has the desired property. A necessary condition for x belonging to K_2 is that $(\sigma^\nu)^T (h_k - T_k x) \leq 0$. There is at most a finite number of such constraints (1.3) because there are only a finite number of optimal bases to the problem (1.5) solved in Step 2. This is no surprise because we already know from Theorem 3.5 that K_2 is polyhedral when ξ is a finite random variable. We thus have proved the following theorem.

Theorem 1. *When ξ is a finite random variable, the L-shaped algorithm finitely converges to an optimal solution when it exists or proves the infeasibility of problem (1.12).*

To illustrate the feasibility cuts, consider Example 4.2:

$$\min \ 3x_1 + 2x_2 = E_{\boldsymbol{\xi}}(15y_1 + 12y_2)$$
$$\text{s.t.} \ 3y_1 + 2y_2 \leq x_1,$$
$$2y_1 + 5y_2 \leq x_2,$$
$$.8\boldsymbol{\xi}_1 \leq y_1 \leq \boldsymbol{\xi}_1,$$
$$.8\boldsymbol{\xi}_2 \leq y_2 \leq \boldsymbol{\xi}_2,$$
$$x, y \geq 0,$$

with $\boldsymbol{\xi}_1 = 4$ or 6 and $\boldsymbol{\xi}_2 = 4$ or 8, independently, with probability $1/2$ each and $\boldsymbol{\xi} = (\boldsymbol{\xi}_1, \boldsymbol{\xi}_2)^T$.

To keep the discussion short, assume the first considered realization of $\boldsymbol{\xi}$ is $(6,8)^T$. If not, many cuts would be needed. Starting from an initial solution $x^0 = (0,0)^T$, a first feasibility cut $3x_1 + x_2 \geq 123.2$ is generated. The first-stage solution is then $x^1 = (41.067, 0)^T$. A second feasibility cut is $x_2 \geq 22.4$. The first-stage solution becomes $x^2 = (33.6, 22.4)^T$. A third feasibility cut $x_2 \geq 41.6$ is generated. The first-stage solution is

$$x^3 = (27.2, \ 41.6)^T,$$

which yields feasible second-stage decisions.

This example also illustrates that generating feasibility cuts by a mere application of Step 2 of the L-shaped method may not be efficient. Indeed, a simple look at the problem reveals that, for feasibility when $\xi_1 = 6$ and $\xi_2 = 8$, it is at least necessary to have $y_1 \geq 4.8$ and $y_2 \geq 6.4$, which in turn implies $x_1 \geq 27.2$ and $x_2 \geq 41.6$. We may then consider the following program as a reasonable initial problem:

$$\min 3x_1 + 2x_2 + \mathcal{Q}(x)$$
$$\text{s.t.} \ x_1 \geq 27.2,$$
$$x_2 \geq 41.6,$$

which immediately appears to be feasible. Such situations frequently occur in practice. More is said about these circumstances in the next section.

5.2   Feasibility

As indicated in the previous section, Step 2 of the L-shaped method consists of determining whether a first-stage decision $x \in K_1$ is also second-stage feasible, i.e., $x \in K_2$. This step may be extremely time-consuming. It may require the solution of up to K phase-one problems of the form (1.5)–(1.6). The process may have to be repeated several times for successive candidate first-stage solutions.

In some cases, Step 2 can be simplified. A first case is when the second stage is always feasible. The stochastic program is then said to have complete recourse. Let, as in (1.1), the second-stage constraint be:

$$Wy = h - Tx, y \geq 0.$$

We repeat here the definition given in Section 3.1d for complete recourse for convenience.

Definition. A stochastic program is said to have *complete recourse* when pos $W = \Re^{m_2}$. It is said to have *relatively complete recourse* when $K_2 \supseteq K_1$, i.e., $x \in K_1$ implies $h - Tx \in$ pos W for any h, T realization of $\mathbf{h}, \mathbf{T}$.

If we consider the farmer's problem in Section 1.1, program (1.1.2) has complete recourse. The second stage just serves as a measure of the cost to the farmer of the decisions taken. Any lack of production can be covered by a purchase. Any production in excess can be sold. If we consider the power generation model (1.3.6), it has complete recourse if there exists at least one technology with zero lead time ($\Delta_i = 0$). If the demand in a given period t exceeds what can be delivered by the available equipment, an investment is made in this (usually expensive) technology to cover the needed demand.

A second case is when it is possible to derive some constraints that have to be satisfied to guarantee second-stage feasibility. These constraints are sometimes called *induced constraints*. They can be obtained from a good understanding of the model. A simple look at the second-stage program in the example of the previous section reveals the conditions for feasibility. Constraints $x_1 \geq 27.2$ and $x_2 \geq 41.6$ are examples of induced constraints. In the power generation model (1.3.6) of Section 1.3, the total possible demand in a given stage t is obtained from (1.3.8) as $\sum_{j=1}^{m} d_j^t$. The maximal possible demand in stage t is thus $D^t = \max_{\xi \in \Xi} \sum_{j=1}^{m} d_j^t$. Stage t feasibility will thus require enough investments in the various technologies to cover the maximal demand, i.e.,

$$\sum_{i=1}^{n} a_i(w_i^{t-\Delta_i} + g_i) \geq D^t .$$

Again, with the introduction of these induced constraints, Step 2 of the L-shaped algorithm can be dropped.

A third case is when Step 2 is not required for all $k = 1, \cdots, K$, but for one single h_k. Assume T is deterministic. Also assume we can transform W so that for all $t \geq 0, t \in$ pos W. This poses no difficulty for inequalities, as it is just a matter of taking the slack variables with a positive coefficient. In the example of the previous section, the following representation of W

satisfies the desired requirement:

$$
\begin{array}{llllll}
3y_1 & + 2y_2 + w_1 & & & & = x_1, \\
2y_1 & + 5y_2 & + w_2 & & & = x_2, \\
y_1 & & + w_3 & & & = d_1, \\
-y_1 & & & + w_4 & & = -.8d_1, \\
& y_2 & & & + w_5 & = d_2, \\
& -y_2 & & & & + w_6 = -0.8d_2.
\end{array}
$$

For any $t \geq 0$, it suffices to take $w = t$ to have a second-stage feasible solution. Assume first some lower bound,

$$ b(x) \leq h_k - T_k x \, , \ k = 1, \cdots, K, $$

exists. Then a sufficient condition for x to be feasible is that the linear system $Wy = b(x), y \geq 0$ is feasible. Indeed, if $Wy = b(x), y \geq 0$ is feasible, then $Wy = b'(x), y \geq 0$ is feasible for any $b'(x) \geq b(x)$ by construction of W.

Theorem 2. *Assume that W is such that $t \in pos\ W$ for all $t \geq 0$. Define $a_i = \min_{k=1,\cdots K} \{h_{ik}\}$ to be the componentwise minimum of h. Also assume there exists one realization $h_\ell, \ell \in \{1, \cdots, K\}$ s.t. $a = h_\ell$. Then, $x \in K_2$ if and only if $Wy = a - Tx, y \geq 0$ is feasible.*

Proof. This is easily checked, as the condition was just seen to be sufficient. It is also necessary because $x \in K_2$ only if $Wy = a - Tx, y \geq 0$ is feasible. $\square$

Again taking the example of the previous section, we observe that, with an appropriate choice of W, the vector $h = (0, 0, \xi_1, -.8\xi_1, \xi_2, -.8\xi_2)^T$. The componentwise minimum is $a = (0, 0, 4, -4.8, 4, -6.4)^T$. Unfortunately, no h coincides with a. The system $\{y | Wy = a - Tx, y \geq 0\}$ is infeasible.

On the other hand, the system is feasible only if $3y_1 + 2y_2 \leq x_1, 2y_1 + 5y_2 \leq x_2, y_1 \geq 0.8\xi_1, y_2 \geq 0.8\xi_2$ is feasible (we just drop the upper bounds on y). This reduced system is feasible if and only if

$$ 3y_1 + 2y_2 \leq x_1, 2y_1 + 5y_2 \leq x_2, y_1 \geq 4.8, y_2 \geq 6.4, $$

i.e., if and only if $x_1 \geq 27.2$ and $x_2 \geq 41.6$, which (as already seen intuitively) is a necessary condition for feasibility. Thus, even if in practice there does not always exist a realization h_ℓ such that $a = h_\ell$, the condition of Theorem 2 may still be helpful.

Exercises

1. Obtain the feasibility cuts for Example 4.1.

2. Feasibility cuts in Benders decomposition have an equivalent in Dantzig-Wolfe decomposition. What is it?

5.3   The Multicut Version

In Step 3 of the L-shaped method, all K realizations of the second-stage program are optimized to obtain their optimal simplex multipliers. These multipliers are then aggregated in (1.10) and (1.11) to generate one cut (1.4). The structure of stochastic programs clearly allows placing several cuts instead of one. In the multicut version, one cut per realization in the second stage is placed. For those familiar with Dantzig-Wolfe decomposition (explored more deeply in Section 6), adding multiple cuts at each iteration corresponds to including several columns in the master program of an inner linearization algorithm (see, e.g., Lasdon [1970] for a general presentation and Birge [1985b] for an analysis of the stochastic case). We first give a presentation of the multicut algorithm, taken from Birge and Louveaux [1988].

The Multicut L-Shaped Algorithm

Step 0. Set $r = \nu = 0$ and $s_k = 0$ for all $k = 1, \cdots, K$.

Step 1. Set $\nu = \nu + 1$. Solve the linear program (3.1)–(3.4)

$$\min z = c^T x + \sum_{k=1}^{K} \theta_k \tag{3.1}$$

$$\text{s.t. } Ax = b \,, \tag{3.2}$$
$$D_\ell x \geq d_\ell, \ell = 1, \cdots, r \,, \tag{3.3}$$
$$E_{\ell(k)} x + \theta_k \geq e_{\ell(k)} \,, \ell(k) = 1, \cdots, s(k), \tag{3.4}$$
$$k = 1, \cdots, K \,,$$
$$x \geq 0 \,,$$

Let $(x^\nu, \theta_1^\nu, \ldots, \theta_K^\nu)$ be an optimal solution of (3.1)–(3.4). If no constraint (3.4) is present for some k, θ_k^ν is set equal to $-\infty$ and is not considered in the computation of x^ν.

Step 2. As before.

Step 3. For $k = 1, \cdots, K$ solve the linear program (1.9). Let π_k^ν be the simplex multipliers associated with the optimal solution of problem k. If

$$\theta_k^\nu < p_k(\pi_k^\nu)^T(h_k - T_k x^\nu), \tag{3.5}$$

define

$$E_{s(k)+1} = p_k(\pi_k^\nu)^T T_k, \tag{3.6}$$

$$e_{s(k)+1} = p_k(\pi_k^\nu)^T h_k, \tag{3.7}$$

and set $s(k) = s(k) + 1$. If (3.5) does not hold for any $k = 1, \cdots, K$, stop; x^ν is an optimal solution. Otherwise, return to Step 1.

We illustrate the multicut L-shaped on Example 1. Starting from $x^1 = 0$, the sequence of iterations is as follows:

Iteration 1:

x^1 is not optimal, send the cuts

$$\theta_1 \geq \frac{1-x}{3} \; ; \; \theta_2 \geq \frac{2-x}{3} \; ; \; \theta_3 \geq \frac{4-x}{3}.$$

Iteration 2:

$x^2 = 10, \theta_1^2 = -3, \theta_2^2 = -8/3, \theta_3^2 = -2$ is not optimal; send the cuts

$$\theta_1 \geq \frac{x-1}{3} \; ; \; \theta_2 \geq \frac{x-2}{3} \; ; \; \theta_3 \geq \frac{x-4}{3}.$$

Iteration 3:

$x^3 = 2, \theta_1^3 = 1/3, \theta_2^3 = 0, \theta_3^3 = 2/3$ is the optimal solution.

Let us define a *major iteration* to consist of the operations performed between returns to Step 1 in both algorithms. By sending multiple cuts, a solution is found in two major iterations instead of four with the single-cut L-shaped method.

A few observations are necessary. By sending disaggregate cuts, more detailed information is given to the first stage. The number of major iterations is expected then to be less than in the single cut method. Because the two methods do not necessarily follow the same path, by chance, the L-shaped method can conceivably do better than the multicut approach. Exercise 1 provides such an example.

In general, however, as numerical experiments reveal, the number of major iterations is reduced. This is done at the expense of a larger first-stage program, because many more cuts are added. The balance between less major iterations but larger first-stage programs is problem-dependent. The results of numerical experiments are available in Birge and Louveaux [1988] and Gassmann [1990]. As a rule of thumb, the multicut approach is expected to be more effective when the number of realizations K is not significantly larger than the number of first-stage constraints m_1.

Finally, some hybrid approach may be worthwhile, where subsets of the realizations are grouped to form a smaller number of combination cuts. Exercise 2 provides such an example.

Exercises

1. Assume $n_1 = 1, m_1 = 0, m_2 = 3, n_2 = 6,$

$$W = \begin{pmatrix} 1 & -1 & -1 & -1 & 0 & 0 \\ 0 & 1 & 0 & 0 & 1 & 0 \\ 0 & 0 & 1 & 0 & 0 & 1 \end{pmatrix},$$

and $K = 2$ realizations of $\boldsymbol{\xi}$ with equal probability $1/2$. These realizations are $\xi^1 = (q^1, h^1, T^1)^T$ and $\xi^2 = (q^2, h^2, T^2)^T$, where $q^1 = (1,0,0,0,0,0)^T, q^2 = (3/2,0,2/7,1,0,0)^T, h^1 = (-1,2,7)^T$, $h^2 = (0,2,7)^T$, and $T^1 = T^2 = (1,0,0)^T$. For the first value of $\xi, Q(x,\xi)$ has two pieces, such that

$$Q_1(x) = \begin{cases} -x - 1 & \text{if } x \leq -1, \\ 0 & \text{if } x \geq -1. \end{cases}$$

For the second value of $\xi, Q(x,\xi)$ has four pieces such that

$$Q_2(x) = \begin{cases} -1.5x & \text{if } x \leq 0, \\ 0 & \text{if } 0 \leq x \leq 2, \\ 2/7(x - 2) & \text{if } 2 \leq x \leq 9, \\ x - 7 & \text{if } x \geq 9. \end{cases}$$

Assume also that x is bounded by $-20 \leq x \leq 20$ and $c = 0$. Starting from any initial point $x^1 \leq -1$, show that one obtains the following sequence of iterate points and cuts for the L-shaped method.

Iteration 1:

$x^1 = -2, \theta^1$ is omitted; new cut $\theta \geq -0.5 - 1.25x$.

Iteration 2:

$x^2 = +20, \theta^2 = -25.5$; new cut $\theta \geq 0.5x - 3.5$.

Iteration 3:

$x^3 = 12/7, \theta^3 = -37/14$; new cut $\theta \geq 0$.

Iteration 4:

$x^4 \in [-2/5, 7], \theta^4 = 0$. If x^4 is chosen to be any value in $[0,2]$, then the algorithm terminates at Iteration 4. The multicut approach would generate the following sequence.

Iteration 1:

$x^1 = -2, \theta^1_1$ and θ^1_2 omitted; new cuts $\theta_1 \geq -0.5x - 0.5, \theta_2 \geq -3/4x$.

Iteration 2:

$x^2 = 20, \theta^2_1 = -10.5, \theta^2_2 = -15$; new cuts $\theta_1 \geq 0, \theta_2 \geq 0.5x - 3.5$.

Iteration 3:

$x^3 = 2.8, \theta^3_1 = 0, \theta^3_1 = -2.1$; new cut $\theta_2 \geq 1/7(x - 2)$.

Iteration 4:

$x^4 = 0.32, \theta_1^4 = 0, \theta_2^4 = -0.24$; new cut $\theta_2 \geq 0$.

Iteration 5:

$x^5 = 0, \theta_1^5 = \theta_2^5 = 0$, stop.

2. Consider Example 1, now with $\boldsymbol{\xi}$ taking values:

    0.5, 1.0, 1.5 with probability 1/9 each,
    2 with probability 1/3,
    3, 4, 5 with probability 1/9 each.

    As can be seen, the expectation of $\boldsymbol{\xi}$ is still 2, and new uncertainty is added around 1 and 4.

    (a) Show that the L-shaped method follows exactly the same path as before ($x^1 = 0, x^2 = 10, x^3 = 7/3, x^4 = 1.5, x^5 = 2$) provided that in Iteration 4, the support is chosen to describe the region $[1.5, 2]$. If it is chosen to describe the region $[1, 1.5]$, one more iteration is needed.

    (b) Show the multicut version also follows the same path as before ($x^1 = 0, x^2 = 10, x^3 = 2$).

    (c) Now consider an intermediate situation, where $\mathcal{Q}(x)$ is approximated by $\frac{1}{3}[\mathcal{Q}_1(x) + \mathcal{Q}_2(x) + \mathcal{Q}_3(x)]$, where $\mathcal{Q}_1(x)$ is the expectation over the three realizations 0.5, 1.0, and 1.5 (conditional on $\boldsymbol{\xi}$ being in the group $\{0.5, 1.0, 1.5\}$), $\mathcal{Q}_2(x) = Q(x, \xi = 2)$, and $\mathcal{Q}_3(x)$ is the (similarly conditional) expectation over the realizations 3, 4, and 5. Thus, the objective becomes $\frac{1}{3}(\theta_1 + \theta_2 + \theta_3)$. Show that in Iteration 1, the cuts at $x^1 = 0$ are $\theta_1 \geq 1 - x, \theta_2 \geq 2 - x$, and $\theta_3 \geq 4 - x$. In Iteration 2, $x^2 = 10$, and the cuts become $\theta_1 \geq x - 1, \theta_2 \geq x - 2$, and $\theta_3 \geq x - 4$. Show, without computations, that only two major iterations are needed. What conclusions can you draw from this example?

5.4 Bunching and Other Efficiencies

One big issue in the efficient implementation of the L-shaped method is in Step 3. The second-stage program (1.9) has to be solved K times to obtain the optimal multipliers, π_k^ν. For a given x^ν and a given realization k, let B be the optimal basis of the second stage. It is well-known from linear programming that B is a square submatrix of W such that $(\pi_k^\nu)^T = q_{k,B}^T B^{-1}, q_k^T - (\pi_k^\nu)^T W \geq 0, B^{-1}(h_k - T_k x^\nu) \geq 0$, where $q_{k,B}$ denotes the restriction of q_k to the selection of columns that define B. Important savings can be obtained in Step 3 when the same basis B is optimal for several realizations of k. This is especially the case when q is deterministic. Then, two different realizations that share the same basis also share the same multipliers π_k^ν. We present the rest of the section, assuming q is deterministic.

To be more precise, define

$$\tau = \{t | t = h_k - T_k x^\nu \text{ for some } k = 1, \cdots, K\} \tag{4.1}$$

as the set of possible right-hand sides in the second stage. Let B be a square submatrix and $\pi^T = q_B^T B^{-1}$. Assume B satisfies the optimality criterion $q^T - \pi^T W \geq 0$. Then define a *bunch* as

$$Bu = \{t \in \tau | B^{-1}t \geq 0\}, \tag{4.2}$$

the set of possible right-hand sides that satisfy the feasibility condition. Thus, π is an optimal dual multiplier for all $t \in Bu$. Note also that, by virtue of Step 2 of the L-shaped method, only feasible first-stage $x^\nu \in K_2$ are considered. This observation means that, by construction,

$$\tau \subseteq \text{pos } W = \{t | t = Wy, y \geq 0\}.$$

We now provide an introduction to possible implementations that use these ideas. For more details, the reader is referred to Gassmann [1990], Wets [1988], or Wets [1983b].

a. Full decomposability

One first possibility is to work out a full decomposition of pos W into component bases. This can only be done for small problems or problems with a well-defined structure. As an example, consider the farming example of Section 1.1. The second-stage representation (1.1.4) is repeated here under the notation of the current chapter:

$$
\begin{aligned}
Q(x,\xi) = \min \quad & -210y_1 + 140y_2 - 150y_3 + 100y_4 - 36y_5 + y_6 \\
\text{s.t.} \quad & y_1 - y_2 - w_1 = 200 - \xi_1 x_1, \\
& y_3 - y_4 - w_2 = 240 - \xi_2 x_2, \\
& y_5 + y_6 + w_3 = \xi_3 x_3, \\
& y_5 + w_4 = 6000, \\
& y, w \geq 0,
\end{aligned}
$$

where w_1 to w_4 are slack variables. This second stage has complete recourse, so pos $W = \Re^4$.

$$\text{The matrix } W = \begin{pmatrix} 1 & -1 & 0 & 0 & 0 & 0 & -1 & 0 & 0 & 0 \\ 0 & 0 & 1 & 0 & 0 & 0 & 0 & -1 & 0 & 0 \\ 0 & 0 & 0 & -1 & 1 & 1 & 0 & 0 & 1 & 0 \\ 0 & 0 & 0 & 0 & 1 & 0 & 0 & 0 & 0 & 1 \end{pmatrix}$$

has 4 rows and 10 columns, so that theoretically $\binom{10}{4} = 210$ bases could be found. However, in practice w_1, w_2, and w_3 are never in the basis, as they are always dominated by y_2, y_4, and y_6, respectively. The matrix where the columns w_1, w_2, and w_3 are removed is sometimes called the *support* of W (see Wallace and Wets [1992]). Also, y_5 is always in the basis (a fact of worldwide importance as it is one of the reasons that created tension between United States and Europe within the GATT negotiations). Moreover y_1 or y_2 and y_3 or y_4 are always basic. In this case, not only is a full decomposition of pos W available, but an immediate analytical expression for the multipliers is also obtained. Thus,

$$\pi_1(\xi) = \begin{cases} 210 & \text{if } \xi_1 x_1 < 200, \\ -140 & \text{otherwise;} \end{cases}$$

$$\pi_2(\xi) = \begin{cases} 150 & \text{if } \xi_2 x_2 < 240, \\ -100 & \text{otherwise;} \end{cases}$$

$$\pi_3(\xi) = \begin{cases} -36 & \text{if } \xi_3 x_3 < 6000, \\ 0 & \text{otherwise;} \end{cases}$$

$$\pi_4(\xi) = \begin{cases} 10 & \text{if } \xi_3 x_3 > 6000, \\ 0 & \text{otherwise.} \end{cases}$$

The dual multipliers are easily obtained because the problem is small and enjoys some form of separability. The decomposition is thus $(1, 3, 5, 6)$, $(1, 3, 5, 10)$, $(1, 4, 5, 6)$, $(1, 4, 5, 10)$, $(2, 3, 5, 6)$, $(2, 3, 5, 10)$, $(2, 4, 5, 6)$, $(2, 4, 5, 10)$, where the four variables in a basis are described by their indices (where the index is $6 + j$ for the jth slack variable). Another example is given in Exercise 1 and Wallace [1986a].

When applicable, full decomposability has proven very efficient. In general, however, it is expected to be applicable only for small problems.

b. Bunching

A relatively simple bunching procedure is as follows. Again let $\tau = \{t | t = h_k - T_k x^\nu$ for some $k = 1, \cdots, K\}$ be the set of possible right-hand sides in the second stage. Consider some k. Denote $t_k = h_k - T_k x^\nu$. It might arbitrarily be $k = 1$, or, if available, a value of k such that $h_k - T_k x^\nu = \bar{t}$, the expectation of all $t_k \in \tau$. Let B_1 be the corresponding optimal basis

172   5. Two-Stage Linear Recourse Problems

and $\pi(1)$ the corresponding vector of simplex multipliers. Then, $Bu(1) = \{t \in \tau | B_1^{-1} t \geq 0\}$. Let $\tau_1 = \tau \backslash Bu(1)$.

We can now repeat the same operations. Some element of τ_1 is chosen. The corresponding optimal basis B_2 and its associated vector of multipliers $\pi(2)$ are formed . Then, $Bu(2) = \{t \in \tau_1 | B_2^{-1} t \geq 0\}$ and $\tau_2 = \tau_1 \backslash Bu(2)$. The process is repeated until all $t_k \in \tau$ are in one of b total bunches. Then, (1.10) and (1.11) are replaced by

$$E_{s+1} = \sum_{\ell=1}^{b} \pi(\ell)^T \sum_{t_k \in Bu(\ell)} p_k T_k \qquad (4.3)$$

and

$$e_{s+1} = \sum_{\ell=1}^{b} \pi(\ell)^T \sum_{t_k \in Bu(\ell)} p_k h_k . \qquad (4.4)$$

This procedure still has some drawbacks. One is that the same $t_k \in \tau$ may be checked many times against different bases. The second is that a new optimization is restarted every time a new bunch is considered. It is obvious here that some savings can be obtained in organizing the work in such a way that the optimal basis in the next bunch is obtained by performing only one (or a few) dual simplex iterations from the previous one. As an example, consider the following second stage:

$$\begin{array}{llllll}
\max & 6y_1 & +5y_2 & +4y_3 & +3y_4 \\
\text{s. t.} & 2y_1 & +y_2 & +y_3 & & \leq \xi_1, \\
& & y_2 & +y_3 & +y_4 & \leq \xi_2, \\
& y_1 & & +y_3 & & \leq x_1, \\
& & 2y_2 & & +y_4 & \leq x_2, \\
& & & y \geq 0.
\end{array}$$

Let $\xi_1 \in \{4,5,6,7,8\}$ with equal probability 0.2 each and $\xi_2 \in \{2,3,4,5,6\}$ with equal probability 0.2 each. There are theoretically $\binom{8}{4} = 70$ different possible bases. In view of the possible realizations of ξ, at most 25 different bases can be optimal.

Let t^1 to t^{25} denote the possible right-hand sides with

$$t^1 = \begin{pmatrix} 4 \\ 2 \\ x_1 \\ x_2 \end{pmatrix}, \ t^2 = \begin{pmatrix} 4 \\ 3 \\ x_1 \\ x_2 \end{pmatrix}, \cdots, t^{25} = \begin{pmatrix} 8 \\ 6 \\ x_1 \\ x_2 \end{pmatrix}.$$

Consider the case where $x_1 = 3.1$ and $x_2 = 4.1$. Let us start from $\xi = \bar{\xi} = (6,4)^T$. Represent a basis again by the variable indices with $4+j$ the index of the jth slack. The optimal basis is $B_1 = \{1,4,7,8\}$ with $y_1 = 3, y_4 = 4, w_3 = 0.1, w_4 = 0.1$, the values of the basic variables.

The optimal dictionary associated with B_1 is

$$z = 3\xi_1 + 3\xi_2 - y_2 - 2y_3 - 3w_1 - 3w_2,$$
$$y_1 = 1/2\xi_1 - 1/2y_2 - 1/2y_3 - 1/2w_1,$$
$$y_4 = \xi_2 - y_2 - y_3 - w_2,$$
$$w_3 = 3.1 - 1/2\xi_1 + 1/2y_2 - 1/2y_3 + 1/2w_1,$$
$$w_4 = 4.1 - \xi_2 - y_2 + y_3 + w_2.$$

This basis is optimal and feasible as long as $\xi_1/2 \le 3.1$ and $\xi_2 \le 4.1$, which in view of the possible values of ξ amounts to $\xi_1 \le 6$ and $\xi_2 \le 4$, so that $Bu(1) = \{t^1, t^2, t^3, t^6, t^7, t^8, t^{11}, t^{12}, t^{13}\}$. Neighboring bases can be obtained by considering either $\xi_1 \ge 7$ or $\xi_2 \ge 5$. Let us start with $\xi_2 \ge 5$. This means that w_4 becomes negative and a dual simplex pivot is required in Row 4. This means that w_4 leaves the basis, and, according to the usual dual simplex rule, y_3 enters the basis.

The new basis is $B_2 = \{1, 3, 4, 7\}$ with an optimal dictionary

$$z = 3\xi_1 + \xi_2 + 8.2 - 3y_2 - 3w_1 - w_2 - 2w_4,$$
$$y_1 = \frac{\xi_1}{2} - \frac{\xi_2}{2} + 2.05 - y_2 - \frac{w_1}{2} + \frac{w_2}{2} - \frac{w_4}{2},$$
$$y_3 = \xi_2 - 4.1 + y_2 - w_2 + w_4,$$
$$y_4 = 4.1 - 2y_2 - w_4,$$
$$w_3 = 5.15 - \frac{\xi_1}{2} - \frac{\xi_2}{2} + \frac{w_1}{2} + \frac{w_2}{2} - \frac{w_4}{2}.$$

The condition $\xi_1 - \xi_2 + 4.1 \ge 0$ always holds. This basis is optimal as long as $\xi_2 \ge 5$ and $\xi_1 + \xi_2 \le 10$, so that $Bu(2) = \{t^4, t^5, t^9\}$.

Neighboring bases are B_1 when $\xi_2 \le 4$ and B_3 obtained when $w_3 < 0$, i.e., $\xi_1 + \xi_2 \ge 11$. This basis corresponds to w_3 leaving the basis and w_2 entering the basis. To keep a long story short, we just summarize the various steps in the following list:

$$B_1 = \{1, 4, 7, 8\} \quad Bu(1) = \{t^1, t^2, t^3, t^6, t^7, t^8, t^{11}, t^{12}, t^{13}\}$$

$$B_2 = \{1, 3, 4, 7\} \quad Bu(2) = \{t^4, t^5, t^9\}$$

$$B_3 = \{1, 3, 4, 6\} \quad Bu(3) = \{t^{10}, t^{14}, t^{15}\}$$

$$B_4 = \{1, 4, 5, 6\} \quad Bu(4) = \{t^{19}, t^{20}, t^{24}, t^{25}\}$$

$$B_5 = \{1, 2, 4, 5\} \quad Bu(5) = \{t^{18}, t^{22}, t^{23}\}$$

$$B_6 = \{1, 2, 4, 8\} \quad Bu(6) = \{t^{16}, t^{17}, t^{21}\}$$
$$B_7 = \{1, 2, 5, 8\} \quad Bu(7) = \emptyset.$$

Several paths are possible, as one may have chosen B_6 instead of B_2 as a second basis. Also, the graph may take the form of a tree, and more elaborate techniques for constructing the graph and recovering the bases can be used, see Gassmann [1988] and Wets [1983b].

Research has also been done to find an appropriate root of the tree (Haugland and Wallace [1988]) and to develop preprocessing techniques (Wallace and Wets [1992]). Other attempts include the sifting procedure, a sort of parametric analysis proposed by Gartska and Rutenberg [1973].

Finally, it is reasonable to expect that parallel processing may be helpful in the search of the optimal multipliers in the second stage. As an example, Ariyawansa and Hudson [1991] have designed a parallel implementation of the L-shaped algorithm, in which the computation of the dual simplex multipliers in Step 3 is parallelized. They report an average speed-up factor of 5.5 on a seven-processor Sequent/Balance, for problems where the number K of realizations in the second stage is large (up to $10,000$).

Exercise

1. Consider the capacity expansion example. Order the equipment in increasing order of utilization cost $q_1 \leq q_2 \leq \ldots$. Observe that it is always optimal to use the equipment in that order. Then obtain a full decomposition of pos W.

5.5   Inner Linearization Methods

As mentioned earlier, the most direct alternative to an outer linearization, or cut generation, approach is an inner linearization or column generation approach (see Geoffrion [1970] for other basic approaches to large-scale problems). In fact, this was the first suggestion of Dantzig and Madansky [1961] for solving stochastic linear programs. They observed that the structure of the dual in Figure 2 fits the prototype for Dantzig-Wolfe decomposition. In fact, we can derive this approach from the L-shaped method by taking duals.

Consider the following dual linear program to (1.2)–(1.4).

$$\max \ \zeta = \rho^T b + \sum_{\ell=1}^{r} \sigma_\ell d_\ell + \sum_{\ell=1}^{s} \pi_\ell e_\ell \tag{5.1}$$

$$\text{s.t.} \ \rho^T A + \sum_{\ell=1}^{r} \sigma_\ell D_\ell + \sum_{\ell=1}^{s} \pi_\ell E_\ell \leq c^T, \tag{5.2}$$

$$\sum_{\ell=1}^{s} \pi_\ell = 1, \sigma_\ell \geq 0, \ell = 1, \ldots, r, \pi_\ell \geq 0, \ell = 1, \ldots, s. \tag{5.3}$$

The linear program in (5.1)–(5.3) includes multipliers σ_ℓ on *extreme rays*, or directions of recession, that cannot be produced with positive combinations of distinct other recession directions, of the duals of the subproblems and multipliers π_ℓ on the expectations of extreme points of the duals of the subproblems. To see this, suppose that (5.1)–(5.3) is solved to obtain a multiplier x^ν on constraint (5.2). Now, consider the following dual to (1.9):

$$\max \quad w = \pi^T(h_k - T_k x^\nu) \quad \text{s.t.} \quad \pi^T W \le q^T. \tag{5.4}$$

If (5.4) is unbounded for any k, we then must have some σ^ν such that $\sigma^{\nu T} W \le 0$ and $\sigma^{\nu T}(h_k - T_k x^\nu) > 0$, or (1.5)–(1.6) has a feasible dual solution (hence optimal primal solution) with a positive value. So, Step 2 of the L-shaped method is equivalent to checking whether (5.4) is unbounded for any k. In this case, we form D_{r+1} and d_{r+1} as in (1.7) and (1.8) of the L-shaped method and add them to (5.1)–(5.3).

Next, note, that if (5.4) is infeasible, the stochastic program is not well-formulated (see Exercise 1). Consider when (5.4) has a finite optimal value for all k. In the L-shaped method, if (1.9) was solvable for all k, then we formed E_{s+1} and e_{s+1} and added them to (1.2)–(1.4). In this case in the inner linearization procedure, we again use (1.10) and (1.11) to form E_{s+1} and e_{s+1} and add them to (5.1)–(5.3).

Solving the duals in Steps 1 to 3 of the L-shaped algorithm then consists of solving (5.1)–(5.3) as a master problem and problems (5.4) as subproblems. Formally, this method is the following inner linearization method.

Inner Linearization Algorithm

Step 0. Set $r = s = \nu = 0$.

Step 1. Set $\nu = \nu + 1$ and solve the linear program in (5.1)–(5.3). Let the solution be $(\rho^\nu, \sigma^\nu, \pi^\nu)$ with a dual solution, (x^ν, θ^ν).

Step 2. For $k = 1, \ldots, K$, solve (5.4). If any infeasible problem (5.4) is found, stop and evaluate the formulation. If an unbounded solution with extreme ray σ^ν is found for any k, then form new columns $(d_{r+1} = (\sigma^\nu)^T h_k, D_{r+1} = (\sigma^\nu)^T T_k)$, set $r = r + 1$, and return to Step 1.

If all problems (5.4) are solvable, then form new columns, E_{s+1} and e_{s+1}, as in (1.10) and (1.11). If $e_{s+1} - E_{s+1} x^\nu - \theta^\nu \le 0$, then stop; $(\rho^\nu, \sigma^\nu, \pi^\nu)$ and (x^ν, θ^ν) are optimal in the original problem (5.1.2).
If $e_{s+1} - E_{s+1} x^\nu - \theta^\nu > 0$, set $s = s + 1$, and return to Step 1.

Clearly, the inner linearization method follows the same steps as the L-shaped method, except that we solve the duals of the problems instead of the primals. Hence, convergence follows directly from the L-shaped method. We could also view this approach directly as in Dantzig-Wolfe decomposition by stating that (5.1)–(5.3) is an inner linearization of the dual of the basic L-shaped problem in (5.1.2) and that the subproblems (5.4) generate new extreme points and rays to add to this inner linearization (see Exercise 2).

If, as in many problems, $n_1 >> m_1$, the primal version has smaller basis matrices, at most of order $m_1 + m_2$, than the $n_1 \times n_1$ bases for the dual. Hence, the L-shaped implementation is usually preferred. Inner linearization can, however, be applied directly to the primal by assuming T is fixed using the form in (3.1.5), which we repeat here:

$$\min z = c^T x + \Psi(\chi) \qquad (5.5)$$
$$\text{s.t. } Ax = b,$$
$$Tx - \chi = 0,$$
$$x \geq 0,$$

where $\Psi(\chi) = E_\omega \psi(\chi, \xi(\omega))$ and $\psi(\chi, \xi(\omega)) = \min\{q(\omega)^T y | Wy = h(\omega) - \chi, y \geq 0\}$. Note that, in this form, we assume that T is fixed but q and h may still be functions of ω. For this reason, we revert to the use of Ψ for the recourse function.

In this case, we wish to build an inner linearization of the function $\Psi(\chi)$ using the generalized programming approach as in Dantzig [1963, Chapter 24]. The basic idea is to replace $\Psi(\chi)$ by the convex hull of points $\Psi(\chi^\ell)$ chosen in each iteration of the algorithm. Each iteration generates a new extreme point of a region of linearity for Ψ, which is polyhedral as we showed in Theorem 3.6. Thus, finite convergence is assured with finite numbers of realizations.

The algorithm follows.

Generalized Programming Method for Two-Stage Stochastic Linear Programs

Step 0. Set $s = t = \nu = 0$.

Step 1. Set $\nu = \nu + 1$ and solve the linear program master problem:

$$\min \ z^\nu = c^T x + \sum_{i=1}^{r} \mu_i \Psi_0^+(\zeta^i) + \sum_{i=1}^{s} \lambda_i \Psi(\chi^i) \qquad (5.6)$$

$$\text{s.t. } Ax = b, \qquad (5.7)$$

$$Tx - \sum_{i=1}^{r} \mu_i \zeta^i - \sum_{i=1}^{s} \lambda_i \chi^i = 0, \qquad (5.8)$$

$$\sum_{i=1}^{r} \lambda_i = 1, \qquad (5.9)$$

$$x, \mu_i \geq 0, i = 1, \dots, r, \lambda_i \geq 0, i = 1, \dots, s.$$

If (5.6)–(5.9) is infeasible or unbounded, stop. Otherwise, let the solution be $(x^\nu, \mu^\nu, \lambda^\nu)$ with associated dual variables, $(\sigma^\nu, \pi^\nu, \rho^\nu)$.

Step 2. Solve the subproblem:

$$\min_{\chi} \Psi(\chi) + (\pi^{\nu})^T \chi - \rho^{\nu}, \qquad (5.10)$$

which we assume has value less than ∞.

If (5.10) is unbounded, a recession direction ζ^{r+1} is obtained, such that for some χ, $\Psi(\chi + \alpha\zeta^{r+1}) + (\pi^{\nu})^T(\chi + \alpha\zeta^{r+1}) \to -\infty$ as $\alpha \to \infty$. In this case, let $\Psi_0^+(\zeta^{r+1}) = \lim_{\alpha \to \infty} \frac{\Psi(\chi + \alpha\zeta^{r+1}) - \Psi(\chi)}{\alpha}$, $r = r + 1$, and return to Step 1.

If (5.10) is solvable, let the solution be χ^{s+1}. If $\Psi(\chi) + (\pi^{\nu})^T \chi - \rho^{\nu} \geq 0$, then stop; $(x^{\nu}, \mu^{\nu}, \lambda^{\nu})$ corresponds to an optimal solution to (5.5). Otherwise, set $s = s + 1$ and return to Step 1.

This algorithm generates columns in (5.6)–(5.9) corresponding to new proposals from the subproblem in (5.10). In the two-stage stochastic linear program form, (5.10) can be recast as:

$$\min \sum_{k=1}^{K} p_k q_k^T y_k + (\pi^{\nu})^T \chi - \rho^{\nu} \qquad (5.11)$$
$$\text{s.t. } Wy_k + \chi = h_k, k = 1, \ldots, K,$$
$$y_k \geq 0, k = 1, \ldots, K.$$

This problem is not generally separable into different subproblems for each k. Hence, for general problems, the L-shaped method has an advantage. In some cases (notably simple recourse), $\Psi(\chi)$ is separable into components for each k, and (5.11) can again be divided into K independent subproblems. We discuss this possibility further in Section 7.

To show that the generalized programming method also converges finitely, we wish to show that an extreme solution in (5.11) is an extreme value of linear regions of $\Psi(\chi)$. We do this for extreme points in the following proposition.

Proposition 3. *Every optimal extreme point, $(\bar{y}_1, \ldots, \bar{y}_K, \bar{\chi})$, of the feasible region in (5.11) corresponds to an extreme point $\bar{\chi}$ of $\{\chi | \Psi(\chi) = \bar{\pi}^T \chi + \theta\}$, where $\bar{\pi} = \sum_{k=1}^{K} \bar{\pi}_k$, and each $\bar{\pi}_k$ is an extreme point of $\{\pi_k | \pi_k^T W \leq q_k^T\}$.*

Proof. Suppose $(\bar{y}_1, \ldots, \bar{y}_K, \bar{\chi})$ is an optimal extreme point in (5.11). In this case, we must have $q_i^T \bar{y}_i \leq q_i^T y_i$ for all $Wy_i = \xi_i - \bar{\chi}$. We must also have that $\bar{y}_i$ is an extreme point of $\{y_i | Wy_i = \xi_i - \bar{\chi}, y_i \geq 0\}$ because, otherwise, we could take $\bar{y}_i = (1/2)(y_i^1 + y_i^2)$ for distinct feasible y_i^1 and y_i^2. So, $\bar{y}_k$ has a complementary dual solution, $\bar{\pi}_k$, that is an extreme point of $\{\pi_k | \pi_k^T W \leq q_k^T\}$ and such that $(q_k^T - \bar{\pi}_k^T W)\bar{y}_k = 0$.

Now, suppose $\bar{\chi}$ is not an extreme point of the linearity region where $\Psi(\chi) = \bar{\pi}^T \chi + \theta$ for $\theta = \Psi(\bar{\chi}) - \bar{\pi}^T \bar{\chi}$ with $\bar{\pi} = \sum_{k=1}^{K} \bar{\pi}_k$. In this case, there exists χ^1 and χ^2 such that $\bar{\chi} = \lambda\chi^1 + (1 - \lambda)\chi^2$ where $0 < \lambda < 1$, for $\Psi(\chi^1) = \bar{\pi}^T\chi^1 + \theta$ and $\Psi(\chi^2) = \bar{\pi}^T\chi^2 + \theta$. We also have that $\Psi(\chi^j) = \sum_{k=1}^{K} q_k^T y_k^j$, where $q_k^T y_k^j = \bar{\pi}_k^T(h_k - \chi^j)$ for $j = 1, 2$, because, by $\bar{\pi}_k^T$ feasible in the kth recourse problem, the only other possibility is $q_k^T y_k^j > \bar{\pi}_k^T(\xi - \chi^j)$, which would imply $\Psi(\chi^j) > \bar{\pi}^T\chi^j + \theta$. This also implies that

$$(\bar{\pi}_k^T W - q_k^T)(\lambda y_k^1 + (1 - \lambda)y_k^2) = 0, \qquad (5.12)$$

which implies that $\lambda y_k^1 + (1 - \lambda)y_k^2 = \bar{y}_k$ because $\bar{y}_k$ is an extreme point of the feasible region in recourse problem k. In this case, $(\bar{y}_1, \ldots, \bar{y}_K, \bar{\chi}) = \lambda(\bar{y}_1^1, \ldots, \bar{y}_K^1, \chi^1) + (1 - \lambda)(\bar{y}_1^2, \ldots, \bar{y}_K^2, \chi^2)$, with both terms feasible in (5.11). This contradicts that $(\bar{y}_1, \ldots, \bar{y}_K, \bar{\chi})$ is an extreme point. $\square$

A similar argument shows that any extreme ray found in solving (5.11) is an extreme ray of a region of linearity of $\Psi(\chi)$ (Exercise 3). Now, we can state the generalized programming finite convergence result.

Theorem 4. *The generalized programming applied to problem (5.5) with subproblem (5.11) solves (5.5) in a finite number of steps.*

Proof: At each solution of (5.11), a new linear region extreme value is generated. First for a new extreme ray, we must have $\Psi_0^+(\zeta^{r+1}) + (\pi^\nu)^T(\zeta^{r+1}) < 0$, while, for $1 \le i \le s$, $\Psi_0^+(\zeta^i) \ge -(\pi^\nu)^T\zeta^i$. For an extreme point, we only add that point if $\Psi(\chi^{s+1}) + (\pi^\nu)^T\chi^{s+1} - \rho^\nu < 0$, while, for $1 \le i \le s$, $\Psi(\chi^s) + (\pi^\nu)^T\chi^s - \rho^\nu \ge 0$. Because the number of such regions is finite and each has a finite number of extreme points and rays, the algorithm converges finitely.

The solution found solves (5.5) because if we reach the termination condition, then

$$\begin{aligned}(\sigma^\nu)^T b + \rho^\nu &\le (\sigma^\nu)^T b + \Psi(\chi) + (\pi^\nu)^T\chi \\ &\le (\sigma^{\nu T}A + (\pi^\nu)^T T)x + \Psi(\chi), (x, \chi) \text{ feasible in } (5.5), \\ &\le c^T x + \Psi(\chi), \end{aligned} \qquad (5.13)$$

for all (x, χ) feasible in (5.5). $\square$

As with the L-shaped method, we can also modify the generalized linear programming approach to consider only active columns so that s and t can be bounded again by m_2. Of course, this approach's greatest potential is in simple recourse problems as we mentioned earlier. It may also be advantageous if an algorithm can take advantage of the special matrix structure in (5.11). The most direct approach in this case is to construct a working basis and to try to perform most linear transformations with submatrices chosen from W. In this case, the procedure becomes quite similar to the procedures for directly attacking (3.1.2) that are given in the next section.

The generalized programming approach is also useful in considering the stochastic program as a procedure for combining *tenders* χ_i (see Nazareth and Wets [1986]) bid from the subproblems. In this case, the method may converge most quickly if the initial set of tenders is chosen well. A method for choosing such an initial set of tenders appears in Birge and Wets [1984].

Exercises

1. Suppose Problem (5.4) is infeasible for some k. What can be said about the original two-stage stochastic linear program? Find examples for these possible situations.

2. Prove directly that the inner linearization method converges to an optimal solution to the two-stage stochastic linear program (3.1.2).

3. Show that any extreme descending ray in (5.11) corresponds to an extreme ray of a linear piece of $\Psi(\chi)$.

5.6 Basis Factorization Methods

As observed earlier in this chapter, the matrices in (1.1) and its dual have a special structure that may allow efficient specific basis factorizations. In this way, the extensive form of the problem may be more efficiently solved by either extreme point or interior point methods. There are similarities with the previous decomposition approaches. We discuss relative advantages and disadvantages at the end of this section.

Basis factorization for extreme point methods has generally been considered the dual structure, although the same ideas apply to either the dual or primal problems. For more details on this approach, we refer to Kall [1979] and Strazicky [1980]. We consider the primal approach because, generally, the number of columns $(n_1 + K n_2)$ is larger than the number of rows $(m_1 + K m_2)$ in the original constraint matrix. In this case, we can write a basic solution as $(x_{I_0}, x_{I_1}, \ldots, x_{I_K}, y_{J_1}, \ldots, y_{J_k})$, where $I_j, j = 0, \ldots, K$, and $J_l, l = 1, \ldots, K$, are index sets that may be altered at each iteration. The constraints are also partitioned according to these index sets so that a basis is:

$$
B = \begin{pmatrix}
A_{I_0} & A_{I_1} & \cdots & A_{I_K} & & & \\
T_{1,I_0} & T_{1,I_1} & \cdots & T_{1,I_K} & W_{J_1} & & \\
\vdots & \vdots & \vdots & \vdots & & W_{J_k} & \\
T_{K,I_0} & T_{K,I_1} & \cdots & T_{K,I_K} & & & W_{J_K}
\end{pmatrix}. \tag{6.1}
$$

The main observation in basis factorization is that we may permute the rows of B to achieve an efficient form. This is the result of the following proposition.

Proposition 5. *A basis matrix, B, for problem (1.1) is equivalent after a row permutation P to*

$$B' = PB = \begin{pmatrix} D & C \\ F & L \end{pmatrix}, \tag{6.2}$$

where D is square invertible and at most $n_1 \times n_1$ and L is an invertible matrix of K invertible blocks of sizes at most $m_2 \times m_2$ each.

Proof. We can perform the required permutation on B in (6.1). First, note that the number of columns in $A_{I_0}, \ldots, A_{I_K}$ is at most n_1 for B to be nonsingular. We must also be able to form a nonsingular submatrix from these columns if B is invertible. Suppose this matrix is composed of $A_{I_0}, \ldots, A_{I_K}$ and rows T_{ku,I_j} from each subproblem $j = 1, \ldots, K$. In this case, we have constructed

$$D = \begin{pmatrix} A_{I_0} & A_{I_1} & \cdots & A_{I_K} \\ T_{1u,I_0} & T_{1u,I_1} & \cdots & T_{1u,I_K} \\ \vdots & \vdots & \vdots & \vdots \\ T_{Ku,I_0} & T_{Ku,I_1} & \cdots & T_{Ku,I_K} \end{pmatrix}.$$

Hence,

$$C = \begin{pmatrix} 0 & 0 & \cdots & 0 & 0 \\ W_{1u,J_1} & 0 & \cdots & 0 & 0 \\ 0 & \ddots & 0 & \cdots & 0 \\ \vdots & & 0 & W_{ku,J_k} & 0 & \vdots \\ 0 & \cdots & 0 & \ddots & 0 \\ 0 & 0 & \cdots & 0 & W_{Ku,J_K} \end{pmatrix}.$$

Next, assume that the remaining rows of T_{k,I_j} are T_{kl,I_j}. We then obtain:

$$F = \begin{pmatrix} T_{1l,I_0} & T_{1l,I_1} & \cdots & T_{1l,I_K} \\ \vdots & \vdots & \vdots & \vdots \\ T_{Kl,I_0} & T_{Kl,I_1} & \cdots & T_{Kl,I_K} \end{pmatrix}$$

and

$$L = \begin{pmatrix} W_{1l,J_1} & 0 & 0 \\ 0 & \cdots W_{kl,J_k} \cdots & 0 \\ 0 & 0 & W_{Kl,J_K} \end{pmatrix}.$$

Because D has rank at least m_1, each W_{kl,J_k} in L has rank at most m_2. This gives the result. $\square$

To show how this result is used, consider the forward transformation to find the basic values of $(x_{I_0}, x_{I_1}, \ldots, x_{I_K}, y_{J_1}, \ldots, y_{J_k})$, which we write as (x_B, y_B), that solve:

$$Dx_B + Cy_B = b'; Fx_B + Ly_B = h', \tag{6.3}$$

where $b' = \begin{pmatrix} b \\ h_u \end{pmatrix}$, $h' = h_l$, h_u corresponds to the components of the right-hand side for rows of T in D, and h_l corresponds to the components with rows in F.

Note that L is invertible, so

$$y_B = L^{-1}(h' - Fx_B). \tag{6.4}$$

Substituting in the first system of equations yields

$$(D - CL^{-1}F)x_B = b' - CL^{-1}h'. \tag{6.5}$$

Hence, we use L to solve for the columns of $L^{-1}F$ and $L^{-1}h'$, then form the working basis, $(D - CL^{-1}F)$, to solve for x_B, and multiply x_B again by $L^{-1}F$ and subtract from $L^{-1}h'$ to obtain y_B. Because most of the work involves just the square block matrices in L and the working basis, substantial effort can be saved in the decomposition procedure (see Exercise 1). The backward transformation can also be performed by taking advantage of this structure (see Exercise 2). The other forward transformation in the simplex method to find the leaving column is, of course, the same as the operations used in (6.4) and (6.5). The entire simplex method then has the following form.

Basic Factorization Simplex Method

Step 0. Suppose that $(x_{B^0}^0, y_{B^{0'}}^0) = (x_{I_0^0}^0, \ldots, x_{I_K^0}^0, y_{J_0^0}^0, \ldots, y_{J_K^0}^0)$ is an initial basic feasible solution for (1.1), with initial indices partitioned according to $B^0 = \{\beta_1^0, \ldots, \beta_{l^0}^0\} = \{I_i^0, i = 0, \ldots, K\}$ and $B^{0'} = \{\beta_{1,1}^{0,'}, \ldots, \beta_{1,l_1'}^{0,'}, \ldots, \beta_{K,1}^{0,'}, \ldots, \beta_{K,l_K'}^{0,'}\} = J_j^0, j = 1, \ldots, K$. Let the initial permutation matrix be P^0, and set $\nu = 0$.

Step 1. Solve $(\rho^T, \pi^T) \begin{pmatrix} D & C \\ F & L \end{pmatrix} = (c_{B^0}^T, \hat{q}_{\beta^0}^T)$, where $\hat{q}_{k,i} = p_k q_{k,i}$.

Step 2. Find $\bar{c}_s = \min_j \{c_j - (\rho^T | \pi^T) P^\nu (A_{\cdot j}^T | T_{1,\cdot j}^T | \cdots | T_{K,\cdot j}^T)^T\}$ and $\bar{q}_{k',s'} = \min_{j,k} \{p_k q_{k,j} - (\rho^T | \pi^T) P^\nu (0 \cdots W_{k,\cdot j} \cdots 0)\}$. If $\bar{c}_s \geq 0$ and $\bar{q}_{k',s'} \geq 0$, then stop; the current solution is optimal. Otherwise, if $\bar{c}_s < \bar{q}_{k',s'}$, go to Step 4. If $\bar{c}_s \geq \bar{q}_{k',s'}$, go to Step 3.

Step 3. Solve for the entering column, $\begin{pmatrix} D & C \\ F & L \end{pmatrix} \bar{W}_{k',\cdot s'} = P^\nu (0 \cdots W_{k',\cdot s'}^T \cdots 0)^T$. Let

$$\theta = x_{B^\nu(r)}^\nu / \bar{W}_{k',rs'} = \min_{\bar{W}_{k',is'} > 0, 1 \leq i \leq l^\nu} \{x_{B^\nu(i)}^\nu / \bar{W}_{k',is'}\} \tag{6.6}$$

and

$$\theta' = y_{B^{\nu'}(r')}^\nu / \bar{W}_{k',r's'} = \min_{\bar{W}_{k',is'} > 0, l^\nu + 1 \leq i \leq m_1 + Km_2} \{y_{B^{\nu'}(i)}^\nu / \bar{W}_{k',is'}\}. \tag{6.7}$$

If no minimum exists in either (6.6) or (6.7), then stop; the problem is unbounded. Otherwise, if $\theta < \theta'$, go to Step 5. If $\theta \geq \theta'$, go to Step 6.

Step 4. Solve for the entering column, $\begin{pmatrix} D & C \\ F & L \end{pmatrix} \bar{A}_{\cdot s'} = P^\nu (A_{\cdot s}^T | T_{1,\cdot s}^T | \cdots$
$|T_{K,\cdot s}^T)^T$. Let

$$\theta = x_{B^\nu(r)}^\nu / \bar{A}_{rs} = \min_{\bar{A}_{is} > 0, 1 \leq i \leq l^\nu} \{ x_{B^\nu(i)}^\nu / \bar{A}_{is'} \} \tag{6.8}$$

and

$$\theta' = y_{B^{\nu'}(r')}^\nu / \bar{A}_{r's} = \min_{\bar{A}_{is'} > 0, l^\nu + 1 \leq i \leq m_1 + K m_2} \{ y_{B^{\nu'}(i)}^\nu / \bar{A}_{is} \}. \tag{6.9}$$

If no minimum exists in either (6.8) or (6.9), then stop; the problem is unbounded. Otherwise, if $\theta < \theta'$, go to Step 5. If $\theta \geq \theta'$, go to Step 6.

Step 5. Let $B^{\nu+1} = B^\nu$, $B^{\nu+1'} = B^{\nu'}$, $I_i^{\nu+1} = I_i^\nu$, and $J^{\nu+1} = J^\nu$. Suppose $B^\nu(r) = I_{j,w}^\nu = t$. If x_s is entering, then let $B^{\nu+1}(r) = I^{\nu+1}(j,w) = s$. If $y_{k's'}$ is entering, then let $B^{\nu+1}(i) = B^\nu(i+1), i \geq r$, $I_{j,i}^{\nu+1} = I_{j,i+1}^\nu, i \geq w$, $J_{k',l'_{k'}+1}^{\nu+1} = s'$, and $l'_{k'} = l'_{k'} + 1$. Update P^ν to $P^{\nu+1}$, the factorization correspondingly, let $\nu = \nu + 1$, and go to Step 1.

Step 6. Let $B^{\nu+1} = B^\nu$, $B^{\nu+1'} = B^{\nu'}$, $I_i^{\nu+1} = I_i^\nu$, and $J^{\nu+1} = J^\nu$. Suppose $B^{\nu'}(r') = J_{k,w}^\nu = t$. If x_s is entering, then let $B^{\nu+1}(\sum_{j=1}^k l_j) = I^{\nu+1}(k, l_k + 1) = s$, $B^{\nu+1}(i) = B^\nu(i-1), i > \sum_{j=1}^k l_j$, $l_k = l_{k+1}$, $J_{k,i}^{\nu+1} = J_{k,i+1}^\nu, i \geq w$. If $y_{k's'}$ is entering, then let $B^{\nu+1}(i) = B^\nu(i+1), i \geq r$, $I_{j,i}^{\nu+1} = I_{j,i+1}^\nu, i \geq w$, $J_{k',l'_{k'}+1}^{\nu+1} = s'$, $J_{k,i}^{\nu+1} = J_{k,i+1}^{\nu+1}, i \geq w$, $l'_k = l'_k - 1$, and $l'_{k'} = l'_{k'} + 1$. Update P^ν to $P^{\nu+1}$, the factorization correspondingly, let $\nu = \nu + 1$, and go to Step 1.

For updating a factorization of the basis as used in (6.4) and (6.5), several cases need to be considered according to the possibilities in Steps 5 and 6 (see Exercise 3). If the entering and leaving variables are both in x, then only D changes. Substantial effort can again be saved. In other cases, only one block of L is altered by any iteration so we can again achieve some savings by only updating the corresponding parts of $L^{-1}F$ and $L^{-1}h$.

As mentioned earlier, this procedure can also apply to the dual of (1.1) and the primal. In this case, the procedure can mimic decomposition procedures and entails essentially the same work per iteration as the L-shaped method (see Birge [1988b]) or the inner linearization method applied to the dual. If choices of entering columns are restricted in a special variant of a decomposition procedure, then factorization and decomposition follow the same path.

In general, decomposition methods have been favored for this class of problems because they offer other paths of solutions, require less overhead, and, by maintaining separate subproblems, allow for parallel computation. The extensive form offers little hope for efficient solution, so it is not surprising that even sophisticated factorizations would not prove beneficial. Because most commercial methods already have substantial capabilities for exploiting general matrix structure, it is difficult to see how substantial gains could be obtained from basis factorization alone for a direct extreme point approach. Combinations of decomposition and factorization approaches may, however, be beneficial, as observed in Birge [1985b].

Factorization schemes also offer substantial promise for interior point methods, where there is much speculation that the solution effort grows linearly in the size of the problem. This observation is supported by the results we present here. For this discussion, we assume that the interior point method follows a standard form version of Karmarkar's projective algorithm as in Anstreicher [1989, 1990], Gay [1987], and Ye [1987]. We choose the standard form because we believe it is more practical for computation than various canonical forms. We also assume an unknown optimal objective value and use Todd and Burrell's [1986] method for updating a lower bound on the optimal objective value. We use an initial lower bound, as is often available in practice. An alternative is Anstreicher's [1989] method to obtain an initial lower bound.

Other interior point methods based on affine scaling (Barnes [1986], Vanderbei, Meketon, and Freedman [1986], Dikin [1967]) also follow the same basic steps with some simplification. They are, however, not provably polynomial methods. We present this polynomial version as a more general case.

We first describe the algorithm for a standard linear program:

$$\min \ c^T x$$
$$\text{s. t. } Ax = b,$$
$$x \geq 0, \tag{6.10}$$

where $x \in \Re^n$, $c \in Z^n$ (i.e., an n-vector of rationals), $b \in Z^m$, $A \in Z^{m \times n}$ with optimal value $c^T x^* = z^*$. In referring to the parameters in (6.10), we use (ext) as a modifier, e.g., $c(ext)$, when necessary to distinguish these parameters in our standard stochastic program form in (1.1).

Suppose we have a strictly interior feasible point x^0 of (6.10), i.e.,

$$Ax^0 = b, x^0 > 0, \tag{6.11}$$

a lower bound β^0 on z^*, and the set of optimal solutions in (6.10) is bounded. Note that if we do not have a feasible solution, we can solve a phase-one problem as in Karmarkar [1984].

The standard form variant of Karmarkar's projective scaling algorithm creates a sequence of points $x^0, x^1, ..., x^k$ by the following steps.

Standard Form Projective Scaling Method

Step 0. Set $\nu = 0$ and lower bound $\beta^0 \leq z^*$.

Step 1. If $c^T x^\nu - \beta^\nu$ is small enough, i.e., less than a given positive number ϵ, then stop. Otherwise, go to Step 2.

Step 2. Let $D = \mathrm{diag}\{x_1^\nu, \ldots, x_n^\nu\}$, $\hat{A} := [AD, -b]$, and let $\Pi_{\hat{A}}$ be the projection onto the null space of $\hat{A}$. Find

$$u = \Pi_{\hat{A}} \begin{pmatrix} Dc \\ 0 \end{pmatrix}, v = \Pi_{\hat{A}} \begin{pmatrix} 0 \\ 1 \end{pmatrix}, \tag{6.12}$$

and let $\mu(\beta^\nu) = \min\{u_i - \beta^\nu v_i : i = 1, \ldots, n+1\}$. If $\mu(\beta^\nu) \leq 0$, let $\beta^{\nu+1} = \beta^\nu$. Otherwise, let $\beta^{\nu+1} = \min\{u_i/v_i : v_i > 0, i = 1, \ldots, n+1\}$. Go to Step 3.

Step 3. Let $c_p = u - \beta^{\nu+1} v - (c^T x^\nu - \beta^{\nu+1}) e/(n+1)$, where $e = (1, \ldots, 1)^T \in \Re^{n+1}$. Let

$$g' = \frac{1}{n+1} e - \alpha \frac{c_p}{\|c_p\|_2}.$$

Let $\bar{g} \in \Re^n$ consist of the first n components of g'. Then $x^{\nu+1} = D\bar{g}/g'_{n+1}$, $\nu = \nu + 1$, go to Step 1.

For the purpose of obtaining a worst-case bound, the step length α in the definition of g' may be set equal to $\frac{1}{3(n+1)}$, as in Gay [1987]. In practice, much better performance is obtained by choosing α using a line search of the "potential functions" (see Anstreicher [1990] and Gay [1987]). We consider the number of arithmetic operations in our complexity analyses. The main computational effort in each iteration of the algorithm is to compute the projections in (6.12), which requires $O(m^2 n)$ arithmetic operations. The algorithm is shown to obtain a solution that can be resolved to an optimal basic feasible solution in $O(nL)$ iterations in Ye and Kojima [1987] and Anstreicher [1989], who also shows how a slightly revised version obtains a complexity of $O(\sqrt{n}L)$ iterations (Anstreicher [1993]). The overall arithmetic complexity for the basic method is $O(m^2 n^2 L)$. Karmarkar [1984] uses a rank–one updating scheme to reduce the complexity to $O(n^{3.5}L)$.

In our case, if we consider (1.1) as in the form of (6.10), then $n = n_0 + Kn_1$, $m = m_0 + Km_1$, $x((ext)) = \begin{pmatrix} x \\ y_1 \\ \vdots \\ y_K \end{pmatrix}$, $c((ext)) = \begin{pmatrix} c \\ p_1 q_1 \\ \vdots \\ p_K q_K \end{pmatrix}$,

$$b((ext)) = \begin{pmatrix} b \\ h_1 \\ \vdots \\ h_K \end{pmatrix}, \text{ and}$$

$$A((ext)) = \begin{pmatrix} A & 0 & \cdots & 0 \\ T_1 & W & \cdots & 0 \\ \vdots & 0 & \ddots & 0 \\ T_K & 0 & \cdots & W \end{pmatrix}. \tag{6.13}$$

The main computational work at each step of Karmarkar's algorithm is to compute the projections in (6.12). The projection can be written as

$$\Pi_{\hat{A}} = (I - \hat{A}^T(\hat{A}\hat{A}^T)^{-1}\hat{A}),$$

where $(\hat{A}\hat{A}^T) = AD^2A^T + bb^T := M + bb^T$. In this case, the work is dominated by computing M^{-1}. The key effort is in solving systems with AD^2A^T for the general A in the formulation in (6.10). Using the specific $A(ext)$ in the stochastic programming matrix as given by (6.13) and letting $D_0 = diag(x^\nu)$, $D_k = diag(y_k^\nu)$, $k = 1, \ldots, K$, we would have

$$M = \begin{pmatrix} AD_0^2A^T & AD_0^2T_1^T & \cdots & AD_0^2T_K^T \\ T_1D_0^2A^T & T_1D_0^2T_1^T + WD_1^2W^T & \cdots & T_1D_0^2T_K^T \\ \vdots & \vdots & \ddots & \vdots \\ T_KD_0^2A^T & T_1D_0^2T_K^T & \cdots & T_KD_0^2T_K^T + WD_K^2W^T \end{pmatrix},$$

$$\tag{6.14}$$

which is clearly much denser than the original constraint matrix in (1.1). In this case, a straightforward implementation of an interior point method that solves systems with M is quite inefficient.

Note, however, that M in (6.14) has a great deal of structure that can be exploited in any solution scheme. This is the object of the factorization scheme given by Birge and Qi [1988]. The following proposition gives the essential characterization of that factorization.

Proposition 6. Let $S_0 = I_2 \in \Re^{m_1 \times m_1}$, $S_l = W_l D_l^2 W_l^T$, $l = 1, \ldots, K$, $S = \text{diag}\{S_0, \ldots, S_K\}$. Then $S^{-1} = \text{diag}\{S_0, S_1^{-1}, \ldots, S_N^{-1}\}$. Let I_1 and I_2 be identity matrices of dimensions n_1 and m_1, respectively. Let

$$G_1 = (D_0)^{-2} + A^T S_0^{-1} A + \sum_{l=1}^{K} T_l^T S_l^{-1} T_l, \quad G_2 = -AG_1^{-1}A^T, \tag{6.15}$$

$$U = \begin{pmatrix} A & I_2 \\ T_1 & 0 \\ \vdots & \vdots \\ T_K & 0 \end{pmatrix}, V = \begin{pmatrix} A & -I_2 \\ T_1 & 0 \\ \vdots & \vdots \\ T_K & 0 \end{pmatrix}.$$

If A, W_k, $k = 1, \ldots, K$ have full row rank, then G_2 and M are invertible and

$$M^{-1} = S^{-1} - S^{-1}U \begin{pmatrix} I_1 & G_1^{-1}A^T \\ 0 & I_2 \end{pmatrix} \begin{pmatrix} I_1 & 0 \\ 0 & G_2^{-1} \end{pmatrix}$$
$$\begin{pmatrix} I_1 & 0 \\ A & I_2 \end{pmatrix} \begin{pmatrix} G_1^{-1} & 0 \\ 0 & -I_2 \end{pmatrix} V^T S^{-1}. \tag{6.16}$$

Proof: Follows Birge and Qi [1988]. See also Birge and Holmes [1992]. $\square$

Following the assumptions and using Karmarkar's complexity result, the number of arithmetic operations using this factorization can be reduced from $O((n_1 + Kn_2)^4)$ as in the general projective scaling method. Birge and Qi show that the effort is, in fact, dominated by $O(K(n_2^3 + n_2^2 n_1 + n_2 n_1^2))$. It is also possible to reduce this bound further as Karmarkar does with a partial rank-one updating scheme. In this case, for $n = n_1 + Kn_2$, the complexity using the factorization in (6.16) becomes $O((n^{0.5}n_2^2 + n \max\{n_1, n_2\} + n_1^3)nL)$ for the entire algorithm, where L represents the size of the data, or, if $K \sim n_1 \sim n_2$, the full arithmetic complexity is $O(n^{2.5}L)$, compared to Karmarkar's general result of $O(n^{3.5}L)$. Thus, the factorization in (6.16) provides an order of magnitude improvement over a general solution scheme if the number of realizations K approaches the number of variables in the first and second stage.

In practice, we would not compute M^{-1} explicitly. The work in (6.16) is dominated by the effort to solve systems of the form

$$Mv = u \tag{6.17}$$

using

$$v = p - r, \tag{6.18}$$

where

$$\begin{aligned} Sp &= u, \\ Gq &= V^T p, \\ Sr &= Uq, \end{aligned} \tag{6.19}$$

where G is the inverse of the matrix between U and V^T in (6.16). The systems in (6.19) require solving systems with S_l, computation of G_1 and G_2, and solving systems with G_1 and G_2. In practice, we find a Cholesky factorization of each S_l, use them to find G_1 and G_2, and find Cholesky factorizations of G_1 and G_2.

Before we describe results using the factorization in (6.16), we consider some other options for interior point methods. These possibilities are

1. Schur complement updates;

2. Column splitting;

3. Solution of the dual.

The Schur complement approach is used in many interior point method implementations (see, e.g., Choi, Monma, and Shanno [1990]). The basic idea is to write M as the sum of a matrix with sparse columns, $A_s D_s^2 A_s^T$, and a matrix with dense columns, $A_d D_d^2 A_d^T$. Using a Cholesky factorization of the sparse matrix, $LL^T = A_s D_s^2 A_s^T$, the method involves solving $Mu = v$ by:

$$\begin{pmatrix} LL^T & -A_d D_d \\ D_d A_d^T & I \end{pmatrix} \begin{pmatrix} v \\ w \end{pmatrix} = \begin{pmatrix} u \\ 0 \end{pmatrix}, \qquad (6.20)$$

which requires solving $[I + D_d A_d^T (LL^T)^{-1} A_d D_d]w = -D_d A_d^T (LL^T)^{-1}b$ and $LL^T v = b + A_d D_d w$, where $I + D_d A_d^T (LL^T)^{-1} A_d D_d$ is a Schur complement.

The Schur complement is thus quite similar to the factorization method given earlier. If every column of x is considered a dense column, then the remaining matrix is quite sparse but rank deficient. The factorization in (6.16) is a method for maintaining an invertible matrix when $A_s D_s^2 A_s^T$ is singular. It can thus be viewed as an extension of the Schur complement to the stochastic linear program. Because of the possible rank deficiency and the size of the Schur complement, the straightforward Schur complement approach in (6.20) is quick but leads to numerical instabilities as Carpenter, Lustig, and Mulvey [1991] report.

Carpenter et al. also propose the column splitting technique. The basic idea is to rewrite problem (1.1) with explicit constraints on nonanticipativity. The formulation then becomes:

$$\min \sum_{k=1}^{K} p_k (c^T x_k + q_k^T y_k) \qquad (6.21)$$
$$\text{s.t. } A x_k = b, \qquad (6.22)$$
$$T_k x_k + W y_k = h_k, \quad k = 1, \dots, K, \qquad (6.23)$$
$$x_k - x_{k+1} = 0, k = 1, \dots, K-1, \qquad (6.24)$$
$$x_k \geq 0, \quad y_k \geq 0, \quad k = 1, \dots, K. \qquad (6.25)$$

The difference now is that the constraints in (6.22) and (6.23) separate into separate subproblems k and constraints (6.24) link the problems together. Alternating constraints, (6.22), (6.23) and (6.24) for each k in sequence, the full constraint matrix has the form:

$$
\bar{A} = \begin{pmatrix}
A & 0 & 0 & 0 & 0 & 0 & 0 & 0 \\
T_1 & W & 0 & 0 & 0 & 0 & 0 & 0 \\
I & 0 & -I & 0 & 0 & 0 & 0 & 0 \\
0 & 0 & A & 0 & 0 & 0 & 0 & 0 \\
0 & 0 & T_2 & W & 0 & 0 & 0 & 0 \\
0 & 0 & I & 0 & -I & 0 & 0 & 0 \\
\vdots & \vdots & 0 & \ddots & \vdots & \ddots & 0 & \vdots \\
0 & 0 & 0 & 0 & I & 0 & -I & 0 \\
0 & 0 & 0 & 0 & 0 & 0 & A & 0 \\
0 & 0 & 0 & 0 & 0 & 0 & T_k & W
\end{pmatrix}.
\qquad (6.26)
$$

If we form $\bar{A}\bar{A}^T$, then we obtain $\bar{A}\bar{A}^T =$

$$
\begin{pmatrix}
AA^T & AT_1^T & A & 0 & 0 & 0 & 0 & 0 \\
T_1A^T & T_1T_1^T+WW^T & T_1 & 0 & 0 & 0 & 0 & 0 \\
A^T & T_1^T & 2I & -A^T & 0 & 0 & 0 & 0 \\
0 & 0 & -A & AA^T & AT_2^T & A & 0 & 0 \\
0 & 0 & T_2A^T & T_2T_2^T+WW^T & T_2 & 0 & 0 & 0 \\
0 & 0 & 0 & T_2^T & 2I & 0 & 0 & 0 \\
\vdots & \vdots & \vdots & \ddots & \ddots & \vdots & 0 & \vdots \\
0 & 0 & 0 & A^T & T_{K-1}^T & 2I & -A^T & 0 \\
0 & 0 & 0 & 0 & 0 & -A & AA^T & AT_K^T \\
0 & 0 & 0 & 0 & 0 & 0 & T_KA^T & T_KT_K^T+WW^T
\end{pmatrix},
$$
$$
\qquad (6.27)
$$

which is clearly much sparser than the original matrix in (6.14). It is, however, larger than the matrix in (6.14) (see Exercise 5) so there is some tradeoff for the reduced density.

The third additional approach is to form the dual of (1.1) and to solve that problem using the same basic interior point method we gave earlier. The dual problem is:

$$
\max \ b^T\rho + \sum_{k=1}^{K} p_k \pi_k^T h_k
\qquad (6.28)
$$

$$
\text{s.t. } A^T\rho + \sum_{k=1}^{K} p_k T_k^T \pi_k \le c, k = 1,\ldots,K,
\qquad (6.29)
$$

$$
W^T\pi_k \le q, k = 1,\ldots,K,
\qquad (6.30)
$$

where the variables are not restricted in sign. For this problem, we can achieve a standard form as in (6.10) by splitting the variables π_k and ρ into differences of non-negative variables and by adding slack variables to constraints (6.29) and (6.30)[1]. In this way the constraint matrix for (6.29) and (6.30) becomes:

$$\begin{pmatrix} A^T & -A^T & T_1^T & -T_1^T & 0 & T_2^T & -T_2^T & \cdots & T_K^T & -T_K^T & 0 & I \\ 0 & 0 & W^T & -W^T & I & 0 & 0 & 0 & 0 & 0 & 0 & 0 \\ 0 & 0 & 0 & 0 & 0 & W^T & -W^T & I & 0 & 0 & 0 & 0 \\ \vdots & \vdots & \vdots & \vdots & \vdots & 0 & \ddots & \ddots & \ddots & 0 & 0 & \vdots \\ 0 & 0 & 0 & 0 & 0 & 0 & 0 & 0 & W^T & -W^T & I & 0 \end{pmatrix}.$$
$$(6.31)$$

The matrix in (6.31) may again be much larger than the matrix in the original, but the gain comes in considering $A'A'^T$ which is now:

$$\begin{pmatrix} 2(A^TA + \sum_{k=1}^{K} T_k^T T_k) + I & 2T_1^T W & 2T_2^T W & \cdots & 2T_k^T W \\ 2W^T T_1 & 2W^T W + I & 0 & 0 & 0 \\ 2W^T T_2 & 0 & 2W^T W + I & 0 & 0 \\ \vdots & 0 & 0 & \ddots & 0 \\ 2W^T T_k & 0 & 0 & 0 & 2W^T W \end{pmatrix},$$
$$(6.32)$$

with an inherent sparsity of which an interior point method can take advantage. In fact, it is not necessary to take the dual to use this alternative factorization form, although we do so in the following computations. As shown in Birge, Freund, and Vanderbei [1992], the Sherman-Morrison-Woodbury formula can be applied to the original problem in (6.10) so that computations with the structure of $A(ext)A(ext)^T$ are replaced by computations with $A(ext)^T A(ext)$. In this way, the matrix in (6.32) replaces the dense matrix in (6.14).

In Carpenter, Lustig, and Mulvey [1991], increasing numbers of scenarios were included into a network formulation of a financial decision problem. They used the variable splitting option with an additional observation that many of the T_k columns were zero and that the corresponding variables need not be split. By splitting only those variables with nonzero T_k entries, they developed a *partial splitting* model that proved most effective. Their results with this partial splitting model show an approximately linear increase in speed-up for using split variables compared to the original primal form with the interior point code OB1 (Lustig, Marsten, and Shanno [1991]). For 18 scenarios, the speed-up was 1.48, while for 72 scenarios of the same problem, the partial splitting speed-up was 11.82.

Carpenter et al. also used the Schur complement approach and achieved additional speed-ups over partial splitting, although the additional speed-up was about the same (essentially two) for all numbers of scenarios. They did, however, report an order of magnitude increase in a measure of primal

---

[1]The dual problem may no longer have a bounded set of optima causing some theoretical difficulties for convergence results. In practice, bounds are placed on the variables to guarantee convergence.

TABLE 1. Speed-ups for the Proposition 6 factorization over the primal and dual
interior point method solutions.

Problem	Number of Scenarios	Speed-Ups Over:	
		Primal	Dual
SC205.02	8	2.156	0.315
SC205.03	16	7.101	0.331
SC205.04	32	26.753	0.379
SC205.05	64	> 30	0.404
SCRS8.02	8	1.571	0.224
SCRS8.03	16	9.396	0.263
SCRS8.04	32	> 30	0.150
SCAGR7.02	8	4.426	0.395
SCAGR7.03	16	15.840	0.439
SCAGR7.04	32	> 30	0.507
SCAGR7.05	64	> 30	0.727
SCTAP.02	8	3.164	0.850
SCTAP.03	16	12.825	1.054
SCTAP.04	32	> 30	1.250
SCSD8.01	4	*	2.691
* Algorithm for primal problem did not converge.			

infeasibility due to numerical instability with the basic Schur complement
approach.

In Birge and Holmes [1992], the dual approach and the primal factorization of Birge and Qi are compared. We repeat the results of some of these
problems in Table 1. The problems are a test set of stochastic programs
based on the multistage problems compiled by Ho and Loute [1981]. They
represent problems in economic development, agriculture, energy modeling, structural design, traffic assignment, and production scheduling. The
problems are solved as multistage stochastic programs in Birge [1985b] with
a small number of scenarios. For the interior point tests reported in Table 1, they were formed as two-stage problems with increasing numbers of
second-stage realizations or scenarios.

Table 1 shows the number of scenarios in each problem and the speed-
up, or the ratio of solution times for solving each problem in the original
primal, and the dual form to the time for solutions using the factorization
in Proposition 6. An affine scaling form (without a centering term) of the
interior point method given earlier was implemented in FORTRAN on an
IBM 3090 at the University of Michigan. The program used SPARSPAK
(Chu et al. [1984]) to solve the systems of equations.

Table 1 shows that the dual form appears to be the most efficient. However, in some cases, this matrix became too large for the MTS operating

system that the machine runs under. This was the case for larger versions of SCSD8, which has seven times as many columns as rows. Also, notice that the factorization approach appears relatively more efficient as the problem size increases. It, therefore, appears that either of these approaches may be used, although many columns would favor the factorization in (6.16), while many rows would favor taking the dual.

The speed-ups over the primal form in Table 1 are especially dramatic. They increase superlinearly in all classes of problems tested. In comparison to Lustig et al.'s results, it appears that either the dual form or the factorization in (6.16) is most efficient because they appear clearly to offer superlinear speed-ups. Of course, some problems may exist where variable splitting is preferred. Other approaches include an augmented system approach (Czyzyk, Fourer, and Mehrotra [1995]) that appears to have efficiency comparable to the factorization in (6.16) and improved stability. Yang and Zenios [1995] have, however, demonstrated substantial speed-ups using the Proposition 6 factorization with parallel processors.

We should note that the solution times for factorization and the dual approach for the problems in Table 1 are still significantly greater than times for using the L-shaped method, although, for the larger problems, they were faster than the MINOS extreme point algorithm. There may, of course, also be some exceptions where these procedures outperform decomposition. The key reason for this decomposition advantage appears to be the ability to solve multiple subproblems quickly after one subproblem is solved because many subproblems have the same optimal basis. Because a basis requires an extreme point, this type of approach does not seem possible for interior point methods. Interior point methods that allow repeated solutions for many subproblems and use the same search direction in these problems may, however, offer similar advantages to decomposition methods. Research is continuing into these alternative interior point approaches.

Exercises

1. Compare the number of operations to solve (6.3) using (6.4) and (6.5) compared to solving (6.3) as an unstructured linear system of equations.

2. Give a similar basis factorization scheme to (6.4) and (6.5) to solve the backward transformation, $(\sigma^T, \pi^T)B = (c_B^T, q_B^T)$, for a basis corresponding to columns B from the constraint matrix of (1.1).

3. Describe an efficient updating procedure for any possible combination of entering and leaving columns in the basis matrix of (6.3) using the factorization scheme in (6.4) and (6.5).

4. Find the number of arithmetic operations for a single step of the interior point method using (6.16). Compare this to the number of arithmetic operations if no special factorization is used.

5. Compare the sizes of the adjacency matrices in (6.14) and (6.26). Assuming that each matrix A, T_k, and W is completely dense, compare the number of nonzero entries in these two matrices.

5.7   Special Cases—Simple Recourse and Network Problems

In many stochastic programming problems, special structure provides additional computational advantages. The most common structures that allow for further efficiencies are simple recourse and network problems. The key features of these problems are separability of any nonlinear objective terms and efficient matrix computations.

Separability is the key to simple recourse computations. In Section 3.1, we described how these problems involve a recourse function that separates into components for each random variable. The stochastic program with simple recourse can then be written as:

$$\min \ z = c^T x + \sum_{i=1}^{m_2} \Psi_i(\chi_i) \tag{7.1}$$

$$\text{s.t. } Ax = b,$$
$$Tx - \chi = 0,$$
$$x \geq 0,$$

where $\Psi_i(\chi_i) = \int_{h_i \leq \chi_i} q^-(\chi_i - h_i) dF(h_i) + \int_{h_i > \chi_i} q^+(h_i - \chi_i) dF(h_i)$. Using this form of the objective in χ, we can substitute in (3.1.9) to obtain:

$$\Psi_i(\chi_i) = q_i^+ \bar{h}_i - (q_i^+ - q_i F_i(\chi_i))\chi_i - q_i \int_{h_i \leq \chi_i} h_i dF(h_i), \tag{7.2}$$

where $\bar{h}_i = E[h_i]$.

The separable objective terms in (7.1) offer advantages for computation. In the next chapter, we discuss nonlinear programming techniques that can apply when the random variables are continuous. Linear programming-based procedures can, however, be used when the random variables have a finite number of values. In this section, we assume that each h_i can take on the values, $h_{i,j}, j = 1, \ldots, K_i$ with probabilities, $p_{i,j}$.

Wets [1983a] gave the basic framework for computation of finitely distributed simple recourse problems as a linear program with upper bounded variables. The idea is to split χ_i into values corresponding to each interval, $[h_{i,j}, h_{i,j+1}]$, so that

$$\chi_i = \sum_{j=0}^{K_i} \chi_{i,j}, \chi_{i,0} \le h_{i,1}, 0 \le \chi_{i,j} \le h_{i,j+1} - h_{i,j}, 0 \le \chi_{i,K_i}. \qquad (7.3)$$

The objective coefficients correspond to the slope of $\Psi(\chi_i)$ in each of these intervals. They are

$$d_{i,0} = -q_i^+, d_{i,j} = -q_i^+ + q_i(\sum_{l=1}^{j} p_{i,l}), j = 1, \ldots, K_i. \qquad (7.4).$$

The piecewise linear program with these objective coefficients and variables is

$$\min z = c^T x + \sum_{i=1}^{m_2}((\sum_{j=0}^{K_i} d_{i,j}\chi_{i,j}) + q_i^+ \bar{h}_i) \qquad (7.5)$$
$$\text{s.t. } Ax = b,$$
$$Tx - \chi = 0,$$
$$x \ge 0 \text{ and } (7.3).$$

The equivalence of (7.1) and (7.5) is given in the following theorem.

Theorem 7. *Problems (7.1) and (7.5) have the same optimal values and sets of optimal solutions, (x^*, χ^*).*

Proof: We first show any solution $(x, \chi_1, \ldots, \chi_{m_2})$ to (7.1) corresponds to a solution $(x, \chi_1, \ldots, \chi_{m_2}, \chi_{1,1}, \ldots, \chi_{m_2,K_{m_2}})$ to (7.5) with the same objective value. We then also show the reverse to complete the proof. Suppose (x, χ) feasible in (1). If $h_{i,j} \le \chi_i < h_{i,j+1}$ for some $1 \le j \le K_i$, then let $\chi_{i,0} = h_{i,1}, \chi_{i,l} = h_{i,l+1} - h_{i,l}, 1 \le l \le j-1, \chi_{i,j} = \chi_i - h_{i,j}$ and $\chi_{i,l} = 0, l \ge j+1$. If $\chi_i < h_{i,0}$, then let $\chi_{i,0} = \chi_i, \chi_{i,l} = 0, l \ge 1$. In this way, we satisfy (7.3). If $\chi_i \ge h_{i,1}$, the variable i objective term in (7.5) with these values is then

$$q_i^+ \bar{h}_i - q_i^+(h_{i,1} + \sum_{l=1}^{j-1}(h_{i,l+1} - h_{i,l}) + (\chi_i - h_{i,j}))$$

$$+ q_i[(\sum_{l=1}^{j-1}[(\sum_{k=1}^{l} p_{i,k})(h_{i,l+1} - h_{i,l})] + \sum_{k=1}^{j} p_{i,k}(\chi_i - h_{i,j})]$$

$$= q_i^+ \bar{h}_i - q_i^+ \chi_i + q_i[((\sum_{k=1}^{j-1} p_{i,k}[\sum_{l=k}^{j-1}(h_{i,l+1} - h_{i,l}) - h_{i,j}]]$$

$$- p_{i,j}h_{i,j} + \sum_{k=1}^{j} p_{i,k}\chi_i]$$

$$= q_i^+ \bar{h}_i - q_i^+ \chi_i - q_i (\sum_{k=1}^{j} p_{i,k} h_{i,k}) + q_i (\sum_{k=1}^{j} p_{i,k}) \chi_i)$$

$$= q_i^+ \bar{h}_i - q_i^+ \chi_i - q_i \int_{\mathbf{h}_i \leq \chi_i} h_i dF(h_i) + q_i F_i(\chi_i) \chi_i$$

$$= \Psi_i(\chi_i), \tag{7.6}$$

where the last equality follows from substitution in (7.2).

If $\chi_i < h_{i,1}$, then the objective term is $q_i^+ \bar{h}_i - q_i^+ \chi_i$ which again agrees with $\Psi_i(\chi_i)$ from (7.2). Hence, any feasible (x, χ) in (7.1) corresponds to a feasible (x, χ) (where χ is extended into the components for each interval) in (7.5).

Suppose now that some (x^*, χ^*) is optimal in (7.5). Because each $q_i > 0$ and $p_{i,j} > 0$, for $h_{i,j} \leq \chi_i^* < h_{i,j+1}$ for some $1 \leq j \leq K_i$, we must have $\chi_{i,0}^* = h_{i,1}$, $\chi_{i,l}^* = h_{i,l+1} - h_{i,l}, 1 \leq l \leq j-1$, $\chi_{i,j}^* = \chi_i^* - h_{i,j}$ and $\chi_{i,l}^* = 0, l \geq j+1$. If not, then $\chi_{i,l}^* < h_{i,l+1} - h_{i,l} - \delta$ for some $l \leq j-1$ and $\chi_{i,\bar{l}}^* > \delta > 0$ for some $\bar{l} \geq j+1$. A feasible change of increasing $\chi_{i,l}^*$ by δ and decreasing $\chi_{i,\bar{l}}^*$ by δ yields an objective decrease of $\delta q_i \sum_{s=l+1}^{\bar{l}} p_{i,s}$ and would contradict optimality. Hence, we must have that the ith objective term in (7.5) is again $\Psi_i(\chi_i^*)$. Similarly, this must be true if $\chi_i^* < h_{i,1}$. Therefore, any optimal solution in (7.1) corresponds to a feasible solution with the same objective value in (7.5), and any optimal solution in (7.5) corresponds to a feasible solution with the same objective value in (7.1). Their optima must then correspond. $\square$

This formulation as an upper bounded variable linear program can lead to significant computational efficiencies. An implementation in Kallberg, White, and Ziemba [1982] uses this approach in a short-term financial planning model with 12 random variables with three realizations, each corresponding to uncertain cash requirements and liquidation costs. They solve the stochastic model with problem (7.5) in approximately 1.5 times the effort to solve the corresponding mean value linear program with expected values substituted for all random variables. This result suggests that stochastic programs with simple recourse can be solved in a time of about the same order of magnitude as a deterministic linear program ignoring randomness.

Further computational advantages for these problems are possible by treating the special structure of the $\chi_{i,j}$ variables as χ_i variables with piecewise, linear convex objective terms. Fourer [1985, 1988] presents an efficient simplex method approach for these problems. This implementation lends further support to the similar mean value problem–stochastic program order of magnitude claim.

Decomposition methods can also be applied to the simple recourse problem with finite distributions, although solution times better than the mean-value linear programming solution would generally be difficult to obtain.

As mentioned in Section 3, the multicut approach offers some advantage for the L-shaped algorithm (in terms of major iterations), but solution times are generally at best comparable with the mean-value linear program time. For generalized programming, because $\Psi(\chi) = \sum_{i=1}^{m_2} \Psi_i(\chi_i)$ and each $\Psi_i(\chi_i)$ is easily evaluated, the subproblem in (5.10) is equivalent to finding χ_i^ν such that

$$-\pi_i^\nu \in \partial \Psi_i(\chi_i^\nu). \tag{7.7}$$

From (7.4) and the argument in Proposition 5.1, $\partial\Psi_i(\chi_i) = \{d_{i,j}\}$ for $h_{i,j} < \chi_i < h_{i,j+1}$ and $\partial\Psi_i(\chi_i) = [d_{i,j-1}, d_{i,j}]$ for $h_{i,j} = \chi_i$. Thus, we can choose $\chi_i^\nu = h_{i,j}$ for $d_{i,j-1} \leq -\pi_i^\nu \leq d_{i,j}, j = 1, \ldots, K_i$. If $\pi_i^\nu < -q_i^+$, then the value in (5.10) is unbounded. The algorithm chooses $\zeta_i^{s+1} = -1$, and $\Psi_{0,i}^+(-1) = q_i^+$. In this way, generalized programming can be implemented easily, but would appear similar to the piecewise linear approach given earlier. The advantage of generalized programming is more apparent, however, when continuous distributions cause nonlinearities as we discuss in the next chapter.

In network problems, the simple recourse formulation can be even more efficiently solved. Suppose, for example, that the random variables $\mathbf{h}_i$ correspond to random demands at m_2 destinations, that the variables x_{st} are flows from s to t, $Ax = b$ corresponds to the network constraints for all source nodes, transshipment nodes, and destinations with known demands, and that Tx represents all the flows entering the destinations with random demand. By adding the constraint,

$$\sum_{i=1}^{m_2}\left(\sum_{j=1}^{l_i} \chi_{i,j}\right) - \sum_{\text{sources } s} \sum_t x_{st} = - \sum_{\text{known demand destinations } r} demand(r),$$

$$\tag{7.8}$$

every variable in (7.5) corresponds to a flow so that (7.5) becomes a network linear program. Hence, efficient network codes can be applied directly to (7.5) in this case.

When T has gains and losses, (7.5) is a generalized network. This problem was one of the first types of practical stochastic linear programs solved when Ferguson and Dantzig [1956] used the generalized network form to give an efficient procedure for allocating aircraft to routes (fleet assignment). We describe this problem to show the possibilities inherent in the stochastic program structure.

The problem includes m_1 aircraft and m_2 routes. The decision variables are x_{sr} aircraft s allocated to route r. The number of aircraft s available is b_s, the passenger capacity of aircraft s on route r is t_{sr}, and the uncertain passenger demand is $\mathbf{h}_r$. Hence, the ith row of $Ax = b$ is $\sum_{r=1}^{m_2} x_{ir} = b_i$. The jth row of $Tx - \chi = 0$ is $\sum_{s=1}^{m_1} t_{sj}x_{sj} - \chi_j = 0$.

The key observation about this problem is that the basis corresponds to a pseudo-rooted spanning forest (see, e.g., Bazaraa, Jarvis, and Sherali [1990]). For this problem, the simplex steps solve with trees and one-trees in an efficient manner. For example, suppose $m_1 = 3$, $m_2 = 3$, $b = (2, 2, 2)$, $t_{1.} = (200, 100, 300)$, $t_{2.} = (300, 100, 200)$, and $t_{3.} = (400, 100, 150)$, $p_{i,j} = 0.5$, and $h_{1,1} = 500$, $h_{1,2} = 700$, $h_{2,1} = 200$, $h_{2,2} = 400$, $h_{3,1} = 200$, $h_{3,2} = 400$. A basic solution is $x_{1,1} = 1$, $x_{1,2} = 1$, $x_{2,1} = 1$, $x_{2,2} = 1$, $x_{3,3} = 4/3$, and $\chi_{3,1} = 100$ with all other variables nonbasic. This basis is illustrated in Figure 4. The forest consists of a cycle and a subtree. Exercises 1, 2, and 3 explore this example in more detail.

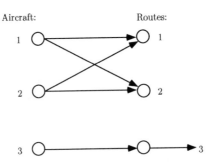

FIGURE 4. Graph of basic arcs for aircraft-route assignment example.

For general network problems, Sun, Qi and Tsai [1990] describe a piecewise linear network method that allows the use of network methods and does not require adding the additional arcs that correspond to the $\chi_{i,j}$ values. Other generalizations for network structured problems allow continuous distributions and apply directly to the nonlinear problem. We discuss these methods in more detail in the next chapter.

The methods all apply to simple recourse problems in which the first-stage variables represent a network. Another class of problems includes network constraints in the second (and following) stages. These problems are called *network recourse* problems. In this case, some computational advantages are again possible.

Most computational experience with solving these problems directly has been with the L-shaped method. The efficiencies occur in constructing feasibility constraints, in generating facets of the polyhedral convex recourse function, and in solving multiple recourse problems using small Schur complement updates of a network basis. These procedures are described in Wallace [1986b]. Other methods for network recourse problems involve nonlinear programming-based procedures. We will also describe these approaches in the next chapter.

Exercises

1. Show that any basis for the aircraft allocation problem consists of a collection of $m_1 + m_2$ basic variables that correspond to a collection of trees and one-trees.

2. Describe a procedure for finding the values of basic variables, multipliers, reduced costs, and entering and leaving basic variables for the structure in the aircraft allocation problem.

3. Solve the aircraft allocation problem using the procedure in (7.2) starting at the basis given with cost data corresponding to $c_{1.} = (300, 200, 100)$, $c_{2.} = (400, 100, 300)$, $c_{3.} = (200, 100, 300)$, $q_i^+ = 25$, $q_i^- = 0$ for all i. You may find it useful to use the graph to compute the appropriate values.

6

Nonlinear Programming Approaches to Two-Stage Recourse Problems

In Chapter 5, we considered methods that were fundamentally large-scale linear programming procedures. When the stochastic program includes nonlinear terms or when continuous random variables are explicitly included, a finite-dimensional linear programming deterministic equivalent no longer exists. In this case, we must use some nonlinear programming types of procedures.

This chapter describes the basic nonlinear programming approaches applied to stochastic programs. In Sections 1 and 2, we begin by describing enhancements of the L-shaped method to include a quadratic regularization term and to allow for quadratic objective terms. Section 3 describes methods based on the stochastic program Lagrangian. Section 4 gives procedures that are specially constructed for simple recourse problems, while Section 5 describes some other basic approaches.

6.1 Regularized Decomposition

Regularized decomposition is a method that combines a multicut approach for the representation of the second-stage value function with the inclusion in the objective of a quadratic regularizing term. This additional term is included to avoid two classical drawbacks of the cutting plane methods. One is that initial iterations are often inefficient. The other is that iterations may become degenerate at the end of the process. Regularized decomposition was introduced by Ruszczyński [1986]. We present a somewhat simplified version of his algorithm using the notation of Section 5.3.

The Regularized Decomposition Method

Step 0. Set $r = \nu = 0, s(k) = 0$ for all $k = 1, \cdots, K$. Select a^1, a feasible solution.

Step 1. Set $\nu = \nu + 1$. Solve the regularized master program

$$\min c^T x + \sum_{k=1}^{K} \theta_k + \frac{1}{2}\|x - a^\nu\|^2 \tag{1.1}$$

$$\text{s.t. } Ax = b,$$
$$D_\ell x \geq d_\ell, \qquad \ell = 1, \ldots, r,$$
$$E_{\ell(k)} x + \theta_k \geq e_{\ell(k)}, \qquad \ell(k) = 1, \ldots, s(k),$$
$$x \geq 0 .$$

Let (x^ν, θ^ν) be an optimal solution to (1.1) where $(\theta^\nu)^T = (\theta_1^\nu, \cdots, \theta_K^\nu)^T$ is the vector of θ_k's. If $s(k) = 0$ for some k, θ_k^ν is ignored in the computation. If $c^T x^\nu + e^T \theta^\nu = c^T a^\nu + Q(a^\nu)$, stop; a^ν is optimal.

Step 2. As before, if a feasibility cut (5.1.3) is generated, set $a^{\nu+1} = a^\nu$ (null infeasible step), and go to Step 1.

Step 3. For $k = 1, \ldots, K$, solve the linear subproblem (5.1.9). Compute $Q_k(x^\nu)$. If (5.3.5) holds, add an optimality cut (5.3.4) using formulas (5.3.6) and (5.3.7). Set $s(k) = s(k + 1)$.

Step 4. If (5.3.5) does not hold for any k, then $a^{\nu+1} = x^\nu$ (exact serious step); go to Step 1.

Step 5. If $c^T x^\nu + Q(x^\nu) \leq c^T a^\nu + Q(a^\nu)$, then $a^{\nu+1} = x^\nu$ (approximate serious step); go to Step 1. Else, $a^{\nu+1} = a^\nu$ (null feasible step), go to Step 1.

Observe that when a serious step is made, the value $Q(a^{\nu+1})$ should be memorized, so that no extra computation is needed in Step 1 for the test of optimality. Note also that a more general regularization would use a term of the form $\alpha\|x - a^\nu\|^2$ with $\alpha > 0$. This would allow tuning of the regularization with the other terms in the objective. As will be illustrated in Exercise 2, regularized decomposition works better when a reasonable starting point is chosen.

Example 1

Consider Exercise 1 of Section 5.3. Take $a^1 = -0.5$ as a starting point. It corresponds to the solution of the problems with $\xi = \bar{\xi}$ with probability 1. We have $Q(a^1) = 3/8$.

Iteration 1: Cuts $\theta_1 \geq 0, \theta_2 \geq -\frac{3}{4}x$ are added. Let $a^2 = a^1$.

Iteration 2: The regularized master is

$$\min \theta_1 + \theta_2 + \frac{1}{2}(x + 0.5)^2$$

$$\text{s.t. } \theta_1 \geq 0, \theta_2 \geq -\frac{3}{4}x$$

with solution $x^2 = 0.25 : \theta_1 = 0, \theta_2 = -3/16$. A cut $\theta_2 \geq 0$ is added. As $Q(0.25) = 0 < Q(a^1), a^3 = 0.25$ (approximate serious step 1).

Iteration 3: The regularized master is

$$\min \theta_1 + \theta_2 + \frac{1}{2}(x - 0.25)^2$$

$$\text{s.t. } \theta_1 \geq 0, \ \theta_2 \geq -\frac{3}{4}x, \theta_2 \geq 0$$

with solution $x^3 = 0.25, \theta_1 = 0, \theta_2 = 0$. Because $\theta^\nu = Q(a^\nu)$, a solution is found.

In Exercise 1, the L-shaped and multicut methods are compared. The value of a starting point is given in Exercise 2.

We now describe the main results needed to prove convergence of the regularized decomposition to an optimal solution when it exists. For notational convenience, we drop the first-stage linear terms $c^T x$ in the rest of the section. This poses no theoretical difficulty, as we may either define $\theta_k = p_k(c^T x + Q_k(x)), k = 1, \cdots, K$ or add a $(K+1)$th term $\theta_{K+1} = c^T x$. With this notation, the original problem can be written as

$$\min Q(x) = \sum_{k=1}^{K} p_k Q_k(x) \tag{1.2}$$

$$\text{s.t. } (5.1.2), x \geq 0,$$

and $Q_k(x) = \min\{q_k^T y | Wy = h_k - T_k x, y \geq 0\}$. This is equivalent to

$$\min e^T \theta = \sum_{k=1}^{K} \theta_k \tag{1.3}$$

$$\text{s.t. } (5.1.2), (5.1.3), (5.3.4), x \geq 0,$$

provided all possible cuts (5.1.3) and (5.3.4) are included.

The regularized master program is

$$\min \eta(x, \theta, a^\nu) = \sum_{k=1}^{K} \theta_k + \frac{1}{2}\|x - a^\nu\|^2 \tag{1.4}$$

$$\text{s.t. } (5.1.2), (5.1.3), (5.3.4), x \geq 0.$$

Note, however, that in the regularized master, only some of the potential cuts (5.1.3) and (5.3.4) are included. We follow the proof in Ruszczyński [1986].

Lemma 1. $e^T \theta^\nu \leq \eta(x^\nu, \theta^\nu, a^\nu) \leq Q(a^\nu)$.

Proof. The first inequality simply comes from $\|x^\nu - a^\nu\|^2 \geq 0$. We then observe that a^ν always satisfies (5.1.2), (5.1.3), as a^1 is feasible and the serious steps always pick feasible a^νs. The solution $(a^\nu, \hat{\theta})$ obtained by choosing $\hat{\theta}_k = p_k Q_k(a^\nu), k = 1, \cdots, K$ necessarily satisfies all constraints (5.3.4) as θ_k is a lower bound on $p_k Q_k(\cdot)$. Thus, $\eta(x^\nu, \theta^\nu, a^\nu) \leq \eta(a^\nu, \hat{\theta}, a^\nu) = Q(a^\nu)$. $\square$

Lemma 2. *If the algorithm stops at Step 1, then a^ν solves the original problem (1.2).*

Proof. By Lemma 1 and the optimality criterion, $e^T \theta^\nu = Q(a^\nu)$ (remember the linear term $c^T x$ has been dropped). It follows that $e^T \theta^\nu = \eta(x^\nu, \theta^\nu, a^\nu)$, which implies $\|x^\nu - a^\nu\|^2 = 0$, hence $x^\nu = a^\nu$. Thus, a^ν solves the regularized master (1.4) with the cuts (5.1.3) and (5.3.4) available at iteration ν. The cone of feasible directions at a^ν does not include any direction of descent of $\eta(x, \theta, a^\nu)$. The cone of feasible directions at x^ν for problem (1.3) is included in the cone of feasible directions at iterations ν of the regularized master ((1.4) contains fewer cuts). Moreover, the gradient of the regularizing term vanishes at a^ν. Thus, the descent directions of the regularized program (1.4) are the same as the descent directions of (1.3). Hence, a^ν solves (1.3), which means a^ν solves the original program (1.2). $\square$

Lemma 3. *If there is a null step at iteration ν, then*

$$\eta(x^{\nu+1}, \theta^{\nu+1}, a^{\nu+1}) > \eta(x^\nu, \theta^\nu, a^\nu) .$$

Proof. Because the objective function of the regularized master is strictly convex, program (1.4) has a unique solution. A null step at iteration ν may be either a null infeasible step or a null feasible step. In the first case, a cut (5.1.3) is added that renders x^ν infeasible. In the second case, a cut (5.3.4) is added that renders (x^ν, θ^ν) infeasible. Thus, as the previous solution becomes infeasible and the solution is unique, the objective function necessarily increases. $\square$

Lemma 4. *If the number of serious steps is finite, the algorithm stops at Step 1.*

Proof. If the number of serious steps is finite, there exists some ν_0 such that $a^\nu = a^{\nu_0}$ for all $\nu \geq \nu_0$. By Lemma 3, this implies the objective function of the regularized master strictly increases at each iteration $\nu, \nu \geq \nu_0$. Because there are only finitely many possible cuts (5.1.3) and (5.3.4), the algorithm must stop. $\square$

Lemma 5. *The number of approximate serious steps is finite.*

Proof: By definition of Step 5, the value of $\mathcal{Q}(\cdot)$ does not increase in an approximate serious step (remember that the term $c^T x$ is dropped here). Approximate serious steps only happen when $\mathcal{Q}(x^\nu) \neq e^T \theta^\nu$. This can only happen finitely many times because the number of cuts (5.3.4) is finite. $\square$

Lemma 6. *If the algorithm does not stop, then either $\mathcal{Q}(a^\nu)$ tends to $-\infty$ as $\nu \to \infty$ or the sequence $\{a^\nu\}$ converges to a solution of the original problem.*

Proof: (i) Let us first consider the case in which the original problem has solution $\hat{x}$. Define $\hat{\theta}$ by $\hat{\theta}_k = p_k Q_k(\hat{x})$. Thus $(\hat{x}, \hat{\theta})$ solves (1.3). Also $(\hat{x}, \hat{\theta})$ must be feasible for the regularized master for all ν. Because (x^ν, θ^ν) is the solution of the regularized master at iteration ν, the derivative of η at (x^ν, θ^ν) in the direction $(\hat{x} - x^\nu, \hat{\theta} - \theta^\nu)$ must be non-negative, i.e.,

$$(x^\nu - a^\nu)^T(\hat{x} - x^\nu) + e^T \hat{\theta} - e^T \theta^\nu \geq 0$$

or

$$(x^\nu - a^\nu)^T(x^\nu - \hat{x}) \leq \mathcal{Q}(\hat{x}) - e^T \theta^\nu, \qquad (1.5)$$

because $e^T \hat{\theta} = \mathcal{Q}(\hat{x})$.

Let S be the set of iterations at which serious steps occur. In view of Lemma 5, without loss of generality, we may consider such a set where all serious steps are exact. Because, for an exact serious step, $e^T \theta^\nu = \mathcal{Q}(x^\nu)$, (5.3.5) does not hold for any k, and $x^\nu = a^{\nu+1}$ by definition of the step, for all $\nu \in S$, (1.5) may be rewritten as

$$(a^{\nu+1} - a^\nu)^T(a^{\nu+1} - \hat{x}) \leq \mathcal{Q}(\hat{x}) - \mathcal{Q}(a^{\nu+1}).$$

By properties of sums of sequences,

$$\|a^{\nu+1} - \hat{x}\|^2 = \|a^\nu - \hat{x}\|^2 + 2(a^{\nu+1} - a^\nu)^T(a^{\nu+1} - \hat{x}) - \|a^{\nu+1} - a^\nu\|^2.$$

By dropping the last terms and using the inequality, for all $\nu \in S$,

$$\|a^{\nu+1} - \hat{x}\|^2 \leq \|a^\nu - \hat{x}\|^2 + 2(a^{\nu+1} - a^\nu)^T(a^{\nu+1} - \hat{x}) \qquad (1.6)$$
$$\leq \|a^\nu - \hat{x}\|^2 + 2(\mathcal{Q}(\hat{x}) - \mathcal{Q}(a^{\nu+1})).$$

Because $\mathcal{Q}(\hat{x}) \leq \mathcal{Q}(a^{\nu+1})$ for all ν, $\|a^{\nu+1} - \hat{x}\| \leq \|a^\nu - \hat{x}\|$, i.e., the sequence $\{a^\nu\}$ is bounded.

Now (1.6) can be rearranged as

$$2(\mathcal{Q}(a^{\nu+1}) - \mathcal{Q}(\hat{x})) \leq \|a^\nu - \hat{x}\|^2 - \|a^{\nu+1} - \hat{x}\|^2.$$

Summing up both sides for $\nu \in S$, it can be seen that

$$\sum_{\nu \in S} \left(\mathcal{Q}(a^{\nu+1}) - \mathcal{Q}(\hat{x}) \right) < \infty,$$

which implies $\mathcal{Q}(a^{\nu+1}) \to \mathcal{Q}(\hat{x})$ for some subsequence $\{a^\nu\}, \nu \in S_1$ where $S_1 \subseteq S$. Therefore, there must exist an accumulation point $\hat{a}$ of $\{a^\nu\}$ with $\mathcal{Q}(\hat{a}) = \mathcal{Q}(\hat{x})$. All a^ν are feasible, hence $\hat{a}$ is feasible and $\hat{a}$ may substitute for $\hat{x}$ in (1.6) implying $\|a^{\nu+1} - \hat{a}\| \leq \|a^\nu - \hat{a}\|$, which shows that $\hat{a}$ is the only accumulation point of $\{a^\nu\}$.

ii) Now assume that the original problem is unbounded but $\{\mathcal{Q}(a^\nu)\}$ is bounded. Thus one can find a feasible $\hat{x}$ and an $\varepsilon > 0$ such that $Q(\hat{x}) \leq Q(a^\nu) - \varepsilon$, $\forall\, \nu$. Then (1.6) gives $\|a^{\nu+1} - \hat{x}\|^2 \leq \|a^\nu - \hat{x}\|^2 - 2\varepsilon$, which yields a contradiction as $\nu \to \infty, \nu \in S$. $\square$

Lemma 7. *If the algorithm does not stop and $\mathcal{Q}\{a^\nu\}$ is bounded, there exists ν_0 such that if a serious step occurs at $\nu \geq \nu_0$, then the solution (x^ν, θ^ν) of (1.4) is also a solution of (1.4) without the regularizing term.*

Proof. Let K_ν denote the set of (x, θ) that satisfy all constraints (5.1.2), (5.1.3), (5.3.4) at iteration ν. The problem (1.4) without the regularizing term is thus:

$$\begin{aligned} &\min\ e^T \theta \\ &\text{s.t. } (x, \theta) \in K_\nu. \end{aligned} \qquad (1.7)$$

Assume Lemma 7 is false. It is thus possible to find an infinite set S such that, for all $\nu \in S$, a serious step occurs and the solution (x^ν, θ^ν) to (1.4) is not optimal for (1.7).

Let K_ν^* denote the normal cone to the cone of feasible directions for K_ν at (x^ν, θ^ν). Nonoptimality of (x^ν, θ^ν) means that the negative gradient of the objective in (1.7), $-d = \begin{bmatrix} 0 \\ -e \end{bmatrix} \notin K_\nu^*$. As this holds for all $\nu \in S$,

$$-d \notin \cup_{\nu \in S} K_\nu^* . \qquad (1.8)$$

Now K_ν is polyhedral. There can only be a finite number of constraints (5.1.2) and cuts (5.1.3) and (5.3.4). Thus, the right-hand-side of (1.8) is the union of a finite number of closed sets and, hence, is closed. There exists an $\varepsilon > 0$ such that

$$\mathcal{B}(-d, \varepsilon) \cap K_\nu^* = \emptyset, \quad \forall\, \nu \in S, \qquad (1.9)$$

where $\mathcal{B}(-d, \varepsilon)$ denotes the ball of radius ε centered at $-d$. On the other hand, (x^ν, θ^ν) solves (1.4); hence,

$$-\nabla \eta(x^\nu, \theta^\nu, a^\nu) \in K_\nu^*, \quad \forall\, \nu \in S . \qquad (1.10)$$

By Lemma 6, $a^\nu \to \hat{x}$. By Lemma 5, there exists a ν_0 such that for $\nu \geq \nu_0$, $e^T \theta^\nu = \mathcal{Q}(a^\nu)$ for all serious steps. Hence, at serious steps $\nu \geq \nu_0$, we have

$$\mathcal{Q}(a^\nu) \geq \eta(x^\nu, \theta^\nu, a^\nu) = \frac{1}{2}\|a^\nu - x^\nu\|^2 + e^T \theta^\nu$$

$$= \frac{1}{2}\|x^\nu - a^\nu\|^2 + \mathcal{Q}(a^\nu).$$

This implies $x^\nu \to a^\nu$, $\forall\, \nu \in S$. Hence,

$$\nabla \eta(x^\nu, \theta^\nu, a^\nu) \to d \ \forall\, \nu \in S,$$

and (1.10) contradicts (1.9). $\square$

Theorem 8. *If the original problem has a solution, then the algorithm stops after a finite number of iterations. Otherwise, it generates a sequence of feasible points $\{a^\nu\}$ such that $Q(a^\nu)$ tends to $-\infty$ as $\nu \to \infty$.*

Proof. By Lemma 2, the algorithm may only stop at a solution. Suppose the original problem has a solution but the algorithm does not stop. By Lemma 6, $\{a^\nu\}$ converges to a solution $\hat{x}$. Lemma 5 implies that for all ν large enough, all serious steps are exact, i.e.,

$$\mathcal{Q}(a^{\nu+1}) = e^T \theta^\nu.$$

By Lemma 7, for ν large enough, x^ν also solves (1.4) without the regularizing term implying

$$e^T \theta^\nu \leq Q(\hat{x}),$$

because problem (1.4) without the regularizing term is a relaxation of the original problem. Because $Q(\hat{x}) \leq Q(a^\nu)$ for all ν, it follows that, for ν large enough, $Q(x^\nu) = Q(\hat{x})$. Thus, no more serious steps are possible, which by Lemma 4 implies finite termination. The unbounded case was proved in Lemma 6. $\square$

Implementation of the regularized decomposition algorithm poses a number of practical questions, such as controlling the size of the master regularized problem and numerical stability. An implementation using a QR factorization and an active set strategy is described in Ruszczyński [1986]. On the problems tested by the author (see also Ruszczyński [1993b]) the regularized decomposition method outperforms all other methods. This includes a regularized version of the L-shaped method, the L-shaped method, or the multicut method. It is confirmed in the experiments made by Kall and Mayer [1996].

Exercises

1. Check that, with the same starting point, both the L-shaped and the multicut methods require five iterations in Example 1.

2. The regularized decomposition only makes sense with a reasonable starting point. To illustrate this, consider the same example taking as starting point a highly negative value, e.g., $a^1 = -20$. At Iteration 1, the cuts $\theta_1 \geq -\frac{x-1}{2}$ and $\theta_2 \geq -\frac{3}{4}x$ are created. Observe that, for many subsequent iterations, no new cuts are generated as the sequence of trial points a^ν move from -20 to $-\frac{75}{4}$, then $-\frac{70}{4}, -\frac{65}{4}, \cdots$ each time by a change of $\frac{5}{4}$, until reaching 0, where new cuts will be generated. Thus a long sequence of approximate serious steps is taken.

3. As we mentioned in the introduction of this section, the regularized decomposition algorithm works with a more general regularizing term of the form $\frac{\alpha}{2}\|x - a^\nu\|^2$.

   (a) Observe that the proof of convergence relies on strict convexity of the objective function (Lemma 3), thus $\alpha > 0$ is needed. It also relies on $\nabla \frac{\alpha}{2}\|x^\nu - a^\nu\|^2 \to 0$ as $x^\nu \to a^\nu$, which is simply obtained by taking a finite α. The algorithm can thus be tuned for any positive α and α can vary within the algorithm.

   (b) Taking the same starting point and data as in Exercise 2, show that by selecting different values of α, any point in $]-20, 20]$ can be obtained as a solution of the regularized master at the second iteration (where 20 is the upper bound on x and the first iteration only consists of adding cuts on θ_1 and θ_2).

   (c) Again taking the same starting point and data as in Exercise 2, how would you take α to reduce the number of iterations? Discuss some alternatives.

   (d) Let $\alpha = 1$ for Iterations 1 and 2. As of Iteration 2, consider the following rule for changing α dynamically. For each null step, α is doubled. At each exact step, α is halved. Show why this would improve the performance of the regularized decomposition in the case of Exercise 2. Consider the starting point $x^1 = -0.5$ as in Example 1 and observe that the same path as before is followed.

6.2 The Piecewise Quadratic Form of the L-Shaped Method

In this section, we consider two-stage quadratic stochastic programs of the form

$$\min z(x) = c^T x + \frac{1}{2}x^T C x + E_\xi[\min[q^T(\omega)y(\omega) + \frac{1}{2}y^T(\omega)D(\omega)y(\omega)]] \quad (2.1)$$

$$\text{s.t. } Ax = b,$$
$$T(\omega)x + Wy(\omega) = h(\omega),$$
$$x \geq 0, \quad y(\omega) \geq 0,$$

where c, C, A, b, and W are fixed matrices of size $n_1 \times 1, n_1 \times n_1, m_1 \times n_1, m_1 \times 1$, and $m_2 \times n_2$, respectively and q, D, T, and h are random matrices of size $n_2 \times 1, n_2 \times n_2, m_2 \times n_1$, and $m_2 \times 1$, respectively. Compared to the linear case defined in (3.1.1), only the objective function is modified. As usual, the random vector ξ is obtained by piecing together the random components of q, D, T, and h. Although more general cases could be studied, we also make the following two assumptions.

Assumption 9. *The random vector ξ has a discrete distribution.*

Recall that an $n \times n$ matrix M is *positive semi-definite* if $x^T M x \geq 0$ for all $x \in \Re^n$ and M is *positive definite* if $x^T M x > 0$ for all $0 \neq x \in \Re^n$.

Assumption 10. *The matrix C is positive semi-definite and the matrices $D(\omega)$ are positive semi-definite for all ω. The matrix W has full row rank.*

The first assumption guarantees the existence of a finite decomposition of the second-stage feasibility set K_2. The second assumption guarantees that the recourse functions are convex and well-defined.

We may again define the recourse function for a given $\xi(\omega)$ by:

$$Q(x, \xi(\omega)) = \min\{q^T(\omega)y(\omega) + \frac{1}{2}y^T(\omega)D(w)y(w)| \tag{2.2}$$
$$T(\omega)x + Wy(\omega) = h(\omega), y(\omega) \geq 0\}, \tag{2.3}$$

which is $-\infty$ or $+\infty$ if the problem is unbounded or infeasible, respectively. The expected recourse function is

$$\mathcal{Q}(x) = E_\xi Q(x, \xi)$$

with the convention $+\infty + (-\infty) = +\infty$.

The definitions of K_1 and K_2 are as in Section 3.4. Theorem 3.32 and Corollaries 3.34 and 3.35 apply, i.e., $\mathcal{Q}(x)$ is a convex function in x and K_2 is convex. Of greater interest to us is the fact that $\mathcal{Q}(x)$ is piecewise quadratic. Loosely stated, this means that K_2 can be decomposed in polyhedral regions called the *cells* of the decomposition and in addition to being convex, $\mathcal{Q}(x)$ is quadratic on each cell.

Example 2

Consider the following quadratic stochastic program

$$\min z(x) = 2x_1 + 3x_2 + E_\xi \min\{-6.5\mathbf{y}_1 - 7\mathbf{y}_2 + \frac{\mathbf{y}_1^2}{2} + \mathbf{y}_1\mathbf{y}_2 + \frac{\mathbf{y}_2^2}{2}\}$$

$$\text{s.t. } 3x_1 + 2x_2 \leq 15, \ y_1 \leq x_1, \ y_2 \leq x_2$$
$$x_1 + 2x_2 \leq 8, y_1 \leq \boldsymbol{\xi}_1, \ y_2 \leq \boldsymbol{\xi}_2$$
$$x_1 + x_2 \geq 0, x_1, x_2 \geq 0, y_1, y_2 \geq 0.$$

This problem consists of finding some product mix (x_1, x_2) that satisfies some first-stage technology requirements. In the second stage, sales cannot exceed the first-stage production and the random demand. In the second stage, the objective is quadratic convex because the prices are decreasing with sales. We might also consider financial problems where minimizing quadratic penalties on deviations from a mean value leads to efficient portfolios.

Assume that $\boldsymbol{\xi}_1$ can take the three values 2, 4, and 6 with probability $1/3$, that $\boldsymbol{\xi}_2$ can take the values 1, 3, and 5 with probability $1/3$, and that $\boldsymbol{\xi}_1$ and $\boldsymbol{\xi}_2$ are independent of each other. For very small values of x_1 and x_2, it always is optimal in the second stage to sell the production, $y_1 = x_1$ and $y_2 = x_2$. More precisely, for $0 \leq x_1 \leq 2$ and $0 \leq x_2 \leq 1, y_1 = x_1, y_2 = x_2$ is the optimal solution of the second stage for all $\boldsymbol{\xi}$. If needed, the reader may check this using the Karush-Kuhn-Tucker conditions.

Thus, $Q(x, \xi) = -6.5x_1 - 7x_2 + \frac{x_1^2}{2} + x_1 x_2 + \frac{x_2^2}{2}$ for all ξ and $\mathcal{Q}(x) = -6.5x_1 - 7x_2 + \frac{x_1^2}{2} + x_1 x_2 + \frac{x_2^2}{2}$. Here, the cell is $\{(x_1, x_2) | 0 \leq x_1 \leq 2, 0 \leq x_2 \leq 1\}$. Within that cell, $\mathcal{Q}(x)$ is quadratic.

Definition 11. A *finite closed convex complex* $\mathcal{K}$ is a finite collection of closed convex sets, called the cells of $\mathcal{K}$, such that the intersection of two distinct cells has an empty interior.

Definition 12. A *piecewise convex program* is a convex program of the form $\inf\{z(x) | x \in S\}$ where f is a convex function on $I\!\!R^n$ and S is a closed convex subset of the effective domain of f with nonempty interior.

Let $\mathcal{K}$ be a finite closed convex complex such that

(a) the n-dimensional cells of $\mathcal{K}$ cover S,

(b) either f is identically $-\infty$ or for each cell C_ν of the complex there exists a convex function $z_\nu(x)$ defined on S and continuously differentiable on an open set containing C_ν which satisfies

  i. $z(x) = z_\nu(x) \ \forall \ x \in C_\nu$, and
  ii. $\nabla z_\nu(x) \in \partial z(x) \ \forall \ x \in C_\nu$.

Definition 13. A *piecewise quadratic function* is a piecewise convex function where on each cell C_ν the function z_ν is a quadratic form.

Taking Example 2, we have both $\mathcal{Q}(x)$ and $z(x)$ piecewise quadratic. On $C_1 = \{0 \le x_1 \le 2, 0 \le x_2 \le 1\}$,

$$\mathcal{Q}_1(x) = -6.5x_1 - 7x_2 + \frac{x_1^2}{2} + x_1 x_2 + \frac{x_2^2}{2}$$

$$\text{and } z_1(x) = -4.5x_1 - 4x_2 + \frac{x_1^2}{2} + x_1 x_2 + \frac{x_2^2}{2}.$$

Defining a polyhedral complex was first done by Walkup and Wets [1967] for the case of stochastic linear programs. Based on this decomposition, Gartska and Wets [1974] proved that the optimal solution of the second stage is a continuous, piecewise linear function of the first-stage decisions and showed that $Q(x, \xi)$ is piecewise quadratic in x. It follows that under Assumption 1, $\mathcal{Q}(x)$ and $z(x)$ are also piecewise quadratic in x.

For the sake of completeness, observe that $z(x)$ is not always $\max_\nu z_\nu(x)$. To this end, consider

$$z(x) = \begin{cases} z_1(x) = \frac{x}{2} & \text{when } 0 \le x \le 2, \\ z_2(x) = (x-1)^2 & \text{when } x \ge 2. \end{cases}$$

This function is easily seen to be piecewise quadratic. On $(0, 1/2), z(x) = z_1(x)$ while $\max\{z_1(x), z_2(x)\} = z_2(x)$.

An algorithm

In this section, we study a finitely convergent algorithm for piecewise quadratic programs (Louveaux [1978]).

Algorithm PQP

Initialization: Let $S_1 = S, x^0 \in S, \nu = 1$.
Iteration ν:

    i. Obtain C_ν, a cell of the decomposition of S containing $x^{\nu-1}$. Let $z_\nu(.)$ be the quadratic form on C_ν.

    ii. Let $x^\nu \in \arg\min\{z_\nu(x)|x \in S_\nu\}$ and $w^\nu \in \arg\min\{z_\nu(x)|x \in C_\nu\}$. If w^ν is the limiting point of a ray on which $z_\nu(x)$ is decreasing to $-\infty$, the original PQP is unbounded and the algorithm terminates.

    iii. If

$$\nabla^T z_\nu(w^\nu)(x^\nu - w^\nu) = 0, \qquad (2.4)$$

    then stop; w^ν is an optimal solution.

    iv. Let $S_{\nu+1} = S_\nu \cap \{x|\nabla^T z_\nu(w^\nu)x \le \nabla^T z_\nu(w^\nu)w^\nu\}$. Let $\nu = \nu + 1$; go to Step i.

Thus, contrary to the L-shaped method in the linear case, the subgradient inequality is not applied at the current iterate point x^ν. Instead, it is applied at w^ν, a point where $z_\nu(.)$ is minimal on C_ν. Under some practical conditions on the constructions of the cells, the algorithm is finitely convergent.

We first prove that the condition,

$$\nabla^T z_\nu(w^\nu)x \le \nabla^T z_\nu(w^\nu)w^\nu, \tag{2.5}$$

is a necessary condition for optimality of x.

Because $\nabla z_\nu(w^\nu) \in \partial z(w^\nu)$, the subgradient inequality applied at w^ν implies that $z(x) \ge z(w^\nu) + \nabla^T z_\nu(w^\nu)(x - w^\nu)$ for all x. Now, x is a minimizer of $z(.)$ only if $z(x) \le z(w^\nu)$. This implies that x is a minimizer of $z(.)$ only if $\nabla^T z_\nu(w^\nu)(x - w^\nu) \le 0$, which is precisely (2.5). Thus, a solution $x \in \arg\min\{z(x)|x \in S_\nu\}$ is also a solution $x \in \arg\min\{z(x)|x \in S\}$.

We next show that any solution $\bar{x} \in \arg\min\{z_\nu(x)|x \in S_\nu\}$ is a solution $\in \arg\min\{z(x)|x \in S_\nu\}$ (and thus by the argument, a solution is in $\arg\min\{z(x)|x \in S\}$) if $\bar{x} \in C_\nu$.

By definition, $\bar{x} \in \arg\min\{z_\nu(x)|x \in S_\nu\}$ is a solution of a quadratic convex program whose objective is continuously differentiable on S_ν; it must satisfy the condition $\nabla^T z_\nu(\bar{x})(x - \bar{x}) \ge 0, \forall x \in S_\nu$. If $\bar{x} \in C_\nu$, then $\nabla z_\nu(\bar{x}) \in \partial z(\bar{x})$. Applying the subgradient inequality for $z(.)$ at $\bar{x}$ implies

$$z(x) \ge z(\bar{x}) + \nabla^T z_\nu(\bar{x})(x - \bar{x}) \ge z(\bar{x}) \ \forall \ x \in S_\nu \ .$$

Thus, if $\bar{x} \in C_\nu$, it is a solution to the original problem.

Finally, if the optimality condition (2.4) holds, applying the gradient inequality to the quadratic convex function $z_\nu(.)$ at w^ν implies

$$z_\nu(x^\nu) \ge z_\nu(w^\nu) + \nabla^T z_\nu(w^\nu)(x^\nu - w^\nu) = z_\nu(w^\nu) \ ,$$

which proves $w^\nu \in \arg\min\{z_\nu(x)|x \in S_\nu\}$. Thus, w^ν is (another) minimizer of $z_\nu(.)$ on S_ν. As $w^\nu \in C_\nu$, the conclusion implies it is a solution to the original problem. A more detailed proof, including properties of the successive sets S_ν and a discussion of the construction of full dimensional cells of a piecewise quadratic program, can be found in Louveaux [1978].

Exercises

1. For Example 2, consider the values $x_1 = 4.5, x_2 = 0$. Check that around these values, $y_2 = x_2$ for all ξ_2, and

$$y_1 = \begin{cases} \xi_1 & \text{if } \xi_1 = 2 \text{ or } 4, \\ x_1 & \text{if } \xi_1 = 6 \end{cases}$$

are the optimal second-stage decisions. Check that the corresponding cell is defined as

$$\{(x_1, x_2)|4 \le x_1 \le 6, 0 \le x_2 \le 1, x_1 + x_2 \le 6.5\}$$

and

$$z(x) = -\frac{29}{3} - \frac{x_1}{6} - 2x_2 + \frac{x_1^2}{6} + \frac{x_1 x_2}{3} + \frac{x_2^2}{2}.$$

2. We now apply the PQP algorithm to the problem of Example 2.

Initialization: $x^0 = (0,0); \nu = 1$
$S_1 = S = \{x | 3x_1 + 2x_2 \le 15, x_1 + 2x_2 \le 8, x_1, x_2 \ge 0\}$.

Iteration 1:
As we saw earlier, $C_1 = \{x | 0 \le x_1 \le 2, 0 \le x_2 \le 1\}$ and $z_1(x) = -4.5x_1 - 4x_2 + \frac{x_1^2}{2} + x_1 x_2 + \frac{x_2^2}{2}$. Using the classical Karush-Kuhn-Tucker condition, we obtain $x^1 = (4.5, 0)^T$ and $w^1 = (2, 1)^T \in C_1$. Hence, $\nabla^T z_1(w^1) = (-1.5, -1)$, $\nabla^T z_1(w^1)(x^1 - w^1) = -2.75 \neq 0$, and

$$S_2 = S \cap \{x | -1.5x_1 - x_2 \le -4\}.$$

Iteration 2:
As we saw in Exercise 1, $x^1 \in C_2 = \{x | 4 \le x_1 \le 6, 0 \le x_2 \le 1, x_1 + x_2 \le 6.5\}$ and

$$z_2(x) = -\frac{29}{3} - \frac{x_1}{6} - 2x_2 + \frac{x_1^2}{6} + \frac{x_1 x_2}{3} + \frac{x_2^2}{2}.$$

We obtain $x^2 = \left(\frac{22}{19}, \frac{43}{19}\right)^T$, a point where the optimality constraint $-1.5x_1 - x_2 \le -4$ is binding. We also obtain $w^2 = \left(4, \frac{2}{3}\right)^T \in C_2$, $\nabla^T z_2(w^2) = (25/18, 0)^T$, and (2.3) does not hold.

$$S_3 = S_2 \cap \left\{x | \frac{25}{18}x_1 \le \frac{100}{18}\right\}.$$

Iteration 3:

  i. We now obtain $x^2 \in C_3 = \{x | 0 \le x_1 \le 2, 1 \le x_2 \le 3\}$. In the second stage, $y_1 = x_1 \vee \xi_1, y_2 = x_2$ when $\xi_2 \ge 3$ and $y_2 = 1$ when $\xi_2 = 1$, so that

$$z_3(x) = -\frac{13}{6} - \frac{25}{6}x_1 - \frac{5}{3}x_2 + \frac{x_1^2}{2} + \frac{2x_1 x_2}{3} + \frac{x_2^2}{3}.$$

  ii. $x^3 = (4, 0)^T; w^3 = w^1 = (2, 1)^T$.
  iii. $S_4 = S_3 \cap \{x | -\frac{3}{2}x_1 + \frac{x_2}{3} \le -\frac{8}{3}\}$.

Iteration 4:

  i. $x^3 \in C_4 = \{x | 2 \le x_1 \le 4, 0 \le x_2 \le 1\}$.
  $z_4(x) = -\frac{11}{3} - \frac{7}{3}x_1 - \frac{10}{3}x_2 + \frac{x_1^2}{3} + \frac{2x_1 x_2}{3} + \frac{x_2^2}{2}$.
  ii. $x^4 \simeq (2.18, 1.81)^T$, a point where $-\frac{3}{2}x_1 + \frac{x_2}{3} = -\frac{8}{3}$.
  $w^4 = (2.5, 1)$.

    iii. $S_5 = S_4 \cap \{x| -\frac{2x_2}{3} \leq -\frac{2}{3}\}$.

Iteration 5:

    i. $x^4 \in C_5 = \{x|2 \leq x_1 \leq 4, 1 \leq x_2 \leq 3\} \cap S$.
        $z_5(x) = -\frac{101}{18} - \frac{19}{9}x_1 - \frac{11}{9}x_2 + \frac{x_1^2}{3} + \frac{4x_1x_2}{9} + \frac{x_2^2}{3}$.

    ii. $x^5 = w^5 = (2.5, 1)^T$ is an optimal solution to the problem.

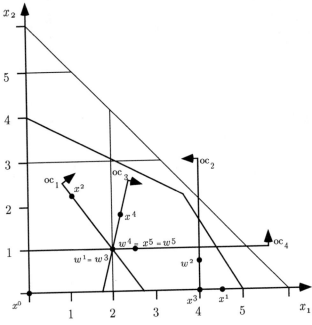

FIGURE 1. The cells and PQP cuts of Example 2.

The PQP iterations for the example are shown in Figure 1. The thinner lines represent the limits of cells and the constraints containing S. The heavier lines give the optimality cuts, OC_ν, for $\nu = 1, 2, 3, 4$. A few comments are in order:

(a) Observe that the objective values of the successive iterate points are not necessarily monotone decreasing. As an example, $z^1(w^1) = -8.5$ and $z^2(w^2) = -\frac{71}{9} > z^1(w^1)$.

(b) A stronger version of (2.4) can be obtained. Let $\bar{z} = \min_\nu\{z(w^\nu)\}$ be the best known solution at iteration ν. Starting from the subgradient inequality at w^ν,

$$z(x) \geq z(w^\nu) + \nabla z_\nu^T(w^\nu)(x - w^\nu)$$

and observing that $z(x) \leq \bar{z}$ is a necessary condition for optimality, we obtain an updated cut,

$$\nabla^T z_\nu(w^\nu)x \leq \nabla^T z_\nu(w^\nu)w^\nu + \bar{z} - z(w^\nu). \tag{2.6}$$

Updating is quite easy, as it only involves the right-hand sides of the cuts. As an example, at Iteration 2, the cut could be updated from

$$\frac{25x_1}{18} \leq \frac{100}{18} \text{ to } \frac{25}{18}x_1 \leq \frac{100}{18} - 8.5 + \frac{71}{9},$$

namely, $\frac{25x_1}{18} \leq \frac{89}{18}$. Similarly, at Iteration 4, $\bar{z}$ becomes $-\frac{103}{12}$ and the right-hand sides of all previously imposed cuts can be modified by $\left(-\frac{103}{12} + 8.5\right)$, i.e., by $-\frac{1}{12}$. In the example, the updating does not change the sequence of iterations.

(c) The number of iterations is strongly dependent on the starting point. In particular, if one cell exists such that the minimizer of its quadratic form over S is in fact within the cell, then starting from that cell would mean that a single iteration would suffice. In Example 2, this is not the case. However, starting from $\{x|2 \leq x_1 \leq 4, 1 \leq x_2 \leq 3\}$ would require only two iterations. This is in fact a reasonable starting cell. Indeed, the intersection of the two nontrivial constraints defining S,

$$3x_1 + 2x_2 \leq 15, x_1 + 2x_2 \leq 8,$$

is the point $(3.5, 2.25)$ that belongs to that cell. (An alternative would be to start from the minimizer of the mean value problem on S.)

(d) If we observe the graphical representation of the cells and of the cuts, we observe that the cuts each time eliminate all points of a cell, except possibly the point w^ν at which they are imposed, and possibly other points on a face of dimension strictly less than n_1. (Working with updated cuts (2.6) sometimes also eliminates the point w^ν at which it is imposed.) The finite termination of the algorithm is precisely based on the elimination of one cell at each iteration. (We leave aside the question of considering cells of full dimension n_1.) There is thus no need at iteration ν to start from a cell containing $x^{\nu-1}$. In fact, any cell not yet considered is a valid candidate. One reasonable candidate could be the cell containing $\frac{x^{\nu-1} + w^{\nu-1}}{2}$, for example, or any convex combination of $x^{\nu-1}$ and $w^{\nu-1}$.

3. Consider the farming example of Section 1.1. As in Exercise 1.1, assume that prices are influenced by quantities. As an individual, the

farmer has little influence on prices, so he may reasonably consider the current solution optimal. If we now consider that all farmers read this book and optimize their choice of crop the same way, increases of sales will occur in parallel for all farmers, bringing large quantities together on the market. Taking things to an extreme, this means that changes in the solution are replicated by all farmers. Assume a decrease in selling prices of $0.03 per ton of grain and of $0.06 per ton of corn brought into the market by each individual farmer. Assume the selling price of beets and purchase prices are not affected by quantities.

Show that the PQP algorithm reaches the solution in one iteration when the starting point is taken as $\{x_1, x_2, x_3 | 80 \leq x_2 \leq 100; 250 \leq x_3 \leq 300; x_1 + x_2 + x_3 = 500\}$. (Remark: Although only one iteration is needed, calculations are rather lengthy. Observe that constant terms are not needed to obtain the optimal solution.)

6.3   Methods Based on the Stochastic Program Lagrangian

Again consider the general nonlinear stochastic program given in (3.4.1), which we repeat here without equality constraints to simplify the following discussion:

$$\begin{array}{ll} \inf z = & f^1(x) + \mathcal{Q}(x) \\ \text{s. t.} & g_i^1(x) \le 0, \quad i = 1, \ldots, m_1, \end{array} \tag{3.1}$$

where $\mathcal{Q}(x) = E_\omega[Q(x, \omega)]$ and

$$Q(x, \omega) = \begin{array}{l} \inf f^2(y(\omega), \omega) \text{ s. t.} \\ b_i^2(x, \omega) + g_i^2(y(\omega), \omega) \end{array} \le 0, i = 1, \ldots, m_2, \tag{3.2}$$

with the continuity assumptions mentioned in Section 3.4.

In general, we can consider a variety of approaches to (3.1) based on available nonlinear programming methods. For example, we may consider gradient projection, reduced gradient methods, and straightforward penalty-type procedures, but these methods all assume that gradients of $\mathcal{Q}$ are available and relatively inexpensive to acquire. Clearly, this is not the case in stochastic programs because each evaluation may involve solving several problems (3.2). Lagrangian approaches have been proposed to avoid this problem.

The basic idea behind the Lagrangian approaches is to place the first- and second-stage links into the objective so that repeated subproblem optimizations are avoided in finding search directions. To see how this approach works, consider writing (3.1) in the following form:

$$\begin{array}{rl} \inf z = f^1(x) + E_\omega[f^2(y(\omega), \omega)] & \\ \text{s. t.} \ g_i^1(x) \le & 0, i = 1, \ldots, m_1, \\ b_i^2(x, \omega) + g_i^2(y(\omega), \omega) \le & 0, i = 1, \ldots, m_2, \text{ a. s.} \end{array} \tag{3.3}$$

If we let (λ, π) be a multiplier vector associated with the constraints, then we can form a dual problem to (3.3) as:

$$\max_{\pi(\omega) \ge 0} w = \theta(\pi), \tag{3.4}$$

where

$$\theta(\pi) =$$

$$\begin{array}{ll} \inf_{x, \mathbf{y}} z = & f^1(x) + E_\omega[f^2(y(\omega), \omega)] \\ & + E_\omega[\sum_{i=1}^{m_2} \pi(\omega)_i (b_i^2(x, \omega) + g_i^2(y(\omega), \omega))] \\ \text{s. t.} & g_i^1(x) \le 0, i = 1, \ldots, m_1. \end{array} \tag{3.5}$$

We show duality in the finite distribution case in the following theorem.

Theorem 14. *Suppose the stochastic nonlinear program (3.1) with all functions convex has a finite optimal value and a point strictly satisfying* *all constraints, and suppose $\Omega = \{1, \ldots, K\}$ with $P\{\omega = i\} = p_i$. Then $z \geq w$ for every feasible $x, y_1, \ldots, y_K$ in (3.1-2) and $\pi_1, \ldots, \pi_K$ feasible in (3.4), and their optimal values coincide, $z^* = w^*$.*

Proof: From the general optimality conditions (see, e.g., Bazaraa and Shetty [1979, Theorem 6.2.1]), the result follows by noting that we may take x satisfying the first-period constraints as a general convex constraint set X so that only the second-period constraints are placed into the dual. We also divide any multipliers on the second-period constraints in (3.3) by p_i if they correspond to $\omega = i$. In this way, the expectation over ω in (3.5) is obtained. $\square$

Now, we can follow a dual ascent procedure in (3.4). This takes the form of a subgradient method. We note that

$$\partial\theta(\bar{\pi}) = co\{(\zeta_1^1, \ldots, \zeta_{m_2}^1)^T, \ldots, (\zeta_1^K, \ldots, \zeta_{m_2}^K)^T\}, \tag{3.6}$$

where *co* denotes the convex hull,

$$\zeta_i^k = b_i^2(\bar{x}, k) + g_i^2(\bar{y}_k, k), \tag{3.7}$$

and $(\bar{x}, \bar{y}_1, \ldots, \bar{y}_K)$ solves the problem in (3.5) given $\pi = \bar{\pi}$. This again follows from standard theory as in, for example, Bazaraa and Shetty [1979, Theorem 6.3.7].

We can now describe a basic gradient method for the dual problem. For our purposes, we assume that (3.5) always has a unique solution.

Basic Lagrangian Dual Ascent Method

Step 0. Set $\pi^0 \geq 0, \nu = 0$ and go to Step 1.

Step 1. Given $\pi = \pi^\nu$ in (3.5), let the solution be $(x^\nu, y_1^\nu, \ldots, y_K^\nu)$. Let $\hat{\pi}_i^k = 0$ if $\pi_i^{\nu,k} = 0$ and $b_i^2(x^\nu, k) + g_i^2(y_k^\nu, k) \leq 0$, and $\hat{\pi}_i^k = b_i^2(x^\nu, k) + g_i^2(y_k^\nu, k)$, otherwise. If $\hat{\pi}^k = 0$ for all k, stop.

Step 2. Let λ^ν minimize $\theta(\pi^\nu + \lambda\hat{\pi})$ over $\pi^\nu + \lambda\hat{\pi} \geq 0, \lambda \geq 0$. Let $\pi^{\nu+1} = \pi^\nu + \lambda^\nu\hat{\pi}$, $\nu = \nu + 1$, and go to Step 1.

Assuming the unique solution property, this algorithm always produces an ascent direction in θ. The algorithm either converges finitely to an optimal solution or, assuming a bounded set of optima, produces an infinite sequence with all limit points optimal (see Exercise 1). For the case of multiple optima for (3.5), some nondifferentiable procedure must be used. In this case, one could consider finding the maximum norm subgradient to be assured of ascent or one could use various bundle-type methods (see Section 4).

The basic hope for computational efficiency in the dual ascent procedure is that the number of dual iterations is small compared to the number of function evaluations that might be required by directly attacking (3.1) and (3.2). Substantial time may be spent solving (3.5) but that should be somewhat easier than solving (3.1) because the linking constraints appear in the objective instead of hard constraints. Overall, however, this type of procedure is generally slow due to our using only a single-point linearization of θ. This observation has led to other types of Lagrangian approaches to (3.1) that use more global or second-order information.

Rockafellar and Wets [1986] suggested one such procedure for a special case of (3.5) where $f^1(x) = c^T x + \frac{1}{2} x^T C x$ and $y(\omega)$ can be eliminated so that the second and third objective terms in (3.5) are:

$$\Phi(\pi, x) = E_\omega[\pi(\omega)^T(h(\omega) - T(\omega)x) - \frac{1}{2}\pi(\omega)^T D(\omega)\pi(\omega)]. \qquad (3.8)$$

In fact, this is always possible if $b^2(x, \omega) = T(\omega)x$, $g^2(y(\omega), \omega) = Wy(\omega) - h(\omega)$, and $f_2(y(\omega), \omega)) = q(\omega)^T y(\omega) + \frac{1}{2} y(\omega)^T D(\omega) y(\omega)$ (Exercise 2). The dual problem in (3.4) is then

$$\max_{\pi \geq 0} \inf_{\{x | g^1(x) \leq 0\}} [c^T x + \frac{1}{2} x^T C x + \Phi(\pi, x)]. \qquad (3.9)$$

In fact, Rockafellar and Wets allowed more general constraints on π that may depend on ω.

Their approach is, however, not to restrict the search to a single search direction but to allow optimization over a low dimensional set in $[\Re_+^{m_2}]^\Omega$. At iteration ν, they replace $[\Re_+^{m_2}]^\Omega$ by Π^ν, and iteratively update this approximation by including not just the gradient direction of the Lagrangian but the best $\pi(\omega)$ for the fixed value of $x = x^\nu$ in $\Phi(\pi, x)$. This point, $\pi^{\nu+1}$, is used to create the new Π^ν. The process repeats as follows.

Lagrangian Finite Generation Method for Linear-Quadratic Stochastic Programs

Step 0. Choose Π^0, set $\nu = 0$, and go to Step 1.

Step 1. Solve (3.9) with the constraint, $\pi \in \Pi^\nu$, in place of $\pi \geq 0$. Let the solution be (π^ν, x^ν) with value, $LB^\nu = \theta(\pi^\nu)$ (a lower bound on the optimal objective value).

Step 2. Find $\pi^{\nu+1}(\omega)$ for each ω to maximize $[\pi(\omega)^T(h(\omega) - T(\omega)x^\nu) - \frac{1}{2}\pi(\omega)^T D(\omega)\pi(\omega)]$ over $\pi(\omega) \geq 0$. (Notice that this optimization finds $\max_{\pi \geq 0} \Phi(\pi, x^\nu)$.) Let $UB^\nu = c^T x^\nu + \frac{1}{2} x^{\nu,T} C x^\nu + \Phi(\pi^{\nu+1}, x^\nu)$ (an upper bound on the optimal objective value). If $UB^\nu - LB^\nu < \epsilon$, then stop with an ϵ-optimal solution.

Step 3. Update Π^ν to $\Pi^{\nu+1}$ by ensuring that $\Pi^{\nu+1}$ includes π^ν and $\pi^{\nu+1}$ (and any other part but not necessarily all of Π^ν). Let $\nu = \nu + 1$, and go to Step 1.

An implementation of this algorithm is described in King [1988a]. The method was used successfully to solve practical water management problems concerning Lake Balaton in Hungary (Somlyódy and Wets [1988]). The algorithm produces a convergent sequence to an optimum provided that the set $\Pi^{\nu+1}$ is sufficiently large and the matrix C is positive definite (enforcing strict convexity). In general, the choice of $\Pi^{\nu+1}$ is similar to those in cutting plane methods. On linear pieces of the objective, restricting $\Pi^{\nu+1}$ too severely may lead to cycling. The alternatives discussed in Eaves and Zangwill [1971] can be used to control this.

The finite generation method is similar to other methods based on inner linearization approaches in nonlinear programming such as the restricted simplicial decomposition approach (Ventura and Hearn [1993]). This procedure essentially replaces the line search in the Topkis-Veinott [1967] feasible direction method with a search over a simplex. The finite generation algorithm is analogously an enhancement over basic Lagrangian dual ascent methods. Both the finite generation and restricted simplicial decomposition methods tend to avoid the zigzagging behavior that often occurs in methods based on single point linearizations.

Another method for speeding convergence is to enforce strictly convex terms in the objective. Rockafellar and Wets discussed methods for adding quadratic terms to the matrices C and $D(\omega)$ so that these matrices become positive definite. In this way, the finite generation method becomes a form of augmented Lagrangian procedure. We next discuss the basic premise behind these procedures.

In an augmented Lagrangian approach, one generally adds a penalty $r\|(b_i^2(\bar{x},k) + g_i^2(\bar{y}_k,k))^+\|^2$ to $\theta(\pi)$ and performs the iterations including this term. The advantage (see the discussion in Dempster [1988]) is that Newton-type steps can be applied because we would obtain a nonsingular Hessian. The result should generally be that convergence becomes superlinear in terms of the dual objective without a significantly greater computational burden over the Lagrangian approach.

The computational experience reported by Dempster suggests that few dual iterations need be used but that a more effective alternative was to include explicit nonanticipative constraints as in (3.4.4) and to place these constraints into the objective instead of the full second-period constraints. In this way, θ becomes

$$\theta'(\rho) = \inf z = f^1(x) + \sum_{k=1}^{K} p_k[f^2(y_k,k)]$$

$$+ \sum_{k=1}^{K} [\rho_k^T (x - x_k) + r/2\|x - x_k\|^2]$$

s. t. $g_i^1(x) \le 0, i = 1, \ldots, m_1,$
$b_i^2(x_k, k) + g_i^2(y_k, k) \le 0, i = 1, \ldots, m_2,$
$k = 1, \ldots, K.$ (3.10)

Notice how in (3.10) the only links between the nonanticipative x decision and the scenario k decisions are in the $(x - x_k)$ objective terms. Dempster suggests solving this problem approximately on each dual iteration by iterating between searches in the x variables and search in the x_k, y_k variables. In this way, the augmented Lagrangian approach of solving (3.10) to find a dual ascent Newton-type direction achieves superlinear convergence in dual iterations. The only problem may come in the time to construct the search directions through solutions of (3.10).

This method also resembles the *progressive hedging algorithm* of Rockafellar and Wets [1991]. Their method achieves a full separation of the separate scenario problems for each iteration and, therefore, has considerably less work at each iteration. However, the number of iterations as we shall see, may be greater. The method appears to offer many computational advantages at least for structured problems as reported by Mulvey and Vladimirou [1991a]. The key to this method's success is that individual subproblem structure is maintained throughout the algorithm. Related implementations by Nielsen and Zenios [1993a, 1993b] on parallel processors demonstrate possibilities for parallelism and the solution of large problems.

The basic progressive hedging method begins with a nonanticipative solution $\hat{x}^\nu$ and a multiplier ρ^ν. The nonanticipative (but not necessarily feasible) solution is used in place of x in (3.10). The first-period constraints are also split into each x_k. In this way, we obtain a subproblem:

$$\inf z = \sum_{k=1}^{K} p_k [f^1(x_k) + f^2(y_k, k) + \rho_k^{\nu,T}(x_k - \hat{x}^\nu) + r/2\|x_k - \hat{x}^\nu\|^2]$$

s. t. $g_i^1(x_k) \le 0, i = 1, \ldots, m_1, k = 1, \ldots, K,$
$b_i^2(x_k, k) + g_i^2(y_k, k) \le 0, i = 1, \ldots, m_2, k = 1, \ldots, K.$ (3.11)

Now (3.11) splits directly into subproblems for each k so these can be treated separately.

Supposing that $(x_k^{\nu+1}, y_k^{\nu+1})$ solves (3.11). We obtain a new nonanticipative decision by taking the expected value of $x^{\nu+1}$ as $\hat{x}^{\nu+1}$ and step in ρ by $\rho^{\nu+1} = \rho^\nu + (x^{\nu+1} - \hat{x}^{\nu+1})$.

The steps then are simply stated as follows.

Progressive Hedging Algorithm

Step 0. Suppose some nonanticipative x^0, some initial multiplier ρ^0, and $r > 0$. Let $\nu = 0$. Go to Step 1.

Step 1. Let $(x_k^{\nu+1}, y_k^{\nu+1})$ for $k = 1, \ldots, K$ solve (3.11). Let $\hat{x}^{\nu+1} = (\hat{x}^{\nu+1,1}, \ldots, \hat{x}^{\nu+1,K})^T$ where $\hat{x}^{\nu+1,k} = \sum_{l=1}^{K} p_l x^{\nu+1,l}$ for all $k = 1, \ldots, K$.

Step 2. Let $\rho^{\nu+1} = \rho^\nu + r(x^{\nu+1,k} - \hat{x}^{\nu+1})$. If $\hat{x}^{\nu+1} = \hat{x}^\nu$ and $\rho^{\nu+1} = \rho^\nu$, then, stop; $\hat{x}^\nu$ and ρ^ν are optimal. Otherwise, let $\nu = \nu + 1$ and go to 1.

The convergence of this method is based on Rockafellar's proximal point method [1976a]. The basis for this approach is not dual ascent but the contraction of the pair, $(\hat{x}^{\nu+1}, \rho^{\nu+1})$, about an optimal point. The key is that the algorithm mapping can be described as $(\Pi x^{\nu+1}, \rho^{\nu+1}/r) = (I - V)^{-1}(\Pi x^\nu, \rho^\nu/r)$, where V is a *maximal monotone operator* and Π is the diagonal matrix of probabilities corresponding to x^k and ρ^k, i.e, where $\Pi_{(k-1)n_1+i,(k-1)n_1+i} = p_k$ for $i = 1, \ldots, n_1$ and $k = 1, \ldots, K$.

To describe this approach we first define a maximal monotone operator at V (see Minty [1961] for more general details) such that for any pairs (w, z) where $z \in V(w)$ and (w', z') for $z' \in V(w')$, we have

$$(w - w')^T V(z - z') \geq 0. \qquad (3.12)$$

The key point here is that if we have a Lagrangian function $l(x, y)$ that is convex in x and concave in y, then the subdifferential set of $l(x, y)$ at $(\bar{x}, \bar{y})$ defined by

$$\{(\zeta, \eta) | \zeta^T(x - \bar{x}) + l(\bar{x}, \bar{y}) \leq l(x, \bar{y}), \forall x; \eta^T(y - \bar{y}) + l(\bar{x}, \bar{y}) \geq l(\bar{x}, y), \forall y\} \qquad (3.13)$$

yields a maximal monotone operator by

$$V(\bar{x}, \bar{y}) = \{(\zeta, \eta)\} \qquad (3.14)$$

for $(\zeta, -\eta) \in \partial l(\bar{x}, \bar{y})$ (Exercise 3).

The second result that follows for maximal monotone operators is that a contraction mapping can be defined on it by taking $(I - V)^{-1}(x, y)$ to obtain (x', y'), or, equivalently, where $(x' - x, y' - y) \in V(x', y')$. The contraction result (Exercise 4) is that, if V is maximal monotone, then, for all $(x', y') = (I - (1/r)V)^{-1}(x, y)$ and $(\bar{x}', \bar{y}') = (I - V)^{-1}(\bar{x}, \bar{y})$,

$$\|(x' - \bar{x}', y' - \bar{y}')\|^2 \leq (x - \bar{x}, y - \bar{y})^T(x' - \bar{x}', y' - \bar{y}'). \qquad (3.15)$$

These results then play the fundamental role in the following proof of convergence.

Theorem 15. *The progressive hedging algorithm, applied to (3.1) with the same conditions as in Theorem 14, converges to an optimal solution, x^*, ρ^*, (or terminates finitely with an optimal solution) and, at each iteration that does not terminate in Step 2,*

$$\|(\Pi\hat{x}^{\nu+1}, \rho^{\nu+1}/r) - (\Pi x^*, \rho^*/r)\| < \|(\Pi\hat{x}^\nu, \rho^\nu/r) - (\Pi x^*, \rho^*)/r\|. \qquad (3.16)$$

Proof. As stated, the key is to find the associated Lagrangian and to show that the iterations follow the mapping as in (3.15). For the Lagrangian, define

$$l(\bar{x}, \bar{\rho}) = \begin{matrix} \inf_x \\ \text{s. t.} \end{matrix} \quad \begin{matrix} (1/r)z(x) + \bar{\rho}^T \Pi x \\ J\Pi x - \bar{x} = 0, \end{matrix} \tag{3.17}$$

where $z(x)$ is defined as $\sum_{k=1}^{K}[f^1(x^k) + Q(x^k, k)]$ for feasible x^k and as $+\infty$ otherwise, Π is defined as the diagonal probability matrix, and J is the matrix corresponding to column sums, $J_{r,s}$ equal one if $r \pmod{n_1} = s \pmod{n_1}$ and zero otherwise. We want to show that $(\Pi(\hat{x}^\nu - \hat{x}^{\nu+1}), (\rho^\nu - \rho^{\nu+1})/r) \in \partial l(\Pi \hat{x}^{\nu+1}, \rho^{\nu+1}/r)$, so we can use the contraction property in (3.15) from the maximal monotone operator defined on $\partial l(\Pi \hat{x}^{\nu+1}, \rho^{\nu+1}/r)$.

Note that, for $\bar{x} = \Pi \hat{x}^\nu$ and $\bar{\rho} = \rho^\nu/r = \sum_{i=1}^\nu (x^i - \hat{x}^i)$, $\bar{x}^T \bar{\rho} = \hat{x}^{\nu,T}$ $\Pi(\sum_{i=1}^\nu (x^i - \hat{x}^i)) = (x')^{\nu,T} J\Pi(\sum_{i=1}^\nu (x^i - \hat{x}^i))$ for $(x')^{\nu,T} = (1/K)\hat{x}^{\nu,T}$. Because $J\Pi x^i = \hat{x}^i$, we have $\bar{x}^T \bar{\rho} = 0$. We can thus add the term, $\bar{x}^T \bar{\rho}$ to the objective in (3.17) without changing the problem. We then obtain:

$$\eta \in \partial_{\bar{\rho}} l(\bar{x}, \bar{\rho}) \Leftrightarrow -\Pi \bar{\rho} \in (1/r)\partial z(\Pi^{-1}(-\eta) + \bar{x}) + \pi^T J\Pi, \tag{3.18}$$

where $J\Pi(\Pi^{-1}(-\eta)) = \bar{x}$ and π is some multiplier. For $\partial_{\bar{x}} l(\bar{x}, \bar{\rho})$, $\zeta = -\pi^T J\Pi$, and some π,

$$\zeta \in \partial_{\bar{x}} l(\bar{x}, \bar{\rho}) \Leftrightarrow \zeta - \Pi \bar{\rho} \in (1/r)\partial Z(x'), \tag{3.19}$$

for some $J\Pi x' = \hat{x}$. We combine (3.18) and (3.19) to obtain that $(\zeta, \eta) \in \partial l(\bar{x}, \bar{\rho})$ if

$$\zeta - \Pi \bar{\rho} \in (1/r)\partial z(\Pi^{-1}(-\eta) + \bar{x}). \tag{3.20}$$

We wish to show that

$$\Pi(\hat{x}^\nu - \hat{x}^{\nu+1}) - \Pi \rho^{\nu+1}/r \in (1/r)\partial z(\Pi^{-1}(\rho^{\nu+1} - \rho^\nu)/r + \hat{x}^{\nu+1}). \tag{3.21}$$

From the algorithm,

$$-\Pi \rho^\nu \in \partial z(x^{\nu+1}) + r\Pi(x^{\nu+1} - \hat{x}^\nu). \tag{3.22}$$

Substituting, $\rho^{\nu+1} = \rho^\nu + r(x^{\nu+1} - \hat{x}^{\nu+1})$, we obtain from (3.22),

$$-\Pi \rho^{\nu+1} + r\Pi(x^{\nu+1} - \hat{x}^{\nu+1}) \in \partial z(x^{\nu+1}) + r\Pi(x^{\nu+1} - \hat{x}^\nu), \tag{3.23}$$

which, after eliminating $r\Pi x^{\nu+1}$ from both sides, coincides with (3.21).

By the nonexpansive property, there exists $(\Pi x^*, \rho^*/r)$, a fixed point of this mapping. By substituting into (3.15), with $(\Pi x^*, \rho^*/r) = (I - V)(\Pi x^*, \rho^*/r)$ and $(\Pi \hat{x}^{\nu+1}, \rho^{\nu+1}/r) = (I - V)(\Pi \hat{x}^\nu, \rho^\nu/r)$, we have (Exercise 5):

$$\|(\Pi \hat{x}^{\nu+1}, \rho^{\nu+1}/r) - (\Pi x^*, \rho^*/r)\| < \|(\Pi \hat{x}^\nu, \rho^{\nu+1}/r) - (\Pi x^*, \rho^*/r)\|. \tag{3.24}$$

Our result follows if (x^*, ρ^*) is indeed a solution of (3.1). Note that in this case, we must have $0 = x^{\nu+1} - \hat{x}^{\nu+1} = x^{\nu+1} - \hat{x}^\nu$, so, from (3.22),

$-\Pi \rho^* \in \partial z(x^*)$. From Theorem 3.2.5, optimality in (3.1) is equivalent to $\rho^T \Pi \in \partial z(x^*)$ for some ρ, where $J\Pi \rho = 0$, which is true because $J\Pi(-\rho^*) = -\sum_\nu J\Pi(x^{\nu+1} - x^\nu) = 0$. Hence, we obtain optimality. The method converges as desired. $\Box$

We note that Rockafellar and Wets obtained these results by defining an inner product as $< \rho, x > = \rho^T \Pi x$ and using appropriate operations with this definition. They also show that, in the linear-quadratic case, the convergence to optimality is geometric.

Variants of this method are possible by considering other inner products and projection operators. For example, we can let $\hat{x}^{\nu+1}$ be the standard orthogonal projection of $x^{\nu+1}$ into the null space of $J\Pi$. This value is the simple average of $x_k^{\nu+1}$ values, so that $\hat{x}_k^{\nu+1}(i) = (1/K)\sum_{k=1}^K x_k^{\nu+1}(i)$ for all $k = 1, \ldots, K$. The multiplier update is then:

$$\rho^{\nu+1} = \rho^\nu + r\Pi^{-1}(x^{\nu+1} - \hat{x}^{\nu+1}). \tag{3.25}$$

One can again obtain the maximal monotone operator property, and, observing that $Jx^{\nu+1} = J\hat{x}^{\nu+1}$, obtain $J\Pi \rho^* = 0$ and optimality.

Example 3

The algorithm's geometric convergence may require many iterations even on small problems as we show in the following small example. Suppose we can invest \$10,000 in either of two investments, A or B. We would like a return of \$25,000, but the investments have different returns according to two future scenarios. In the first scenario, A returns just the initial investment while B returns 3 times the initial investment. In the second scenario, A returns 4 times the initial investment and B returns twice the initial investment.The two scenarios are considered equally likely. To reflect our goal of achieving \$25,000, we use an objective that squares any return less than \$25,000. The overall formulation is then:

$$
\begin{aligned}
\min z = \quad & 0.5(y_1^2 + y_2^2) \\
\text{s. t.} \quad & x_A + x_B \le 10, \\
& x_A + 3x_B + y_1 \ge 25, \\
& 4x_A + 2x_B + y_2 \ge 25, \\
& x_A, x_B, y_1, y_2 \ge 0.
\end{aligned} \tag{3.26}
$$

Clearly, this problem has an optimal solution at $x_A^* = 2.5$ and $x_B^* = 7.5$ with an objective value $z^* = 0$. A single iteration of Step 1 in the basic Lagrangian method is all that would be required to solve this problem for any positive π value. A single iteration is also all that would be necessary in the augmented Lagrangian problem in (3.10). The price for this efficiency is, however, the incorporation of all subproblems into a single master problem. Progressive hedging on the other hand maintains completely separate subproblems. We will follow the first two iterations of PHA for $r = 2$ here.

Iteration 0:

Step 0. Begin with a multiplier vector of $\rho^0 = 0$, and let $x_1^0 = (x_{1A}^0, x_{1B}^0) = (0, 10)^T$ and let $x_2^0 = (x_{2A}^0, x_{2B}^0) = (10, 0)^T$. The initial value of $\hat{x}^0 = (5, 5)^T$.

Step 1. We wish to solve:

$$\min(1/2)[y_1^2 + y_2^2 + (x_{1A}^1 - 5)^2 + (x_{1B}^1 - 5)^2 + (x_{2A}^1 - 5)^2 + (x_{2B}^1 - 5)^2]$$
$$\text{s. t. } x_{1A}^1 + x_{1B}^1 \leq 10,$$
$$x_{2A}^1 + x_{2B}^1 \leq 10,$$
$$x_{1A}^1 + 3x_{1B}^1 - y_1 \geq 25,$$
$$4x_{2A}^1 + 2x_{2B}^1 - y_2 \geq 25,$$
$$x_{1A}^1, x_{1B}^1, x_{2A}^1, x_{2B}^1, y_1, y_2 \geq 0. \tag{3.27}$$

This problem splits into separate subproblems for x_{1A}^1, x_{1B}^1, y_1 and x_{2A}^1, x_{2B}^1, y_2, as mentioned earlier. For x_{1A}^1, x_{1B}^1, y_1 feasible in (3.27), the K-K-T conditions are that there exist $\lambda_1 \geq 0, \lambda_2 \geq 0$ such that

$$2(x_{1A}^1 - 5) + \lambda_1 - \lambda_2 \geq 0,$$
$$2(x_{1B}^1 - 5) + \lambda_1 - 3\lambda_2 \geq 0,$$
$$2y_1 - \lambda_2 \geq 0,$$
$$(2(x_{1A}^1 - 5) + \lambda_1 - \lambda_2)x_{1A}^1 = 0,$$
$$(2(x_{1B}^1 - 5) + \lambda_1 - 3\lambda_2)x_{1B}^1 = 0,$$
$$(2y_1 - \lambda_2)y_1 = 0,$$
$$(x_{1A}^1 + x_{1B}^1 - 10)\lambda_1 = 0,$$
$$(x_{1A}^1 + 3x_{1B}^1 - y_1 - 25)\lambda_2 = 0, \tag{3.28}$$

which has a solution of $(x_{1A}^1, x_{1B}^1, y_1) = (10/3, 20/3, 5/3)$ and $(\lambda_1, \lambda_2) = (20/3, 10/3)$. Similar conditions exist for the second subproblem, which has a solution $(x_{2A}^1, x_{2B}^1, y_2) = (5, 5, 0)$. We then let $(\hat{x}_{iA}^1, \hat{x}_{iB}^1) = (4\frac{1}{6}, 5\frac{5}{6})$ for $i = 1, 2$.

Step 2. The new multiplier is $\rho^1 = (\rho_{1A}^1, \rho_{1B}^1, \rho_{2A}^1, \rho_{2B}^1)^T = 2((10/3 - 25/6), (20/3 - 35/6), (5 - 25/6), (5 - 35/6))^T = (-5/3, 5/3, 5/3, -5/3)^T$.

Iteration 2:

Step 1. The first subproblem is now

$$\min y_1^2 - (5/3)(x_{1A}^2 - 25/6) + (5/3)(x_{1B}^2 - 35/6) + (x_{1A}^2 - 25/6)^2$$
$$+ (x_{1B}^2 - 35/6)^2$$
$$\text{s. t. } x_{1A}^2 + x_{1B}^2 \leq 10,$$
$$x_{1A}^2 + 3x_{1B}^2 - y_1 \geq 25,$$
$$x_{1A}^2, x_{1B}^2, y_1 \geq 0, \tag{3.29}$$

which again has an optimal solution, $(x_{1A}^2, x_{1B}^2, y_1^2) = (10/3, 20/3, 5/3)$. Curiously, we also have the second subproblem solution of $(x_{2A}^2, x_{2B}^2, y_2^2) = (10/3, 20/3, 0)$. In this case, $(\hat{x}_{iA}^2, \hat{x}_{iB}^2) = (10/3, 20/3)$ for $i = 1, 2$.

TABLE 1. PHA iterations for Example 3.

k	$\hat{x}_A^k$	$\hat{x}_B^k$	ρ_{1A}^k $= -\rho_{2A}^k$	ρ_{1B}^k $= -\rho_{2B}^k$	x_{1A}^k	x_{1B}^k	x_{2A}^k	x_{2B}^k
0	5.0	5.0	0.0	0.0	3.33	6.67	5.0	5.0
1	4.17	5.83	-1.67	1.67	3.33	6.67	3.33	6.67
2	3.33	6.67	-1.67	1.67	3.06	6.94	2.50	7.50
3	2.78	7.22	-1.11	1.11	2.78	7.22	2.41	7.59
4	2.59	7.41	-0.74	0.74	2.65	7.35	2.41	7.59
5	2.53	7.47	-0.49	0.49	2.59	7.41	2.43	7.57
6	2.50	7.50	-0.33	0.33	2.56	7.44	2.45	7.55
7	2.50	7.50	-0.22	0.22	2.54	7.46	2.46	7.54
8	2.50	7.50	-0.15	0.15	2.53	7.48	2.48	7.52
9	2.50	7.50	-0.10	0.10	2.52	7.48	2.48	7.52
10	2.50	7.50	-0.07	0.07	2.51	7.49	2.49	7.51
11	2.50	7.50	-0.04	0.04	2.51	7.49	2.49	7.51
12	2.50	7.50	-0.03	0.03	2.50	7.50	2.50	7.50

Step 2. Because the subproblems returned the same solution, $\rho^2 = \rho^1$. We continue because the x values changed, even though we took no multiplier step.

The full iteration values are given in Table 1. Notice how the method achieves convergence in the x values before the ρ values have converged. Also, notice how the convergence appears to be geometric. This type of performance appears to be typical of PHA. It should be noted again, however, that the iterations are quite simple and that little overhead is required.

Exercises

1. Show that the basic dual ascent method converges to an optimal solution under the conditions given.

2. Show that (3.4) can be reduced to (3.9) when $b^2(x, \omega) = T(\omega)x$, $g^2(y(\omega), \omega) = Wy(\omega) - h(\omega)$, $f^2(y(\omega), \omega)) = q(\omega)^T y(\omega) + \frac{1}{2}y(\omega)^T D(\omega)y(\omega)$, and D is positive definite.

3. Show that V as defined in (3.14) is a maximal monotone operator.

4. Prove the contraction property in (3.15).

5. Use (3.15) to obtain (3.24).

6. Apply the dual ascent method and the augmented Lagrangian method with problem (3.10) to the example in (3.26). Start with

zero multipliers (ρ), positive π, and positive penalty r. Show that each obtains an optimal solution in at most one iteration.

6.4   Nonlinear Programming in Simple Recourse Problems

The previous sections considered basically nonlinear problems that could be modeled with deterministic equivalents when the number of random variable realizations was finite. As mentioned in Chapter 5, the simple recourse problem may allow computation even when the underlying distribution is continuous. Recall that the simple recourse problem has the form:

$$\min z = c^T x + \sum_{i=1}^{m_2} \Psi_i(\chi_i) \tag{4.1}$$
$$\text{s.t. } Ax = b,$$
$$Tx - \chi = 0,$$
$$x \geq 0,$$

where $\Psi_i(\chi_i) = \int_{h_i \leq \chi_i} q^-(\chi_i - h_i) dF(h_i) + \int_{h_i > \chi_i} q^+(h_i - \chi_i) dF(h_i)$. Using this form of the objective in χ, we again substitute in (3.1.9) to obtain:

$$\Psi_i(\chi_i) = q_i^+ \bar{h}_i - (q_i^+ - q_i F_i(\chi_i))\chi_i - q_i \int_{h_i \leq \chi_i} h_i dF_i(h_i). \tag{4.2}$$

The most direct methods for solving (4.1) are to use standard nonlinear programming techniques. We briefly describe some of the alternatives here. The most common procedures applied here are single-point linearization approaches, such as the Frank-Wolfe method, multiple-point linearization, such as generalized linear programming, and active set or reduced variable methods, similar to simplex method extensions. Other methods are described in Nazareth and Wets [1986].

The Frank-Wolfe method for simple recourse problems appears in Wets [1966] and Ziemba [1970]. The basic procedure is to approximate the objective using the gradient and to solve a linear program to find a search direction. The algorithm contains the following basic steps. We assume that each random variable $\mathbf{h}_i$ has an absolutely continuous distribution function F_i so that each Ψ_i is differentiable. In this case, the gradient of $\Psi(Tx)$ is easily calculated as $\nabla \Psi(Tx) = (q^+ - q)^T(\bar{F})T$, where $\bar{F} = diag\{F_i(T_i.x)\}$, the diagonal matrix of the probability that $\mathbf{h}_i$ is below $T_i.x$.

Frank-Wolfe Method for Simple Recourse Problems

Step 0. Suppose a feasible solution x^0 to (4.1). Let $\nu = 0$. Go to Step 1.

Step 1. Let $\hat{x}^\nu$ solve:

$$\min z = (c^T + (q^+ - q)^T(\bar{F}^\nu)T)x \qquad (4.3)$$
$$\text{s.t. } Ax = b,$$
$$x \geq 0,$$

where $\bar{F}^\nu = diag\{F_i(T_i.x^\nu)\}$.

Step 2. Find $x^{\nu+1}$ to minimize $c^T(x^\nu + \lambda(\hat{x}^\nu - x^\nu)) + \sum_{i=1}^{m_2} \Psi_i(T(x^\nu + \lambda(\hat{x}^\nu - x^\nu)))$ over $0 \leq \lambda \leq 1$. If $x^{\nu+1} = x^\nu$, stop with an optimal solution. Otherwise, let $\nu = \nu + 1$ and return to Step 1.

The basis for this approach is that x^* is optimal in (4.1) if and only if x^* solves (4.3) with $x^* = x^\nu$. If x^ν is not a solution of (4.1), then $x^{\nu+1} \neq x^\nu$, and descent occurs along $\hat{x}^\nu - x^\nu$. Exercise 1 asks for the details of this convergence result.

The L-shaped method and generalized linear programming can be considered extensions of the linearization approach that use multiple points of linearization. We have already considered the L-shaped method in some detail in the previous chapter. For generalized programming, the key advantage is that $\Psi(\chi)$ is separable. Williams [1966] and Beale [1961] observed the advantage of this property and gave generalized programming procedures for specific problems. In the case of the general problem in (4.1), the master problem of (3.5.6)–(3.5.9) becomes

$$\min \; z^\nu = c^T x + \sum_{j=1}^{m_2}(\sum_{i=1}^{r_j}\mu_{ji}\Psi 0_j^+(\zeta_{ji}) + \sum_{i=1}^{s_j}\lambda_{ji}\Psi_j(\chi_{ji})) \qquad (4.4)$$
$$\text{s.t. } Ax = b, \qquad (4.5)$$
$$T_i.x - \sum_{i=1}^{r_j}\mu_{ji}\zeta_{ji} - \sum_{i=1}^{s_j}\lambda_{ji}\chi_{ji} = 0, j = 1,\ldots,m_2, \qquad (4.6)$$
$$\sum_{i=1}^{s_j}\lambda_{ji} = 1, \qquad (4.7)$$
$$x, \mu_{ji} \geq 0, i = 1,\ldots,r_j; \lambda_{ji} \geq 0, i = 1,\ldots,s_j, j = 1,\ldots,m_2,$$

where we can divide the components of χ in the constraints because of the separability.

We then have a subproblem of the form in (3.5.10) for each j:

$$\min_{\chi_j} \Psi_j(\chi_j) + \pi_j^\nu \chi_j - \rho_j^\nu. \qquad (4.8)$$

We can create an entering column whenever any of the values in (4.8) is negative. If all are non-negative, then the algorithm again terminates with an optimal value.

Example 4

As an example of generalized programming, suppose the following situation. We have \$400 to buy boxes of blueberries (\$5 per box) and cherries (\$7 per box) from a farmer. We take the berries to the town market where we hope to sell them (\$11 per blueberry box and \$15 per cherry box). Any unsold berries at the end of the market day can be sold to a local baker (\$3 per blueberry box and \$5 per cherry box).

The demand for berries is stochastic. We assume that blueberry demand during market hours is uniformly distributed between 10 and 30 boxes and that cherry demand is uniformly distributed between 20 and 40 boxes. In the simple recourse problem, the correlation between these demands does not affect the recourse function value, so we only need this marginal information.

The initial decisions are x_1, the number of boxes of blueberries to buy, and x_2, the number of boxes of cherries to buy. The full problem is then to find x^*, χ^* to

$$
\begin{aligned}
\min\ z = {}& 2x_1 + 2x_2 + \Psi_1(\chi_1) + \Psi_2(\chi_2) && (4.9) \\
\text{s.t. } & 5x_1 + 7x_2 \le 400, \\
& x_1 - \chi_1 = 0, \\
& x_2 - \chi_2 = 0, \\
& x_1, x_2 \ge 0,
\end{aligned}
$$

where

$$
\Psi_1(\chi_1) = \begin{cases} -8\chi_1 & \text{if } \chi_1 \le 10, \\ \frac{1}{5}\chi_1^2 - 12\chi_1 + 20 & \text{if } 10 \le \chi_1 \le 30, \\ -160 & \text{if } \chi_1 \ge 30, \end{cases}
$$

$$
\nabla\Psi_1(\chi_1) = \begin{cases} -8 & \text{if } \chi_1 \le 10, \\ \frac{2}{5}\chi_1 - 12 & \text{if } 10 \le \chi_1 \le 30, \\ 0 & \text{if } \chi_1 \ge 30, \end{cases}
$$

$$
\Psi_2(\chi_2) = \begin{cases} -10\chi_2 & \text{if } \chi_2 \le 20, \\ \frac{1}{4}\chi_2^2 - 20\chi_1 + 100 & \text{if } 20 \le \chi_2 \le 40, \\ -300 & \text{if } \chi_2 \ge 40, \end{cases}
$$

and

$$
\nabla\Psi_2(\chi_2) = \begin{cases} -10 & \text{if } \chi_2 \le 20, \\ \frac{1}{2}\chi_2 - 20 & \text{if } 20 \le \chi_2 \le 40, \\ 0 & \text{if } \chi_2 \ge 40. \end{cases}
$$

The generalized programming method follows these iterations.

Iteration 0:

Step 0. We start with (4.4)–(4.7) with $\nu = r^j = s^j = 0$.

Step 1. The obvious solution is $x^0 = (0,0)^T$ with multipliers, $\pi^0 = \rho^0 = (0,0)^T$.

Step 2. Setting $\pi_i^0 = -\nabla\Psi_i(\chi_{11})$, we obtain $\chi_{11} = 30$ and $\chi_{21} = 40$ with $\Psi_1(\chi_{11}) = -160$ and $\Psi_2(\chi_{21}) = -300$ and clearly $\Psi_j(\chi_{j,s_j+1}) + \pi_j^\nu \chi_{j,s_j+1} - \rho_j^\nu < 0$ for each $j = 1,2$. Now, $s_1 = s_2 = 1$, $\nu = 1$ and we repeat.

Iteration 1:

Step 1. We assume that we can dispose of berries (to avoid creating an infeasibility in (4.4)–(4.7)). The master problem then has the form:

$$\min z = 2x_1 + 2x_2 - 160\lambda_{11} - 300\lambda_{21} \tag{4.10}$$
$$\text{s.t. } 5x_1 + 7x_2 \le 400,$$
$$x_1 - 30\lambda_{11} \ge 0,$$
$$x_2 - 40\lambda_{21} \ge 0,$$
$$\lambda_{11} = 1,$$
$$\lambda_{21} = 1,$$
$$x_1, x_2, \lambda_{11}, \lambda_{21} \ge 0.$$

The solution is $z^1 = -300$, $x^1 = (24,40)^T$, $\lambda_{11} = 0.8, \lambda_{21} = 1.0, \pi^1 = (5.333, 6.667)^T$ and $\rho^1 = (0, -33.333)^T$.

Step 2. Setting $\pi_i^0 = -\nabla\Psi_i(\chi_{11})$, we obtain $\chi_{12} = 16.667$ and $\chi_{22} = 26.667$ with $\Psi_1(\chi_{11}) = -124.4$ and $\Psi_2(\chi_{22}) = -255.55$. Again, $\Psi_j(\chi_{j,s_j+1}) + \pi_j^\nu \chi_{j,s_j+1} - \rho_j^\nu < 0$ for each $j = 1,2$ with $\Psi(\chi_{12}) + \pi_1^1 \chi_{12} - \rho_1^1 = -35.5$ and $\Psi(\chi_{22}) + \pi_2^1 \chi_{22} - \rho_2^1 = -44.4$. Now, $s_1 = s_2 = 2$, $\nu = 2$.

Iteration 2:

Step 1. The new master problem is:

$$\min z = 2x_1 + 2x_2 - 160\lambda_{11} - 124.4\lambda_{12}$$
$$-300\lambda_{21} - 255.55\lambda_{22} \tag{4.11}$$
$$\text{s.t. } 5x_1 + 7x_2 \le 400,$$
$$x_1 - 30\lambda_{11} - 16.667\lambda_{12} \ge 0,$$
$$x_2 - 40\lambda_{21} - 26.667\lambda_{22} \ge 0,$$
$$\lambda_{11} + \lambda_{12} = 1,$$
$$\lambda_{21} + \lambda_{22} = 1,$$
$$x_1, x_2, \lambda_{11}, \lambda_{12}, \lambda_{21}, \lambda_{22} \ge 0.$$

The solution is $z^2 = -316.0$, $x^2 = (24,40)^T$, $\lambda_{11}^2 = 0.55, \lambda_{12}^2 = 0.45, \lambda_{21}^2 = 1.0$, $\pi^2 = (2.667, 2.934)^T$ and $\rho^2 = (-80.0, -182.6)^T$.

Step 2. Setting $\pi_i^2 = -\nabla\Psi_i(\chi_{i,s_i+1})$, we obtain $\chi_{13} = 23.33$ and $\chi_{23} = 34.13$ with $\Psi_1(\chi_{13}) = -151.1$ and $\Psi_2(\chi_{23}) = -291.4$. Here, $\Psi_1(\chi_{13}) + \pi_1^2 \chi_{13} - \rho_1^2 = -8.88$ and $\Psi_2(\chi_{23}) + \pi_2^2 \chi_{23} - \rho_2^2 = -8.61$. Now, $s_1 = s_2 = 3$, $\nu = 3$.

Iteration 3:

Step 1. The new master problem is:

$$\min z = 2x_1 + 2x_2 - 160\lambda_{11} - 124.4\lambda_{12} - 151.1\lambda_{13}$$
$$\quad\quad -300\lambda_{21} - 255.55\lambda_{22} - 291.4\lambda_{23} \quad\quad (4.12)$$
$$\text{s.t. } 5x_1 + 7x_2 \le 400,$$
$$x_1 - 30\lambda_{11} - 16.667\lambda_{12} - 23.333\lambda_{13} \ge 0,$$
$$x_2 - 40\lambda_{21} - 26.667\lambda_{22} - 34.133\lambda_{23} \ge 0,$$
$$\lambda_{11} + \lambda_{12} + \lambda_{13} = 1,$$
$$\lambda_{21} + \lambda_{22} + \lambda_{23} = 1,$$
$$x_1, x_2, \lambda_{ij} \ge 0.$$

The solution is $z^3 = -327.57$, $x^3 = (23.333, 34.133)^T$, $\lambda_{13}^3 = 1.00$, $\lambda_{23}^3 = 1.0$, $\pi^3 = (2.0, 2.0)^T$ and $\rho^3 = (-104.44, -223.13)^T$.

Step 2. Setting $\pi_i^3 = -\nabla\Psi_i(\chi_{i,s_i+1})$, we obtain $\chi_{14} = 25$ and $\chi_{24} = 36$ with $\Psi_1(\chi_{14}) = -155$ and $\Psi_2(\chi_{24}) = -296$. Here, $\Psi_1(\chi_{14}) + \pi_1^3\chi_{14} - \rho_1^3 = -0.56$ and $\Psi_2(\chi_{24}) + \pi_2^3\chi_{24} - \rho_2^3 = -0.87$. Now, $s_1 = s_2 = 4$, $\nu = 3$.

Iteration 4:

Step 1. We add λ_{14} and λ_{24} with their objective and constraint entries to (4.12) to obtain the same form of the master problem. The solution is now $z^4 = -329$, $x^4 = (25, 36)^T$, $\lambda_{14}^4 = 1.00$, $\lambda_{24}^4 = 1.0$, $\pi^4 = (2.0, 2.0)^T$ and $\rho^4 = (-105, -224)^T$.

Step 2. Because $\pi^4 = \pi^3$, we obtain $\chi_{i5} = \chi_{i4}$, and $\Psi_i(\chi_{i5}) + \pi_i^4\chi_{i5} - \rho_i^4 = 0$ for $i = 1, 2$. Hence, no columns can be added. We stop with the optimal solution, $x^* = (25, 36)^T$ with objective value $z^* = 329$.

Notice that in this example the budget constraint is not binding. We only spend \$377 of the total possible, \$400. If we had solved this problem as separate news vendor problems in each type of berry, we would have obtained the same solution. In fact, this is one of the suggestions for initial tenders to start the generalized programming process (see Birge and Wets [1984] and Nazareth and Wets [1986]). In this case, we would terminate on the first step with this initial offer.

Notice also that the algorithm appears to converge quite quickly here. In general, the retention of information about gradients at many points should improve convergence over techniques that use only local information. Second-order information is also valuable, assuming twice differentiable functions. This is the motivation behind Beale's [1961] approach of quadratic approximation. His method is another form of the generalized programming approach for convex separable functions.

The other procedures specifically used on the simple recourse problem concern some form of active set or simplex based strategy. Wets [1966] and

Ziemba [1970] give the basic reduced gradient or convex simplex method procedure. This method consists of computing a search direction corresponding to a change in the value of a nonbasic variable (assuming only basic variables change concomitantly). The basis is changed if the line search implies that basic variable becomes zero. Otherwise, the nonbasic variable's value is updated and other nonbasic variables are checked for possible descent.

A different approach is given by Qi [1986], who suggests alternating between the solution of a linear program with χ fixed and the solution of a reduced variable convex program. The linear program is to find

$$\min_x c^T x + \Psi(\chi^\nu)$$
$$\text{s. t. } Ax = b,$$
$$Tx = \chi^\nu,$$
$$x \geq 0, \tag{4.13}$$

to obtain $x^{\nu+1} = (x_B^\nu, x_N^\nu)$, where $x_N^\nu = 0$. Then solve the reduced convex program:

$$\min_{x,\chi} c^T x + \Psi(\chi)$$
$$\text{s. t. } Ax = b,$$
$$Tx = \chi,$$
$$x_B \geq 0, x_N = 0 \tag{4.14}$$

to obtain $\hat{x}^{\nu+1}, \chi^{\nu+1}$. The algorithm is the following.

Alternating Algorithm for Simple Recourse Problems

Step 0. Let $\nu = 0$, choose a feasible solution x^0 to (4.13) and let χ^0 be part of a solution to (4.14) with N defined according to x^0. Go to Step 1.

Step 1. Solve (4.13). Let $X^{\nu+1} = \{x \text{ optimal in (4.13) }\}$. Choose $x^{\nu+1} \in X^{\nu+1}$ such that $c^T x^{\nu+1} + \Psi(Tx^{\nu+1}) < c^T x^\nu + \Psi(Tx^\nu)$. If none exists, then stop. Otherwise, go to Step 2.

Step 2. Solve (4.14) with N defined for $x^{\nu+1}$ to obtain $\chi^{\nu+1}$. Let $\nu = \nu + 1$ and return to 1.

The algorithm converges to an optimal solution because $x^{\nu+1}$ can always be found with $c^T x^{\nu+1} + \Psi(Tx^{\nu+1}) < c^T x^\nu + \Psi(Tx^\nu)$ whenever x^ν is not optimal (Exercise 5). Of course, the algorithm's advantage is when the number of first-period variables n_1 is much greater than the number of second-period random variables m_2, so that problem (4.14) does represent a computation savings over solving (4.1) directly.

This algorithm (and indeed the convex simplex method) raises the possibility for multiple optima of the linear program (degeneracy). In this case, many solutions may be searched before improvement is found. In tests of partitioning in discretely distributed general stochastic linear programming problems (Birge [1985b]), this problem was found to overcome computational advantages of reducing the working problem size. The approach has, therefore, not been followed extensively in practice although it may, of course, offer efficient computation on some problems.

Other methods for simple recourse have built on the special structure. For transportation constraints, Qi [1985] gives a method based on using the forest structure of the basis to obtain a search direction and improved forest solution. This method only requires the solution of one-dimensional monotone equations apart from standard tree solutions. Piecewise linear techniques as in Sun, Qi, and Tsai [1990] can also be adapted here to general network structures and used in conjunction with Qi's forest procedure to produce a convergent algorithm.

Exercises

1. Show that the Frank-Wolfe method for the simplex recourse problem converges to an optimal solution (assuming that one exists).

2. Solve the example in (4.9) using the L-shaped method.

3. Solve the example in (4.9) using the Frank-Wolfe method.

4. In the general stochastic linear programming model (with fixed T, (3.1.5)), show that solving (4.13) with $\chi^\nu = \chi^*$ yields an optimal solution x^*. Use this to show that there always exists a solution to (3.1.5) with at most $m_1 + m_2$ nonzero variables (Murty [1968]). What does this imply for retaining cuts in the L-shaped method?

5. Show that the alternating algorithm for simple recourse problems converges to an optimal solution assuming that the support of $\mathbf{h}$ is compact. (Hint:From any x^ν, consider a path to x^*, use the convexity of Ψ, and consider the solution as x^ν is approached from x^*.)

6.5 Other Nonlinear Programming–Based Methods

In the previous sections, we considered cutting plane methods and Lagrangian methods for problems with discrete random variables and simple recourse-based techniques for problems with continuous random variables. Other nonlinear programming procedures can also be applied to stochastic programs, although these other procedures have not received as much attention in stochastic programming problems. A notable exception is Noël and Smeers' [1987] multistage combined inner linearization and augmented

Lagrangian procedure, which we will describe in more detail in the next chapter.

The difficulty with discrete random variables is that Ψ or Q generally loses differentiability. In this case, derivative-based methods cannot apply. As we saw, the L-shaped method and other cutting plane approaches are a standard approach that requires only subgradient information. We also saw that augmented Lagrangian techniques can smooth nondifferentiable functions.

Explicit nondifferentiable methods include the nonmonotonic reduced subgradient procedure considered by Ermoliev [1983]. Another possibility is to use bundles of subgradients as in Lemaréchal [1978] and Kiwiel [1983]. In fact, results by Plambeck et al. [1996] show good performance for bundle methods in practical stochastic programs.

Nonsmooth generalizations of the Frank-Wolfe procedure are also possible. These and other options are described in detail in Demyanov and Vasiliev [1981].

With general continuous random variables or with large numbers of discrete random vector realizations, direct nonlinear programming procedures generally break down because of difficulties in evaluating function and derivative values. In these cases, one must rely on approximation. These approximations either take the form of bounds on the actual function values or are in some sense statistical estimates of the actual function values. We present these approaches in Chapters 9 to 11.

7
Multistage Stochastic Programs

As the Chapter 1 examples demonstrate, many operational and planning problems involve sequences of decisions over time. The decisions can respond to realizations of outcomes that are not known a priori. The resulting model for optimal decision making is then a multistage stochastic program. In Section 3.5, we gave some of the basic properties of multistage problems. In this chapter, we explore the variety of solution procedures that have been proposed specifically for multistage stochastic programs.

In general, the methods for two-stage problems generalize to the multistage case but include additional complications. Because of these difficulties, we will describe only those methods that have shown some success in implementations.

As stated in Section 3.5, the multistage stochastic linear program with a finite number of possible future scenarios still has a deterministic equivalent linear program. However, as the graph in Figure 3.4 began to suggest, the structure of this problem is somewhat more complex than that of the two-stage problem. The extensive form does not appear readily accessible to manipulations such as the factorizations for extreme or interior point methods that were described in Chapter 5. The overhead for these procedures appears difficult to overcome (see Birge [1980] for a discussion of multistage basis factorization and its requirements).

The methods that appear most promising are again based on decompositions, some form of Lagrangian relaxation, and uses of separability. In Section 1, we describe the basic nested decomposition procedures for multistage stochastic linear programs. Section 2 describes approaches for multistage nonlinear problems again based on nested decomposition and the progressive hedging algorithm or Lagrangian approach. Section 3 considers the use of block separability in these problems and describes the special case of simple recourse or full separability.

7.1 Nested Decomposition Procedures

Nested decomposition procedures were proposed for deterministic models by Ho and Manne [1974] and Glassey [1973]. These approaches are essentially inner linearizations that treat all previous periods as subproblems to a current period master problem. The previous periods generate columns that can be used by the current-period master problem.

A difficulty with these primal nested decomposition or inner linearization methods is that the set of inputs may be fundamentally different for different last period realizations. Because the number of last period realizations is the total number of scenarios in the problem, these procedures are not well adapted to the bunching procedures described in Section 5.4. Some success has been achieved, however, by Noël and Smeers [1987], as we will describe, by applying inner linearization to the dual, which is again outer linearization of the primal problem.

The general primal approach is, therefore, to use an outer linearization built on the two-stage L-shaped method. Louveaux [1980] first performed this generalization for multistage quadratic problems, as we discuss in Section 2. Birge [1985b] extended the two-stage method in the linear case as in the following description. The approach also appears in Pereira and Pinto [1985].

The basic idea of the nested L-shaped or Benders decomposition method is to place cuts on $Q^{t+1}(x^t)$ in (3.5.3) and to add other cuts to achieve an x^t that has a feasible completion in all descendant scenarios. The cuts represent successive linear approximations of Q^{t+1}. Due to the polyhedral structure of Q^{t+1}, this process converges to an optimal solution in a finite number of steps.

In general, for every stage $t = 1, \ldots, H-1$ and each scenario at that stage, $k = 1, \ldots, N^t$, we have the following master problem, which generates cuts to stage t and proposals for stage $t + 1$:

$$\min (c_k^t)^T x_k^t + \theta_k^t \tag{1.1}$$
$$\text{s. t. } W^t x_k^t = h_k^t - T_k^{t-1} x_{a(k)}^{t-1}, \tag{1.2}$$
$$D_{k,j}^t x_k^t \geq d_{k,j}^t, j = 1, \ldots, r_{k,j}^t, \tag{1.3}$$
$$E_{k,j}^t x_k^t + \theta_k^t \geq e_{k,j}^t, j = 1, \ldots, s_{k,j}^t, \tag{1.4}$$
$$x_k^t \geq 0, \tag{1.5}$$

where $a(k)$ is the ancestor scenario of k at stage $t - 1$, $x_{a(k)}^{t-1}$ is the current solution from that scenario, and where for $t = 1$, we interpret $b = h^1 - T^0 x^0$

as initial conditions of the problem. We may refer also to the stage H problem in which θ_k^H and constraints (1.3) and (1.4) are not present. To designate the period and scenario of the problem in (1.1)–(1.5), we also denote this subproblem, $NLDS(t,k)$.

We first describe a basic algorithm for iterating among these stages. We then discuss some enhancements of this basic approach. In the following, $\mathcal{D}^t(j)$, denotes the period t descendants of a scenario j at period $t-1$. We assume that all variables in (3.5.1) have finite upper bounds to avoid complications presented by unbounded solutions (although, again, these can be treated as in Van Slyke and Wets [1969]).

Nested L-Shaped Method for Multistage Stochastic Linear Programs

Step 0. Set $t = 1$, $k = 1$, $r_k^t = s_k^t = 0$, add the constraint $\theta_k^t = 0$ to (1.1)–(1.5) for all t and k, and let $DIR = FORE$. Go to Step 1.

Step 1. Solve the current problem, $NLDS(t,k)$. If infeasible and $t = 1$, then stop; problem (3.5.1) is infeasible. If infeasible and $t > 1$, then let $r_{a(k)}^{t-1} = r_{a(k)}^{t-1} + 1$, let $DIR = BACK$. Let the infeasibility condition be obtained by a dual basic solution, $\pi_k^t, \rho_k^t \geq 0$, such that $(\pi_k^t)^T W^t + (\rho_k^t)^T D_k^t \leq 0$ but $(\pi_k^t)^T(h_k^t - T_k^{t-1}x_{a(k)}^{t-1}) + (\rho_k^t)^T d_k^t > 0$. Let $D_{a(k),r_{a(k)}^{t-1}}^{t-1} = (\pi_k^t)^T T_k^{t-1}$, $d_{a(k),r_{a(k)}^{t-1}}^{t-1} = \pi_k^t h_k^t + (\rho_k^t)^T d_k^t$. Let $t = t - 1$, $k = a(k)$ and return to Step 1.

If feasible, update the values of x_k^t, θ_k^t, and store the value of the complementary basic dual multipliers on constraints (1.2)–(1.4) as $(\pi_k^t, \rho_k^t, \sigma_k^t)$, respectively. If $k < K^t$, let $k = k + 1$, and return to (1.1). Otherwise, ($k = K^t$), if $DIR = FORE$ and $t < H$, let $t = t + 1$ and return. If $t = H$, let $DIR = BACK$. Go to Step 2.

Step 2. For all scenarios $j = 1, \ldots, K^{t-1}$ at $t - 1$, compute

$$E_j^{t-1} = \sum_{k \in \mathcal{D}^t(j)} \frac{p_k^t}{p_j^{t-1}} (\pi_k^t)^T T_k^{t-1}$$

and

$$e_j^{t-1} = \sum_{k \in \mathcal{D}^t(j)} \frac{p_k^t}{p_j^{t-1}} [(\pi_k^t)^T h_k^t + \sum_{i=1}^{r_k} (\rho_{ki}^t)^T d_{ki}^t + \sum_{i=1}^{s_k} (\sigma_{ki}^t)^T e_{ki}^t].$$

The current conditional expected value of all scenario problems in $\mathcal{D}^t(j)$ is then $\bar{\theta}_j^{t-1} = e_j^{t-1} - E_j^{t-1}x_j^{t-1}$. If the constraint $\theta_j^{t-1} = 0$ appears in $NLDS(t-1,j)$, then remove it, let $s_j^{t-1} = 1$, and add a constraint (1.4) with E_j^{t-1} and e_j^{t-1} to $NLDS(t-1,j)$.

If $\bar{\theta}_j^{t-1} > \theta_j^{t-1}$, then let $s_j^{t-1} = s_j^{t-1} + 1$ and add a constraint (1.4) with E_j^{t-1} and e_j^{t-1} to $NLDS(t-1, j)$. If $t = 2$ and no constraints are added to $NLDS(1)$ $(j = K^1 = 1)$, then stop with x_1^1 optimal. Otherwise, let $t = t - 1$, $k = 1$. If $t = 1$, let $DIR = FORE$. Go to Step 1.

Many alternative strategies are possible in this algorithm in terms of determining the next subproblem (1.1)–(1.5) to solve. For feasible solutions, the preceding description explores all scenarios at t before deciding to move to $t - 1$ or $t + 1$. For feasible iterations, the algorithm proceeds from t in the direction of DIR until it can proceed no further in that direction. This is the "fast-forward-fast-back" procedure proposed by Wittrock [1983] for deterministic problems and implemented with success by Gassmann [1990] for stochastic problems. One may alternatively enforce a move from t to $t - 1$ ("fast-back") or from t to $t + 1$ ("fast-forward") whenever it is possible. From experiments conducted by Gassmann [1990], the fast-forward-fast-back sequencing protocol seems to work better than either of these alternatives. Morton [1994] compares with other alternative protocols and reaches much the same conclusion.

For infeasible solutions at some stage, this algorithm immediately returns to the ancestor problem to see whether a feasible solution can be generated. This alternative appears practical because subsequent iterations with a currently infeasible solution do not seem worthwhile.

We note that much of this algorithm can also run in parallel. We refer to Ruszczyński [1993a] who describes parallel procedures in detail. Again, one should pay attention in parallel implementations to the possible additional work for solving similar subproblems as we mentioned in Chapter 5. The convergence of this method is relatively straightforward, as given in the following.

Theorem 1. *If all Ξ^t are finite and all x^t have finite upper bounds, then the nested L-shaped method converges finitely to an optimal solution of (3.5.1).*

Proof: First, we wish to demonstrate that all cuts generated by the algorithm are valid outer linearizations of the feasible regions and objectives in (3.5.3). By induction on t, suppose that all feasible cuts (1.3) generated by the algorithm for periods t or greater are valid. For $t = H$, no cuts are present so this is true for the last period. In this case, for any $\pi_k^t, \rho_k^t \geq 0$ such that $(\pi_k^t)^T W^t + (\rho_k^t)^T D_k^t \leq 0$, we must have $(\pi_k^t)^T(h_k^t - T_k^{t-1}x_{a(k)}^{t-1}) + (\rho_k^t)^T d_k^t \leq 0$ to maintain feasibility. Because this is the cut added, these cuts are valid for $t - 1$. Thus, the induction is proved.

Now, suppose the cuts in (1.1, 1.4) are an outer linearization of $Q_k^{t+1}(x_k^t)$ for t or greater and all k. In this case, for any $(\pi_k^t, \rho_k^t, \sigma_k^t)$ feasible in (1.1)–(1.5) for t and k, $(\pi_k^t)^T(h_k^t - T_k^t x_{a(k)}^{t-1}) + \sum_{i=1}^{r_k}(\rho_{ki}^t)^T d_{ki}^t + \sum_{i=1}^{s_k}(\sigma_{ki}^t)^T e_{ki}^t$ is a lower bound on $Q_{a(k)}^t(x_{a(k)}^{t-1}, k)$ for any $x_{a(k)}^{t-1}$, each k, and $a(k)$. Thus, we must have

$$Q_{a(k)}^t(x_{a(k)}^{t-1}) \geq \sum_{k \in \mathcal{D}^t(a(k))} (\frac{p_k^t}{p_{a(k)}^{t-1}})((\pi_k^t)^T(h_k^t - T_k^t x_{a(k)}^{t-1})$$
$$+ \sum_{i=1}^{r_k} (\rho_{ki}^t)^T d_{ki}^t + \sum_{i=1}^{s_k} (\sigma_{ki}^t)^T e_{ki}^t), \tag{1.6}$$

which says that $\theta_k^{t-1} \geq -E_{a(k)}^{t-1} x_{a(k)}^{t-1} + e_{a(k)}^{t-1}$, as found in the algorithm. Thus, again, we achieve a valid cut on $Q_{a(k)}^{t-1}$ for any $a(k)$, completing the induction.

Now, suppose that the algorithm terminates. This can only happen if (1.1)–(1.5) is infeasible for $t = 1$ or if each subproblem for $t = 2$ has been solved and no cuts are generated. In the former case, the problem is infeasible, because the cuts (1.1,1.3) are all outer linearizations of the feasible region. In the latter case, we must have $\theta^1 = Q^2(x^1)$, the condition for optimality.

For finiteness, proceed by induction. Suppose that at stage t, at most a finite number of cuts from stage $t + 1$ to H can be generated for each k at t. For H, this is again trivially true. Because at most a finite number of cuts are possible at each k, at most a finite number of basic solutions, $(\pi_k^t, \rho_k^t, \sigma_k^t)$, can be generated to form cuts for $a(k)$. Thus, at most a finite number cuts can be generated for all $a(k)$ at $t - 1$, again completing the induction.

The proof is complete by noting that every iteration of Step 1 or 2 produces a new cut. Because there is only a finite number of possible cuts, the procedure stops finitely. $\Box$

The nested L-Shaped method has many features in common with the standard two-stage L-shaped algorithm. There are, however, peculiarities about the multistage method. We consider the following example in some detail to illustrate these features. In particular, we should note that the two-stage method always produces cuts that are supports of the function Q if the subproblem is solved to optimality. In the multistage case, with the sequencing protocol just given, we may not actually generate a true support so that the cut may lie strictly below the function being approximated.

Example 1

Suppose we are planning production of air conditioners over a three month period. In each month, we can produce 200 air conditioners at a cost of $100 each. We may also use overtime workers to produce additional air conditioners if demand is heavy, but the cost is then $300 per unit. We have a one-month lead time with our customers, so that we know that in Month

1, we should meet a demand of 100. Orders for Months 2 and 3 are, however, random, depending heavily on relatively unpredictable weather patterns. We assume this gives an equal likelihood in each month of generating orders for 100 or 300 units.

We can store units from one month for delivery in a subsequent month, but we assume a cost of \$50 per unit per month for storage. We assume also that all demand must be met. Our overall objective is to minimize the expected cost of meeting demand over the next three months. (We assume that the season ends at that point and that we have no salvage value or disposal cost for any leftover items. This resolves the end-of-horizon problem here.)

Let x_k^t be the regular-time production in scenario k at month t, let y_k^t be the number of units stored from scenario k at month t, let w_k^t be the overtime production in scenario k at month t, and let d_k^t be the demand for month t under scenario k. The multistage stochastic program in deterministic equivalent form is:

$$\min x^1 + 3.0w^1 + 0.5y^1 + \sum_{k=1}^{2} p_k^2 (x_k^2 + 3.0w_k^2 + 0.5y_k^2)$$

$$+ \sum_{k=1}^{4} p_k^3 (x_k^3 + 3.0w_k^3)$$

$$\text{s. t. } x^1 \le 2,$$
$$x^1 + w^1 - y^1 = 1,$$
$$y^1 + x_k^2 + w_k^2 - y_k^2 = d_k^2, \qquad (1.7)$$
$$x_k^2 \le 2, k = 1, 2,$$
$$y_{a(k)}^2 + x_k^3 + w_k^3 - y_k^3 = d_k^3,$$
$$x_k^3 \le 2, k = 1, \dots, 4,$$
$$x_k^t, y_k^t, w_k^t \ge 0, k = 1, \dots, K^t, t = 1, 2, 3,$$

where $a(k) = 1$, if $k = 1, 2$ at period 3, $a(k) = 2$ if $k = 3, 4$ at period 3, $p_k^2 = 0.5$, $k = 1, 2$, $p_k^3 = 0.25$, $k = 1, \dots, 4$, $d_1^2 = 1$, $d_2^2 = 3$, and $d^3 = (1, 3, 1, 3)^T$.

The nested L-shaped method applied to (1.7) follows these steps for the first two iterations. We list an iteration at each change of DIR.

Step 0. All subproblems $NLDS(t, k)$ have the explicit $\theta_k^t = 0$ constraint. $DIR = FORE$.

Iteration 1:

Step 1. Here $t = 1$, $k = 1$. The subproblem $NLDS(1,1)$ is:

$$\min x^1 + 3w^1 + 0.5y^1 + \theta^1$$
$$\text{s. t. } x^1 \leq 2,$$
$$x^1 + w^1 - y^1 = 1, \tag{1.8}$$
$$x^1, w^1, y^1 \geq 0,$$
$$\theta^1 = 0,$$

which has the solution $x^1 = 1$; other variables are zero.

Step 1. Now, $t = 2$, $k = 1$, and $NLDS(2,1)$ is

$$\min x_1^2 + 3w_1^2 + 0.5y_1^2 + \theta_1^2$$
$$\text{s. t. } x_1^2 \leq 2,$$
$$x_1^2 + w_1^2 - y_1^2 = 1, \tag{1.9}$$
$$x_1^2, w_1^2, y_1^2 \geq 0,$$
$$\theta_1^2 = 0,$$

which has the solution, $x_1^2 = 1$; other variables are zero.

Step 1. Here, $t = 2$, $k = 2$, and $NLDS(2,2)$ is

$$\min x_2^2 + 3w_2^2 + 0.5y_2^2 + \theta_2^2$$
$$\text{s. t. } x_2^2 \leq 2,$$
$$x_2^2 + w_2^2 - y_2^2 = 3, \tag{1.10}$$
$$x_2^2, w_2^2, y_2^2 \geq 0,$$
$$\theta_2^2 = 0,$$

which has the solution, $x_2^2 = 2$, $w_2^2 = 1$; other variables are zero.

Step 1. Next, $t = 3$, $k = 1$. $NLDS(3,1)$ is

$$\min x_1^3 + 3w_1^3 + 0.5y_1^3 + \theta_1^3$$
$$\text{s. t. } x_1^3 \leq 2,$$
$$x_1^3 + w_1^3 - y_1^3 = 1, \tag{1.11}$$
$$x_1^3, w_1^3, y_1^3 \geq 0,$$
$$\theta_1^3 = 0,$$

which has the solution, $x_1^3 = 1$; other primal variables are zero. The complementary basic dual solution is $\pi_1^3 = (0,1)^T$.

Step 1. Next, $t = 3$, $k = 2$. $NLDS(3,2)$ has the same form as $NLDS(3,1)$, except we replace the second constraint with $x_2^3 + w_2^3 - y_2^3 = 3$. It has the solution, $x_2^3 = 2$, $w_2^3 = 1$; other primal variables are zero. The complementary basic dual solution is $\pi_2^3 = (-2,3)^T$.

Step 1. For $t = 3$, $k = 3$, we have the same subproblem and solution as $t = 3$, $k = 1$, so $x_3^3 = 1$; other primal variables are zero. The complementary basic dual solution is $\pi_3^3 = (0,1)^T$.

Step 1. For $t = 3$, $k = 4$, we have the same subproblem and solution as $t = 3$, $k = 2$, $x_4^3 = 2$, $w_4^3 = 1$; other primal variables are zero. The complementary basic dual solution is $\pi_4^3 = (-2,3)^T$. Now, $DIR = BACK$, and we go to Step 2.

Iteration 2:

Step 2. For scenario $j = 1$ and $t - 1 = 2$, we have

$$E_{11}^2 = (\frac{0.25}{0.5})(\pi_1^3 T_1^3 + \pi_2^3 T_2^3)$$
$$= (0.5)(0 \quad 1)\begin{pmatrix} 0 & 0 & 0 \\ 0 & 0 & 1 \end{pmatrix} + (0.5)(-2 \quad 3)\begin{pmatrix} 0 & 0 & 0 \\ 0 & 0 & 1 \end{pmatrix} \qquad (1.12)$$
$$= (0 \quad 0 \quad 2)$$

and

$$e_{11}^2 = (\frac{0.25}{0.5})(\pi_1^3 h_1^3 + \pi_2^3 h_2^3)$$
$$= (0.5)(0 \quad 1)\begin{pmatrix} 2 \\ 1 \end{pmatrix} + (0.5)(-2 \quad 3)\begin{pmatrix} 2 \\ 3 \end{pmatrix} \qquad (1.13)$$
$$= 3,$$

which yields the constraint, $2y_1^2 + \theta_1^2 \geq 5$, to add to $NLDS(2,1)$.
For scenario $j = 2$ at $t - 1 = 2$, we have the same, $E_{21}^2 = (0 \quad 0 \quad 2)$, $e_{21}^2 = 3$. Now $t = 2$ and $k = 1$.

Step 1. $NLDS(2,1)$ is now:

$$\min x_1^2 + 3w_1^2 + 0.5y_1^2 + \theta_1^2$$
$$\text{s. t. } x_1^2 \leq 2,$$
$$x_1^2 + w_1^2 - y_1^2 = 1, \qquad (1.14)$$
$$2y_1^2 + \theta_1^2 \geq 3,$$
$$x_1^2, w_1^2, y_1^2 \geq 0,$$

which has an optimal basic feasible solution, $x_1^2 = 2$, $y_1^2 = 1$, $\theta_1^2 = 1$, $w_1^2 = 0$, with complementary dual values, $\pi_1^2 = (-0.5, 1.5)^T$, $\sigma_{11}^2 = 1$.

Step 1. $NLDS(2,2)$ has the same form as (1.14) except that the demand constraint is $x_2^2 + w_2^2 - y_2^2 = 3$. The optimal basic feasible solution found to this problem is $x_2^2 = 2$, $w_2^2 = 1$, $\theta_2^2 = 3$, $y_2^2 = 0$, with complementary dual values, $\pi_2^2 = (-2, 3)^T$, $\sigma_{11}^2 = 1$. We continue in $DIR = BACK$ to Step 2.

Step 2. For scenario $t - 1 = 1$, we have

$$E_1^1 = (0.5)(\pi_1^2 T_1^2 + \pi_2^2 T_2^2)$$

$$= (0.5)\,(-0.5 \quad 1.5)\begin{pmatrix} 0 & 0 & 0 \\ 0 & 0 & 1 \end{pmatrix} + (0.5)\,(-2 \quad 3)\begin{pmatrix} 0 & 0 & 0 \\ 0 & 0 & 1 \end{pmatrix}$$

$$= (0 \quad 0 \quad 2.25)$$

and

$$(1.15)$$

$$e_1^1 = (0.5)(\pi_1^2 h_1^2 + \pi_2^2 h_2^2) + (0.5)(\sigma_{11}^2 e_{11}^2 + \sigma_{21}^2 e_{21}^2)$$

$$= (0.5)\,(-0.5 \quad 1.5)\begin{pmatrix} 2 \\ 1 \end{pmatrix} + (0.5)\,(-2 \quad 3)\begin{pmatrix} 2 \\ 3 \end{pmatrix} + (0.5)((1)(3) + (1)3)$$

$$= (0.5)(0.5 + 5 + 6) = 5.75,$$

$$(1.16)$$

which yields the constraint, $2.25y^1 + \theta^1 \geq 5.75$, to add to $NLDS(1)$.

Step 1. $NLDS(1)$ is now:

$$\min x^1 + 3w^1 + 0.5y^1 + \theta^1$$

$$\text{s. t. } x^1 \leq 2,$$

$$x^1 + w^1 - y^1 = 1, \qquad (1.17)$$

$$2.25y^1 + \theta^1 \geq 5.75,$$

$$x^1, w^1, y^1 \geq 0,$$

with optimal basis feasible solution, $x^1 = 2$, $y^2 = 1$, $w^1 = 0$, $\theta^1 = 3.5$. $DIR = FORE$.

This procedure continues through six total iterations to solve the problem. At the last iteration, we obtain $\bar{\theta}^1 = 3.75 = \theta^1$, so no new cuts are generated for Period 1. We stop with a current solution as optimal, $x^{1*} = 2$, $y^{1*} = 1$, $z^* = 2.5 + 3.75 = 6.25$. In Exercise 1, we ask the reader to generate each of the cuts.

Following the nested L-shaped method completely takes many steps in this example, six iterations or changes of direction corresponding to three forward passes and three backward passes. Figure 1 illustrates the process and provide some insight into nested decomposition performance.

In Figure 1, the solid line gives the objective value in (1.7) as a function of total production $prod^1 = x^1 + w^1$ in the first period. The dashed lines correspond to the cuts made by the algorithm (*Cut 1,2*). The first cut was $2.25y^1 + \theta \geq 5.75$ from (1.15)–(1.16) on Iteration 2. Because $y^1 = x^1 + w^1 - 1$, we can substitute for y^1 to obtain, $2.25x^1 + 2.25w^1 + \theta \geq 8$. The objective in (1.17) is $z^1 = x^1 + 3w^1 + 0.5y^1 + \theta$, so, combined with $1 \leq x^1 \leq 2$, we can substitute $\theta \geq 8 - 2.25(prod^1)$ to obtain $z^1(prod^1) = 7.5 + (1.5)\min\{2, prod^1\} + 3.5(prod^1 - 2)^+ - 2.25prod^1$, where $prod^1 \geq 1$. This can also be written as:

$$z^1(prod^1) = \begin{cases} 7.5 - 0.75prod^1 & \text{if } prod^1 \leq 2, \\ 3.5 + 1.25prod^1 & \text{if } prod^1 > 2, \end{cases} \tag{1.18}$$

which corresponds to the wide dashed line (*Cut 1*) in Figure 1.

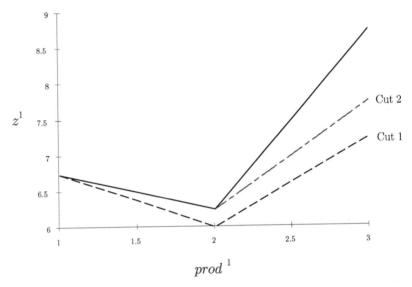

FIGURE 1. The first period objective function (solid line) for the example and cuts (dashed lines) generated by the nested L-shaped method.

The second cut occurs on Iteration 4 (verify this in Exercise 1) as $2x^1 + 2w^1 + \theta \geq 7.75$, which yields $z^1(prod^1) = x^1 + 3w^1 + 0.5y^2 + \theta \geq 7.25 + (1.5)\min\{2, prod^1\} + 3.5(prod^1 - 2)^+ - 2prod^1$ or

$$z^1(prod^1) \geq \begin{cases} 7.25 - 0.5prod^1 & \text{if } prod^1 \leq 2, \\ 3.25 + 1.5prod^1 & \text{if } prod^1 > 2. \end{cases} \tag{1.19}$$

This cut corresponds to the narrow width dashed line (*Cut 2*) in Figure 1.

The optimal value and solution in terms of $prod^1$ can be read from Figure 1 as each cut is added. With only Cut 1, the lowest value of z^1 occurs when $prod^1 = 2$. With Cuts 1 and 2, the minimum is also achieved at $prod^1 = 2$. Note that the first cut is not, however, a facet of the objective function's graph. The cuts meet the objective at $prod^1 = 1$ and $prod^1 = 2$, respectively, but they need not even do this, as we mentioned earlier (see Exercise 2). The other parts of the Period 1 cuts are generated from bounds on Q_2^2.

This example illustrates some of the features of the nested L-shaped method. Besides our not being guaranteed of obtaining a support of the function at each step, another possible source of delay in the algorithm's convergence is in degeneracy. As the example illustrates, the solutions at each step occur at the links of the piecewise linear pieces generated by the method (Exercises 4 and 5). At these places, many bases may be optimal so that several bases may be repeated. Some remedies are possible, as in Birge [1980] and, for deterministic problems, Abrahamson [1983].

As with the standard two-stage L-shaped method, the nested L-shaped method acquires its greatest gains by combining the solutions of many subproblems through bunching (or sifting). Birge [1985b] reported order of magnitude speed-ups over MINOS solutions of the deterministic equivalent problems. Gassmann [1990] reported even more significant speed-ups by taking advantage of the fast-forward-fast-back procedure and implementing bunching in an extremely efficient manner.

We note that multicuts may also be valuable here, although the full potential of this approach has not been explored in a computational study. Infanger [1991, 1994] has also suggested the uses of generating many cuts simultaneously when future scenarios all have similar structure. This procedure may make bunching efficient for periods other than H by making every constraint matrix identical for all scenarios in a given period. In this way, only objective and right-hand side constraint coefficients vary among the different scenarios.

In terms of primal decomposition, we mentioned the work of Noël and Smeers at the outset of this chapter. They apply nested Dantzig-Wolfe decomposition to the dual of the original problem. As we saw in Chapter 5, this is equivalent to applying outer linearization to the primal problem. The only difference is that they allow for some nonlinear terms in their constraints, which would correspond to a nonlinear objective in the primal model. Because the problems are still convex, nonlinearity does not really alter the algorithm. The only problem may be in the finiteness of convergence.

The advantage of a primal or dual implementation generally rests in the problem structure, although primal or dual simplex may be used in either method, making them indistinguishable. Gassmann presents some indication that dual iterations may be preferred in bunching. In general, many primal columns and few rows would tend to favor a primal approach (outer linearization as in the L-shaped method) while few columns and many rows would tend to favor a dual approach. In any case, the form of the algorithm and all proofs of convergence apply to either form.

Exercises

1. Continue Example 1 with the nested L-shaped method until you obtain an optimal solution.

2. Construct a multistage example in which a cut generated by the second period in following in the nested L-shaped method does not meet $Q^1(x^1)$ for any value of x^1, i.e., $-E_1^1 x^1 + e_1^1 < Q(x^1)$.

3. Show that the situation in (1.1) is not possible if the fast-forward protocol is always followed.

4. Suppose a feasibility cut (1.3) is active for x_k^t for any t and k. Show that every basic feasible solution of $NLDS(t+1,j)$ with input x_k^t for some scenario $j \in \mathcal{D}^{t+1}(k)$ must be degenerate.

5. Suppose two optimality cuts (1.4) are active for (x_k^t, θ_k^t) for any t and k. Show that either the subproblems generate a new cut with $\bar{\theta}_k^t > \theta_k^t$ or an optimal solution of $NLDS(t+1,j)$ with input x_k^t for some scenario $j \in \mathcal{D}^{t+1}(k)$ must be degenerate.

6. Using four processors, what efficiency can be gained by solving the preceding example in parallel? Find the utilization of each processor and the speed-up of elapsed time, assuming each subproblem requires the same solution time.

7. Suppose θ^1 is broken into separate components for Q_1^2 and Q_2^2 as in the two-stage multicut approach. How does that alter the solution of the example?

7.2 Quadratic Nested Decomposition

Decomposition techniques for multistage nonlinear programs are available for the case in which the objective function is quadratic convex, the constraint set polyhedral, and the random variables discrete. For the sake of clarity, we repeat the recursive definition of the deterministic equivalent program, already given in Section 3.5.

$$(MQSP) \quad \min \; z_1(x^1) = (c^1)^T \cdot x^1 + (x^1)^T D^1 x^1 + Q^2(x^1) \qquad (2.1)$$
$$\text{s.t. } W^1 x^1 = h^1$$
$$x^1 \geq 0,$$

where $Q^t(x^{t-1}, \xi^t(\omega)) =$

$$\min (c^t(\omega))^T x^t(\omega) + (x^t(\omega))^T D^t(\omega) x^t(\omega) + Q^{t+1}(x^{t+1})$$
$$\text{s.t. } W^t x^t(\omega) = h^t(\omega) - T^{t-1}(\omega) x^{t-1}$$
$$x^t(\omega) \geq 0, \qquad (2.2)$$

$$Q^{t+1}(x^t) = E_{\xi^{t+1}} Q^{t+1}(x^t, \xi^{t+1}(\omega)) \, , \ t = 1, \cdots, H-1, \quad (2.3)$$

and $\quad Q^H(x^{H-1}) = 0.$ $\hspace{9cm}$ (2.4)

In $MQSP$, D^t is an $n_t \times n_t$ matrix. All other matrices have the dimensions defined in the linear case. The random vector, $\xi^t(\omega)$, is formed by the elements of $c^t(\omega)$, $h^t(\omega)$, $T^{t-1}(\omega)$, and $D^t(\omega)$. We keep the notation that ξ^t is an N_t-vector on (Ω, A^t, P), with support Ξ^t. Finally, we again define

$$K^t = \{x^t | Q^{t+1}(x^t) < \infty\}.$$

We also define $z^t(x^t) = (c^t)^T x^t + (x^t)^T D^t x^t + Q^{t+1}(x^t)$.

Theorem 2. *If the matrices $D^t(\omega)$ are positive semi-definite for all $\omega \in \Omega$ and $t = 1, \cdots, H$, then the sets K^t and the functions $Q^{t+1}(x^t)$ are convex for $t = 1, \cdots, H-1$. If Ξ^t is also finite for $t = 2, \cdots, H$, then K^t is polyhedral. Moreover $z^t(x^t)$ is either identically $-\infty$ or there exists a decomposition of K^t into a polyhedral complex such that the tth-stage deterministic equivalent program (2.2) is a piecewise quadratic program.*

Proof. The piecewise quadratic property of (2.2) is obtained by inductively applying to each cell of the polyhedral complex of K^t the result that if $z^t(.)$ is a finite positive semi-definite quadratic form, there exists a piecewise affine continuous optimal decision rule for (2.2). All others results were given in Section 3.5. $\square$

We now describe a nested decomposition algorithm for $MQSP$ first presented in Louveaux [1980]. For simplicity in the presentation of the algorithms, we assume relatively complete recourse. This means that we skip the step that consists of generating feasibility cuts. If needed, those cuts are generated exactly as in the multistage linear case. We keep the notation of $a(k)$ for the ancestor scenario of k at stage $t - 1$. As in Section 7.1, c_k^t, D_k^t, and Q_k^{t+1} represent realizations of c^t, D^t, and Q^{t+1} for scenario k and x_k^t is the corresponding decision vector. In Stage 1, we use the notations, z_1 and z_1^1 and x_1 and x_1^1, as equivalent.

Nested PQP Algorithm for $MQSP$

Step 0. Set $t = 1, k = 1, C_1 = S_1 = K_1$. Choose $x_1^1 \in K_1$.

Step 1. If $t = H$, go to Step 2. For $i = t + 1, \cdots, H$, let $k = 1$, $z_1^i(x_1^i) = (c_1^i)^T x_1^i + (x_1^i)^T D_1^i x_1^i$ and $C_1^i(x_{a(1)}^{i-1}) = S_1^i(x_{a(1)}^{i-1}) = K_i(x_{a(1)}^{i-1})$. Choose $x_1^i \in K_i(x_{a(1)}^{i-1})$. Set $t = H$.

Step 2. Find $v \in \arg\min\{z_k^t(x_k^t) | x_k^t \in S_k^t(x_{a(k)}^{t-1})\}$. Find $w \in \arg\min\{z_k^t(x_k^t) \mid x_k^t \in C_k^t(x_{a(k)}^{t-1})\}$. If w is the limiting point on a ray on which $z_k^t(.)$ is decreasing to $-\infty$, then $(DEP)_k^t$ is unbounded and the algorithm terminates.

Step 3. If $\nabla^T z_k^t(w)(v - w) = 0$, go to Step 4. Otherwise, redefine

$$S_k^t(x_{a(k)}^{t-1}) \leftarrow S_k^t(x_{a(k)}^{t-1}) \cap \{x_k^t | \nabla^T z_k^t(w)(x_k^t - w) \le 0\}.$$

Let $x_k^t = v, z_k^t = (c_k^t)^T x_k^t + (x_k^t)^T D_k^t x_k^t$ and $C_k^t = K_t(x_{a(k)}^{t-1})$. Go to Step 1.

Step 4. If $t = 1$, stop; w is an optimal first-period decision. Otherwise, find the cell $G_k^t(x_{a(k)}^{t-1})$ containing w and the corresponding quadratic form $Q_k^t(x_{a(k)}^{t-1})$. Redefine

$$z_{a(k)}^{t-1}(x_{a(k)}^{t-1}) \leftarrow z_{a(k)}^{t-1}(x_{a(k)}^{t-1}) + p_k^t Q_k^t(x_{a(k)}^{t-1})$$
$$C_{a(k)}^{t-1}(x_{a(k)}^{t-1}) \leftarrow C_{a(k)}^{t-1}(x_{a(k)}^{t-1}) \cap G_{a(k)}^t(x_k^{t-1}) \,.$$

If $k = K^t$, let $t \leftarrow t - 1$, go to Step 2. Otherwise, let $k \leftarrow k + 1, z_k^t(x_k^t) = (c_k^t)^T x_k^t + (x_k^t)^T D_k^t x_k^t, C_k^t = S_k^t(x_{a(k)}^{t-1}) = K_t(x_{a(k)}^{t-1})$. Choose $x_k^t \in S_k^t(x_{a(k)}^{t-1})$. Go to Step 1.

Theorem 3. *The nested PQP algorithm terminates in a finite number of steps by either detecting an unbounded solution or finding an optimal solution of the multistage quadratic stochastic program with relatively complete recourse.*

Proof. The proof of the finite convergence of the PQP algorithm in Section 6.2 amounts to showing that Step 2 of the algorithm can be performed at most a finite number of times. The same result holds for a given piecewise quadratic program (2.2) in the nested sequence. The theorem follows from the observations that there is only a finite number of different problems (2.2) and that all other steps of the algorithm are finite. $\square$

Numerical experiments are reported in Louveaux [1980]. It should be noted that the $MQSP$ easily extends to the multistage piecewise convex case. The limit there is that the objective function and the description of the cell are usually much more difficult to obtain. One simple example is proposed in Exercise 3.

It is interesting to observe that $MQSP$ has a tendency to require few iterations when the quadratic terms play a significant role and a good starting point is chosen. (This probably relates to the good behavior of regularized decomposition.)

Example 1 (continued)

Assume that the cost of overtime is now quadratic (for example, larger increases of salary are needed to convince more people to work overtime). We replace everywhere $3.0w_k^t$ by $2.0w_k^t + (w_k^t)^2$. Assume all other data are unchanged. Take as the starting point a situation where $0 \le y^1 \le 1, 0 \le y_k^2 \le 1, k = 1, 2$. (It is relatively easy to see what the corresponding values for the other first- and second-stage variables should be.) We now proceed backward. Let $t = 3$.

i) $t = 3, k = 1$. We solve

$$\min x_1^3 + 2w_1^3 + (w_1^3)^2$$
$$\text{s.t. } y_1^2 + x_1^3 + w_1^3 = 1, x_1^3 \leq 2,$$
$$x_1^3, w_1^3 \geq 0,$$

where inventory at the end of Period 3 has been omitted for simplicity. The solution is easily seen to be $x_1^3 = 1 - y_1^2, w_1^3 = 0$ and is valid for $0 \leq y_1^2 \leq 1$. It follows that

$$Q_1^3(y_1^2) = 1 - y_1^2.$$

ii) $t = 3, k = 2$. We solve

$$\min x_2^3 + 2w_2^3 + (w_2^3)^2$$
$$\text{s.t. } y_1^2 + x_2^3 + w_2^3 = 3, x_2^3 \leq 2,$$
$$x_2^3, w_2^3 \geq 0.$$

The solution is now $x_2^3 = 2, w_2^3 = 1 - y_1^2$, valid for $0 \leq y_1^2 \leq 1$. It yields $Q_2^3(y_1^2) = 4 - 2y_1^2 + (1 - y_1^2)^2$.

Combining (i) and (ii), we obtain

$$Q_1^2(y_1^2) = \frac{1}{2}Q_1^3(y_1^2) + \frac{1}{2}Q_2^3(y_1^2) = \frac{5}{2} - \frac{3}{2}y_1^2 + \frac{(1 - y_1^2)^2}{2}$$

and

$$C_1^2(y_1^2) = \{y_1^2 | 0 \leq y_1^2 \leq 1\}.$$

iii) and iv) Because the randomness is only in the right-hand side, we conclude that cases (iii) and (iv) are identical to (i) and (ii), respectively. Hence,

$$Q_2^2(y_2^2) = \frac{5}{2} - \frac{3}{2}y_2^2 + \frac{(1 - y_2^2)^2}{2} \text{ and } C_2^2(y_2^2) = \{y_2^2 | 0 \leq y_2^2 \leq 1\}.$$

$t = 2$

i) $t = 2, k = 1$. The objective z_1^2 is computed as

$$z_1^2 = x_1^2 + 2w_1^2 + (w_1^2)^2 + 0.5y_1^2 + \frac{5}{2} - \frac{3}{2}y_1^2 + \frac{(1 - y_1^2)^2}{2},$$

i.e.,

$$z_1^2 = \frac{5}{2} + x_1^2 + 2w_1^2 + (w_1^2)^2 - y_1^2 + \frac{(1 - y_1^2)^2}{2}.$$

The constraint sets are

$$S_1^2 = \{x_1^2, w_1^2, y_1^2 | y^1 + x_1^2 + w_1^2 - y_1^2 = 1, 0 \leq x_1^2 \leq 2, x_1^2, w_1^2, y_1^2 \geq 0\}$$

and
$$C_1^2 = S_1^2 \cap \{0 \leq y_1^2 \leq 1\}.$$

The solution v of minimizing $z_1^2(.)$ over S_1^2 is

$$y_1^2 = 1, x_1^2 = 2 - y^1.$$

Because the solution belongs to C_1^2, we can take $w = v$. (Beware that w without superscript and subscript corresponds to the optimal solution on a cell defined in Step 2, while w with superscript and subscript corresponds to overtime.) Thus, this point satisfies the optimality criterion in Step 3. It yields

$$Q_1^2(y^1) = \frac{5}{2} + 2 - y^1 - 1 = \frac{7}{2} - y^1 \text{ and}$$
$$C_1^2(y^1) = \{y^1 | 0 \leq y^1 \leq 2\}.$$

ii) $t = 2, k = 2$. The objective z_2^2 is similarly computed as

$$z_2^2 = \frac{5}{2} + x_2^2 + 2w_2^2 + (w_2^2)^2 - y_2^2 + \frac{(1 - y_2^2)^2}{2}.$$

The constraint set

$$S_2^2 = \{x_2^2, w_2^2, y_2^2 | y^1 + x_2^2 + w_2^2 - y_2^2 = 3, 0 \leq x_2^2 \leq 2, x_2^2, w_2^2, y_2^2 \geq 0\}$$

only differs in the right-hand side of the inventory constraint with

$$C_2^2 = S_2^2 \cap \{0 \leq y_2^2 \leq 1\}.$$

The solution v is now $x_2^2 = 2, w_2^2 = 1 - y^1, y_2^2 = 0$. Again $v \in C_2^2$, so that we have $w = v$, which satisfies the optimality criterion in Step 3. It yields

$$Q_2^2(y^1) = \frac{5}{2} + 2 + 2(1 - y^1) + (1 - y^1)^2 + \frac{1}{2} = 7 - 2y^1 + (1 - y^1)^2$$
$$C_2^2(y^1) = \{y^1 | 0 \leq y^1 \leq 1\}.$$

$t = 1$.

The current objective function is computed as

$$z_1 = 21/4 - y^1 + \frac{(1 - y^1)^2}{2} + x^1 + 2w^1 + (w^1)^2.$$

The constraint sets are
$$S_1^1 = \{x^1, w^1, y^1 | x^1 + w^1 - y^1 = 1, x^1 \leq 2, x^1, w^1, y^1 \geq 0\},$$
$$C_1^1 = S_1^1 \cap \{0 \leq y^1 \leq 1\}.$$

The solution v of minimizing z_1 over S_1^1 is

$$x^1 = 2, y^1 = 1, w^1 = 0,$$

with objective value $z_1 = \frac{25}{4}$. Because this solution belongs to C^1, it is the optimal solution of the problem. Thus, no cut has ever been needed to optimize the problem.

Block Separability

The definition of block separability was given in Section 3.5. It permits a separate calculation of the recourse functions for the aggregate level decisions and the detailed level decisions. This is an advantage in terms of the number of variables and constraints, but often it makes the computation of the recourse functions and of the cells of the decomposition much easier in the case of a quadratic multistage program. This has been exploited in Louveaux [1986] and Louveaux and Smeers [1997].

We will illustrate a further benefit. It also consists of separating the random vectors. Consider the production of one single product. Now, assume the product cannot be stored (as in the case of a perishable good) or that the policy of the firm is to use a just-in-time system of production so that only a fixed safety stock is kept at the end of each period.

Assume that units are such that one worker produces exactly one product per stage. Two elements are uncertain: labor cost and demand. Labor cost is currently 2 per period. Next period, labor cost may be 2 or 3, with equal probability. Current revenue is 5 per product in normal time and 4 in overtime. Overtime is possible for up to 50% of normal time. Demand is a uniform continuous random variable within $(0,200)$ and $(0,100)$, respectively, for the next two periods. The original workforce is 50. Hiring and firing is possible once a period, at the cost of one unit each. Clearly, the labor decision is the aggregate level decision.

To keep notation in line with Section 3.5, we consider a three-stage model. In Stage 1, the decision about labor is made, say for Year 1. Stage 2 consists of production of Year 1 and decision about labor for Year 2. Stage 3 only consists of production of Year 2. Let ξ_1^t be labor cost in stage t, while ξ_2^t is the demand in stage t. Let w^t be the workforce in stage t. Then,

$$Q_w^t(w^{t-1}, \xi_1^t) = \min |w^t - w^{t-1}| + \xi_1^t w^t + Q^{t+1}(w^t), \tag{2.5}$$
$$Q^{t+1}(w^t) = E_{\xi^{t+1}}[Q_w^{t+1}(w^t, \xi_1^{t+1}) + Q_y^{t+1}(w^t, \xi_2^{t+1})], \tag{2.6}$$

and $Q_y^{t+1}(w^t, \xi_1^{t+1})$ is minus the expected revenue of production in stage $t+1$ given a workforce w^t and a demand scenario ξ_2^{t+1}. It is obtained as follows.

Let D^t represent the maximal demand in stage t (200 for $t = 2$, 100 for $t = 3$). Observe that the expectation of ξ_2^t is $D^t/2$ because ξ_2^t is uniformly continuous over $[0, D^t]$. If $w^t \geq D^t$, all demand can be satisfied with normal time. If $w^t \leq D^t \leq 1.5w^t$, demand up to w^t is satisfied with normal time, the rest in overtime. Finally, if $D^t \geq 1.5w^t$, normal time is possible up to a demand of w^t, overtime from w^t to $1.5w^t$, and extra demand is lost. Taking expectations over these cases, we obtain

$$Q_y^{t+1}(w^t) =$$

$$E_{\xi^{t+1}}[Q_y^{t+1}(w^t, \xi_2^{t+1})] = \begin{cases} -2.5D^t & \text{if } w^t \geq D^t, \\ \frac{(w^t)^2}{2D^t} - w^t - 2D^t & \text{if } w^t \leq D^t \leq 1.5w^t, \\ \frac{5(w^t)^2}{D^t} - 7w^t & \text{if } 1.5w^t \leq D^t. \end{cases}$$

This problem can now be solved with the $MQSP$ algorithm. Assume $w^0 = 50, w^1 \geq 50$.

Let Stage (2,1) represent the first labor scenario in Stage 2, i.e., $\xi_1^2 = 2$. The problem consists of finding

$$\min |w^2 - w^1| + 2w^2 + Q^3(w^2)$$

$$\text{s.t. } w^2 \geq 0.$$

We compute $Q^3(w^2) = Q_y^3(w^2) = \frac{5(w^2)^2}{100} - 7w^2$, for $w^2 \leq \frac{200}{3}$, because $D^3 = 100$. We also replace $|w^2 - w^1|$ by an explicit expression in terms of hiring (h^2) and firing (f^2). The problem in Stage (2,1) now reads:

$$Q_w^2(w^1, 1) = \min \ h^2 + f^2 - 5w^2 + \frac{5(w^2)^2}{100}$$

$$\text{s.t. } w^2 - h^2 + f^2 = w^1$$

$$w^2 \geq 0, h^2 \geq 0, f^2 \geq 0.$$

Under this form, the problem is clearly quadratic convex (remember w^2 is w in Stage 2, not the square of w). Classical Karush-Kuhn-Tucker conditions give the optimal solution $w^2 = w^1$, as long as $40 \leq w^1 \leq 60$. Then

$$Q_w^2(w^1, 1) = -5w^1 + \frac{5(w^1)^2}{100}.$$

Similarly, in Scenario (2,2) where $\xi_1^2 = 3$, the solution of

$$\min |w^2 - w^1| + 3w^2 + Q^3(w^2)$$

$$\text{s.t. } w^2 \geq 0$$

is $w^2 = 50, f^2 = w^1 - 50$, as long as $w^1 \geq 50$. Then

$$Q_w^2(w^1, 2) = w^1 - 125,$$

and

$$Q_w^2(w^1) = -\frac{125}{2} - 2w^1 + \frac{2.5(w^1)^2}{100},$$

which is valid within $C^2 = \{50 \leq w^1 \leq 60\}$.

The Stage 1 objective is $\min h^1 + f^1 + 2w^1 + Q_y^2(w^1) + Q_w^2(w^1)$, so that the Stage 1 problem reads

$$\min h^1 + f^1 - 7w^1 + \frac{(w^1)^2}{20} - \frac{125}{2}$$

$$\text{s.t. } w^1 - h^1 + f^1 = 50,$$

$$w^1, h^1, f^1 \geq 0.$$

Its optimal solution, $w^1 = 60, h^1 = 10$, belongs to C^2 and is thus also the optimal solution of the global problem with objective value -292.5.

Exercises

1. Consider Example 1 with quadratic terms as in this section and take $1 \le y^1 \le 2, 1 \le y_1^2 \le 2, 0 \le y_2^2 \le 1$ as a starting point. Show that the following steps are generated. Obtain $0.5Q_1^3(y_1^2) + 0.5Q_2^3(y_1^2) = \frac{5}{4} - \frac{1}{4}y_1^2$. In $t = 2, k = 1$, solution v is $x_1^2 = 0, y_1^2 = y^1 - 1$ while w is $y_1^2 = 1, x_1^2 = 2 - y^1$, both with $w_1^2 = 0$. A cut $x_1^2 + 2w_1^2 + \frac{1}{4}y_1^2 \le \frac{9}{4} - y^1$ is added. The new starting point is v, which corresponds to $0 \le y_1^2 \le 1$. Then the case $t = 2, k = 1$ is as in the text, yielding

$$Q_1^2(y^1) = \frac{7}{2} - y^1 \text{ and } C_1^2(y^1) = \{0 \le y^1 \le 2\}.$$

   In $t = 2, k = 2$, the calculations appear in the text, we obtain $Q_2^2(y^1) = 6 - y^1$ and $C_2(y^1) = \{1 \le y^1 \le 3\}$. Thus, in $t = 1, z_1 = x^1 + 2w^1 + (w^1)^2 + \frac{19}{4} - y^1/2$ and $C = \{1 \le y^1 \le 2\}$. Again, the solution $v : x^1 = 1, y^1 = 0, w^1 = 0$ does not coincide with $w : x^1 = 2, y^1 = 1, w^1 = 0$. A cut $x^1 - \frac{y^1}{2} + w^1 \le 3/2$ is generated. The new starting point now coincides with the one in the text and the solution is obtained in one more iteration.

2. Does the block separable property depend on having a single product? To help answer this question, take the example in the block separability paragraph and assume a second product with revenue 0.6 in normal time and 0.3 in overtime. One worker produces 10 such products in one stage. Obtain $Q_y^{t+1}(w^t)$,

   (a) if demand in Period t is known to be 400;

   (b) if demand in Period t is uniform continuous within $[0, 500]$ and $[0, 100]$, respectively, for the two periods.

3. In the case of one product, obtain $Q_y^{t+1}(w^t)$ if demand follows a negative exponential distribution with known parameter λ. Based on Louveaux [1978], extend the $MQSP$ to the piecewise convex case, then solve the problem with $\lambda = 0.01$ and 0.02 for the two periods.

7.3 Other Approaches to Multiple Stages

Many two-stage methods may also be enhanced for multiple stages using some form of block separability. One such approach assumes deviations from some mean value can be corrected by a penalty only relating to the current period. This method basically applies a simple recourse strategy in every period. For example, in Kallberg, White and Ziemba [1982] and Kusy and Ziemba [1986], penalties are imposed to meet financial requirements in each period of a short-term financial planning model. With this type of penalty, the various simple recourse methods may be applied to achieve efficient computation.

Other methods that seem particularly adapted for multiple stages include some strict separation of solutions for different scenarios by imposing nonanticipativity constraints explicitly. For example, the progressive hedging algorithm (PHA) is easily adapted by simply defining the projection, Π, to project onto the space of nonanticipative solutions by defining it as the conditional expectation of all solutions at time t that correspond to the same history up to t.

The PHA is particularly well-adapted for problems, such as networks, where maintaining the original problem structure in each scenario problem leads to efficiency. Computational experience on such models (see Mulvey and Vladimirou [1991b]) indicates that PHA may perform well in these circumstances.

Another approach for multistage problems that appears to have particular potential for nonlinear problems is a method from Mulvey and Ruszczyński [1995] called diagonal quadratic approximation (DQA). This method approximates quadratic penalty terms in a Lagrangian type of objective so that each subproblem is again easy to solve and can be spread across a wide array of distributed processors. DQA appears to have particular advantages for nonlinear problems because it requires few additional assumptions. Computational results suggest it is even useful for linear problems.

Exercise

1. Show how to implement PHA on Example 1. Follow three iterations of the algorithm.

8
Stochastic Integer Programs

As seen in Section 3.3, properties for stochastic integer programs are scarce. The absence of general efficient methods reflects this difficulty. Some techniques have been proposed that address specific problems or use a particular property. Clearly, much work needs to be done to solve integer stochastic programs efficiently. The available techniques described here provide elements of answers to start solving some of the integer programs. This field is expected to evolve a great deal in the future.

8.1 Integer L-Shaped Method

In this section, we present a general scheme for solving stochastic integer programs. For the sake of completeness, we recall the definition of a stochastic integer program as

$$(SIP) \qquad \min_{x \in X} z = c^T x + E_{\xi} \min \{q(\omega)^T y | Wy = h(\omega) - T(\omega)x, y \in Y\}$$

$$\text{s. t. } Ax = b,$$

where the definitions of c, b, ξ, A, W, T, and h are as before. However, X and/or Y contains some integrality or binary restrictions on x and/or y. We may again define a deterministic equivalent program of the form:

$$(DEP) \qquad \min_{x \in X} z = c^T x + \mathcal{Q}(x)$$

$$\text{s. t. } Ax = b,$$

with $\mathcal{Q}(x)$ the expected value of the second stage defined as in Section 3.1. Let $\bar{X}$ be the polytope defined by the set of constraints in X other than those defining the first-stage variable type. Hence, with continuous first-stage variables, $X = \bar{X} \cap \Re_+^{n_1}$, and for problems with integer first-stage variables, $X = \bar{X} \cap Z_+^{n_1}$. At a given stage of the algorithm, we consider the *current problem* (CP):

$$(CP) \qquad \min c^T x + \theta \tag{1.1}$$
$$\text{s. t. } Ax = b, \tag{1.2}$$
$$D_l x \geq d_l, \quad l = 1, \cdots, r, \tag{1.3}$$
$$E_l x + \theta \geq e_l, \quad l = 1, \cdots, s, \tag{1.4}$$
$$x \geq 0, \theta \in \Re. \tag{1.5}$$

The current problem is obtained from DEP by three relaxations: the integrality restrictions are relaxed in $x \geq 0$, the restriction $x \in \bar{X}$ is relaxed in a number of constraints (1.3) called *feasibility cuts*, and the exact definition of $\mathcal{Q}(x)$ is relaxed in a polyhedral representation by θ and the constraints (1.4) called *optimality cuts*.

Relaxing the integrality requirement is the basis of any *branch and cut* or *branch and bound* scheme. Integrality is recovered by the handling scheme which creates a number of nodes we call *pendant* as long as they have not been fully examined. The principle of branch and bound can be found in, for example, Taha [1992].

First-stage constraints include $Ax = b$ and $x \in \bar{X}$. Constraints put in $\bar{X}$ are precisely those that are typically relaxed. This is the case when they are not known in advance, as for the induced constraints. It is also the case for constraints that are well known but so numerous that it would be unrealistic to impose all of them from the beginning. Relaxing these constraints is thus a matter of reducing their number so that only those needed along the way are imposed. Such is the case for subtour elimination constraints in vehicle routing problems or for valid inequalities derived by polyhedral combinatoric arguments (see, e.g., Nemhauser and Wolsey [1988]).

Definition 1. A set of feasibility cuts is said to be *valid* at x if there exists a finite r such that

$$x \in \{x | D_l x \geq d_l, l = 1, \cdots, r\} \text{ implies } x \in \bar{X}.$$

Note that weaker forms of validity may exist. In particular, if $X = \bar{X} \cap Z_+^{n_1}$, $x \in \{x | D_l x \geq d_l, l = 1, \cdots, r\} \cap Z_+^{n_1}$ implies $x \in X$ is a sufficient condition for validity, which may be weaker than Definition 1.

Relaxing the exact representation of $\mathcal{Q}(x)$ by a polyhedral representation is exactly what is done in the L-shaped or Benders decomposition method. That it extends to the stochastic integer case will be indicated later in this section.

Definition 2. A set of optimality cuts is said to be *valid* at $x \in X$ if there exists a finite s such that

$$(x, \theta) \in \{(x, \theta) | E_l x + \theta \geq e_l, l = 1, \cdots, s\}$$

implies $\theta \geq \mathcal{Q}(x)$.

Assumption 3. *For fixed x, $\mathcal{Q}(x)$ is computable in a finite number of steps.*

Integer L-shaped Method

Step 0. Set $r = s = \nu = 0, \bar{z} = \infty$. The value of θ is set to $-\infty$ or to an appropriate lower bound and is ignored in the computation. A list is created that contains only a single pendant node corresponding to the initial subproblem.

Step 1. Select some pendant node in the list as the current problem; if none exists, stop.

Step 2. Set $\nu = \nu + 1$. Solve the current problem. If the current problem has no feasible solution, fathom the current node; go to Step 1. Otherwise, let (x^ν, θ^ν) be an optimal solution.

Step 3. Check for any relaxed constraint violation. If one exists, add one feasibility cut (1.3), set $r = r + 1$, and go to Step 2. If $c^T x^\nu + \theta^\nu > \bar{z}$, fathom the current problem and go to Step 1.

Step 4. Check for integrality restrictions. If a restriction is violated, create two new branches following the usual branch and cut procedure. Append the new nodes to the list of pendant nodes, and go to Step 1.

Step 5. Compute $\mathcal{Q}(x^\nu)$ and $z^\nu = c^T x^\nu + \mathcal{Q}(x^\nu)$. If $z^\nu < \bar{z}$, update $\bar{z} = z^\nu$.

Step 6. If $\theta^\nu \geq \mathcal{Q}(x^\nu)$, then fathom the current node and return to Step 1. Otherwise, impose one optimality cut (1.4), set $s = s + 1$, and return to Step 2.

Proposition 4. *Under Assumption 3, for any problem for which a valid set of feasibility cuts and a valid set of optimality cuts exist, the integer L-shaped method yields an optimal solution (when one exists) in a finite number of steps.*

Proof. Finiteness comes directly from the fact that each of the three relaxations can be recovered in a finite way and that Step 5 is finite under Assumption 3. $\square$

As we will see next, a multicut approach is often preferred. It suffices then to replace the objective (1.1) and the optimality cuts (1.4) by their multicut equivalents (5.3.1) and (5.3.4). Also, a valid set of optimality cuts must exist for each realization of the random vector. This addition creates no extra theoretical difficulty.

A set of valid feasibility cuts and optimality cuts is known to exist in the continuous case and forms the basis of the classical L-shaped method (Section 5.1). These cuts are based on duality theory in linear programming. They can also be used in the case where only the first-stage variables contain some integrality restrictions. This offers an interesting alternative to the cutting plane method used in Wollmer [1980]. The first application of the integer L-shaped method was proposed by Laporte and Louveaux [1993] for the case of binary first- and second-stage decision variables. A full characterization of the integer L-shaped method based on general duality theory can be found in Carøe and Tind [1996]. A stochastic version of the branch and cut method using statistical estimation of the recourse function instead of exact evaluation can be found in Norkin, Ermoliev, and Ruszczyński [1997]. We now present a simplified version of Carøe and Tind [1996] for feasibility and optimality cuts based on second-stage branch and bound decomposition when the random variable is discrete. We also include a number of observations from Laporte and Louveaux [1993].

a. Feasibility cuts

As usual, let $K_2(\xi)$ denote the second-stage feasibility set for a given ξ and $K_2 = \cap_{\xi \in \Xi} K_2(\xi)$. Let also $C_2(\xi)$ denote the set of first-stage decisions that are feasible for the continuous relaxation of the second stage, i.e.,

$$C_2(\xi) = \{x | \exists y \text{ s. t. } Wy = h(\omega) - T(\omega)x, y \geq 0\}.$$

Clearly, $K_2(\xi) \subset C_2(\xi)$, and any induced constraint valid for $C_2(\xi)$ is also valid for $K_2(\xi)$. Also, detecting that some point $x \in C_2(\xi)$ does not belong to $K_2(\xi)$ is relatively easy, as only a phase one problem is needed:

$$
\begin{aligned}
(P1) \qquad w(x,\xi) = &\min e^T v^+ + e^T v^- \\
&\text{s. t. } Wy + v^+ - v^- = h(\omega) - T(\omega)x, \\
&\qquad\qquad y \in Z_+^{n_2}, v^+, v^- \geq 0. \qquad (1.6)
\end{aligned}
$$

As usual, $x \in K_2(\xi)$ if and only if $w(x,\xi) = 0$. If $x \notin K_2(\xi)$, we would like to generate a feasibility cut. Let (y, v^+, v^-) be a solution to (P1), and because $x \notin K_2(\xi)$, we have $w(x,\xi) = e^T v^+ + e^T v^- > 0$. If $y \in Z_+^{n_2}$, then a cut of the form (5.1.3) can be generated. If $y \notin Z_+^{n_2}$, then some of the components of y are not integer. A branch and bound algorithm can be applied to (P1). This will generate a branching tree where, at each node, additional simple upper or lower bounds are imposed on some variables.

Let $\rho = 1, \dots, R$ index all *terminal nodes*, i.e., nodes that have no successors, of the second-stage branching tree. Let Y^ρ be the corresponding subregions. They form a partition of $\Re_+^{n_2}$, i.e., $\Re_+^{n_2} = \cup_{\rho=1,\dots,R} Y^\rho$ and

$Y^\rho \cap Y^\sigma = \emptyset, \rho \neq \sigma$. Now, $x \in K_2(\xi)$ if and only if $x \in \cup_{\rho=1,\ldots,R} K_2^\rho(\xi)$, where

$$K_2^\rho(\xi) = \{x | \exists y \in Y^\rho \text{ s. t. } Wy \leq h(\omega) - T(\omega)x, y \geq 0\}.$$

However, because Y^ρ is obtained from $\Re_+^{n_2}$ by some branching process, it is defined by adding a number of bounds to some components of y. Thus, $K_2^\rho(\xi)$ is a polyhedron for which linear cuts are obtained through a classical separation or duality argument. It follows that $x \in K_2(\xi)$ if and only if at least one among R sets of cuts is satisfied.

In practice, one constructs the branching tree of the second stage associated with one particular $\bar{x}$ and generates one cut per terminal node of the restricted tree. This means that one first-stage feasibility cut (8.1.3) corresponds to the requirement that one out of R cuts is satisfied. As expected, this takes the form of a *Gomory function*. It can be embedded in a linear programming scheme by the addition of extra binary variables, one for each of the R cuts, as follows. Assume the ρth cut is represented by $u_\rho^T x \leq d_\rho$. One introduces R binary variables, $\delta_1, \ldots, \delta_R$. The requirement that at least one of the R cuts is satisfied is equivalent to

$$u_\rho^T x \leq d_\rho + M_\rho(1 - \delta_\rho), \rho = 1, \ldots, R,$$

$$\sum_{\rho=1}^R \delta_\rho \geq 1,$$

$$\delta_\rho \in \{0, 1\}, \rho = 1, \ldots, R,$$

where M_ρ is a large number such that $u_\rho^T x \leq d_\rho + M_\rho, \forall x \in K_1$.

Finally, observe that $x \in K_2$ if and only $x \in K_2(\xi), \forall \xi \in \Xi$. As in the continuous case (Section 5.2), it is sometimes enough to consider $x \in K_2(\xi)$ for one particular ξ.

Example 1

Consider again Example 3.3, when the second stage is defined as

$$-y_1 + y_2 \leq \xi - x_1,$$
$$y_1 + y_2 \leq 2 - x_2,$$
$$y_1, y_2 \geq 0 \text{ and integer,}$$

where ξ takes on the values 1 and 2 with equal probability 0.5. It suffices here to consider $x \in K_2(1)$ because $K_2(1) \subset K_2(2)$. First, consider $x = (2, 2)^T$. From Section 5.2, we find a violated continuous induced constraint:

$$x_1 + x_2 \leq 3.$$

Next, consider $x = (1.4, 1.6)^T$. Problem (P1) is

$$\min v_1 + v_2$$
$$\text{s. t.} \ -y_1 + y_2 - v_1 \leq -0.4,$$
$$y_1 + y_2 - v_2 \leq 0.4,$$
$$y_1, y_2 \geq 0 \text{ and integer,}$$

where v_1 and v_2 correspond to v^- in (1.6) and v^+ is not needed due to the inequality form of the constraints. The optimal solution of the continuous relaxation of (P1) is given by the following dictionary:

$$w = v_1 + v_2,$$
$$y_1 = 0.4 + y_2 + s_1 - v_1,$$
$$s_2 = 0 - 2y_2 - s_1 + v_1 + v_2.$$

Its solution is $w = 0$, which implies $x \in C_2(1)$. However, y_1 is not integer. Branching creates two nodes, $y_1 \leq 0$ and $y_1 \geq 1$, respectively. In the first branch, the bound $y_1 \leq 0$ creates the additional constraint $y_1 + s_3 = 0$. After one dual iteration, the following optimal dictionary is obtained:

$$w = 0.4 + y_2 + s_1 + s_3 + v_2,$$
$$y_1 = 0 - s_3,$$
$$s_2 = 0.4 - y_2 + s_3 + v_2,$$
$$v_1 = 0.4 + y_2 + s_1 + s_3.$$

Associating the dual variables $(-1, 0, -1)$ with the right-hand sides $(1 - x_1, 2 - x_2, 0)$, one obtains the feasibility cut, $x_1 - 1 \leq 0$, for this branch.

Similarly, in the second branch, the bound $y_1 \geq 1$ creates a constraint $y_1 - s_3 = 1$. After two dual iterations, the optimal dictionary is:

$$w = 0.6 + y_2 + s_2 + s_3 + v_1,$$
$$y_1 = 1 + s_3,$$
$$v_2 = 0.6 + y_2 + s_2 + s_3,$$
$$s_1 = 0.6 - y_2 + s_3 + v_1.$$

Associating the dual variables $(0, -1, 1)$ to the right-hand sides $(1 - x_1, 2 - x_2, 1)$, one obtains the feasibility cut, $x_2 - 1 \leq 0$, for the second branch. Thus, $R = 2$, as the solutions in the two nodes satisfy the integrality requirement and are thus terminal. The feasibility cut is thus, as required, that either $x_1 - 1 \leq 0$ or $x_2 - 1 \leq 0$ must be satisfied. Because we also have $x_1 \leq 2$ and $x_2 \leq 2$, we may take $M_1 = M_2 = 1$ so that we have to impose the following set of conditions:

$$x_1 \leq 2 - \delta_1,$$
$$x_2 \leq 2 - \delta_2,$$
$$\delta_1 + \delta_2 \geq 1,$$
$$\delta_1, \delta_2 \in \{0, 1\}.$$

b. Optimality cuts

We consider here a multicut approach,

$$\theta = \sum_{k=1}^{K} \theta_k,$$

where, as usual, K denotes the cardinality of Ξ. We search for optimality cuts on a given θ_k. Based on branching on the second-stage problem, one obtains a partition of $\Re_+^{n_2}$ into R terminal nodes $Y^\rho = \{y | a^\rho \leq y \leq b^\rho\}$, $\rho = 1, \ldots, R$. The objective value of the second-stage program over Y^ρ is

$$Q^\rho(x^\nu, \xi_k) = \min\{q^T y | Wy = h(\xi^k) - T(\xi^k)x^\nu, a^\rho \leq y \leq b^\rho\}.$$

It is the solution of a linear program that by classical duality theory is also

$$Q^\rho(x^\nu, \xi_k) = (\pi^\rho)^T (h(\xi^k) - T(\xi^k)x^\nu) + (\underline{\pi}^\rho)^T a^\rho + (\bar{\pi}^\rho)^T b^\rho\},$$

where π^ρ, $\underline{\pi}^\rho$, and $\bar{\pi}^\rho$ are the dual variables associated with the original constraints, lower and upper bounds on $y \in Y^\rho$, respectively.

To simplify notation, we represent this expression as:

$$Q^\rho(x^\nu, \xi_k) = (\sigma_k^\rho)^T x^\nu + \tau_k^\rho,$$

with $(\sigma_k^\rho)^T = -(\pi^\rho)^T T(\xi^k)$ and $\tau_k^\rho = (\pi^\rho)^T h(\xi^k) + (\underline{\pi}^\rho)^T a^\rho + (\bar{\pi}^\rho)^T b^\rho$. By duality theory, we know that $Q^\rho(x, \xi^k) \geq (\sigma_k^\rho)^T x^\nu + \tau_k^\rho$. Moreover, $Q(x, \xi^k) = \min_{\rho=1,\ldots,R} Q^\rho(x, \xi^k)$. Thus,

$$\theta_k \geq p_k \big(\min_{\rho=1,\ldots,R} (\sigma_k^\rho)^T x^\nu + \tau_k^\rho \big). \tag{1.7}$$

Note that some of the terminal nodes may be infeasible, in which case their dual solutions contain unbounded rays with dual objective values going to ∞, so that the minimum is in practice restricted to the feasible terminal nodes.

This expression takes the form of a Gomory function, as expected. Again, it unfortunately requires R extra binary variables to be included in a mixed integer linear representation.

Example 2

Consider the second-stage program

$$E_{\xi} \min\{-8y_1 - 9y_2 \text{ s. t. } 3y_1 + 2y_2 \leq \xi, -y_1 + y_2 \leq x_1, y_2 \leq x_2, y \geq 0, \text{ integer}\}.$$

Consider the value $\xi_1 = 8$ and $\bar{x} = (0,6)^T$. The optimal dictionary of the continuous relaxation of the second-stage program is:

$$z = -136/5 + 17s_1/5 + 11s_2/5,$$
$$y_1 = 8/5 - s_1/5 + 2s_2/5,$$
$$y_2 = 8/5 - s_1/5 - 3s_2/5,$$
$$s_3 = 22/5 + s_1/5 + 3s_2/5,$$

where s_1, s_2, and s_3 are the slack variables of the three constraints. Branching on y_1 gives two nodes, $y_1 \leq 1$ and $y_1 \geq 2$, which turn out to be the only two terminal nodes. For the first node, adding the constraint $y_1 + s_4 = 1$ yields the following dictionary after one dual iteration:

$$z = -17 + 9s_2 + 17s_4,$$
$$s_1 = 3 + 2s_2 + 5s_4,$$
$$y_2 = 1 - s_2 - s_4,$$
$$s_3 = 5 + s_2 + s_4,$$
$$y_1 = 1 - s_4.$$

We thus have dual variables $(0, -9, 0)$ associated with the right-hand side $(8, x_1, x_2)$ of the constraints and -17 associated with the bound 1 on y_1. Hence, $Q^1(\bar{x}, 8) = -9x_1 - 17$.

Similarly, we add $y_1 - s_4 = 2$ for the second node. We obtain:

$$z = -25 + 9/2s_1 + 11/2s_4,$$
$$y_1 = 2 + s_4,$$
$$y_2 = 1 - s_1/2 - 3/2s_4,$$
$$s_3 = 5 + s_1/2 + 3/2s_4,$$
$$s_2 = 1 + s_1/2 + 5/2s_4.$$

We now have dual variables, $(-9/2, 0, 0)$, associated with the right-hand side $(8, x_1, x_2)$ of the constraints and $11/2$ associated with the lower bound 2 on y_1. Hence, $Q^2(\bar{x}, 8) = -25$. Applying (1.7), we conclude that

$$\theta_1 \geq p_1 \min(-9x_1 - 17, -25),$$

where p_1 is the probability of $\xi = \xi_1$.

c. Lower bounding functionals

Lower bounding functionals are simply valid inequalities that apply to $\mathcal{Q}(x)$. As finiteness of the integer L-shaped method is obtained through the optimality cuts (1.4) or their variants, no special requirement is needed for lower bounding functionals.

It is not even required that they be binding at at least one point. On the other hand, they should be effective in describing a valid bound other than the existing ones in a sufficiently large region. Lower bounding functionals can often be derived from the context of the problem. Such is the case, for example, in a priori optimization problems (Laporte, Louveaux, and Mercure [1994]) and in a location problem with stochastic demands (Laporte, Louveaux, and Van Hamme [1994]).

For problems whose second stage is best described as a mathematical program (rather than an implicit or closed form expression), lower bounding functionals are easily derived from the related continuous problem. As before, let

$$Q(x,\xi(\omega))= \min_{y}\{q(\omega)^T y | Wy = h(\omega) - T(\omega)x, y \geq 0 \text{ and integer}\} \quad (1.8)$$
$$\text{with } \mathcal{Q}(x)= E_{\boldsymbol{\xi}}Q(x,\boldsymbol{\xi}). \quad (1.9)$$

We introduce the following notation for the solution to the continuous relaxation of the second-stage program,

$$C(x,\xi(\omega))= \min_{y}\{q(\omega)^T y | Wy = h(\omega) - T(\omega)x, y \geq 0\} \quad (1.10)$$
$$\text{with } C(x)= E_{\boldsymbol{\xi}}C(x,\boldsymbol{\xi}), \quad (1.11)$$

with the usual conventions for infeasible and unbounded cases and $+(\infty)+(-\infty) = +(\infty)$ for the expectation.

Proposition 5. *Any continuous L-shaped optimality cut is a valid lower bound on $\mathcal{Q}(x)$.*

Proof. First note that $Q(x,\xi) \geq C(x,\xi)$ for all x and all ξ, where this result also holds if some problems are unbounded or infeasible. Taking expectations implies $\mathcal{Q}(x) \geq C(x)$. The conclusion follows by applying the continuous L-shaped optimality cut to $C(x)$, namely,

$$\mathcal{Q}(x) \geq C(x) \geq C(x^\nu) + \partial C(x^\nu)^T(x - x^\nu), \quad (1.12)$$

where x^ν is any first-stage feasible solution. $\square$

Corollary 6. *The following is a valid lower bounding functional on $\mathcal{Q}(x)$:*

$$\theta \geq C(x^\nu) + \partial C(x^\nu)(x - x^\nu). \quad (1.13)$$

Example 2 (continued)

From the optimal continuous relaxation of the second stage for ξ_1, we obtain a continuous optimality cut (1.13),

$$\theta_1 \geq p_1(-136/5 - 11/5x_1),$$

which is more restrictive than $p_1 \min(-9x_1 - 17, -25)$ for $x_1 \geq 1.5$.

8.2   Simple Integer Recourse

As seen in Section 3.3, a two-stage stochastic program with simple integer recourse can be transformed into

$$\min c^T x + \sum_{i=1}^{m} \Psi_i(\chi_i)$$
$$\text{s. t. } Ax = b, Tx = \chi, x \in X \subset Z_+^{n_1}, \tag{2.1}$$

where

$$\Psi_i(\chi_i) = q_i^+ u_i(\chi_i) + q_i^- v_i(\chi_i) \tag{2.2}$$

with

$$u_i(\chi_i) = E\lceil \xi_i - \chi_i \rceil^+, \tag{2.3}$$

defined as the expected shortage, and

$$v_i(\chi_i) = E\lceil \chi_i - \xi_i \rceil^+, \tag{2.4}$$

defined as the expected surplus. As before, we assume

$$q_i^+ \geq 0, q_i^- \geq 0.$$

Also from Section 3.3, we know that the values of the expected shortage and surplus can be computed in finitely many steps, either exactly or within a prespecified tolerance ε.

Before turning to algorithms, we still need some results concerning the functions Ψ_i; for simplicity in the exposition we omit the index i. As we also know from Section 3.3, the function Ψ is generally not convex and is even discontinuous when ξ has a discrete distribution. It turns out, however, that some form of convexity exists between function values evaluated in (not necessarily integer) points that are integer length apart. Thus, let $x^0 \in \Re$ be an arbitrary point. Let $i \in Z$ be some integer.

Define $x^1 = x^0 + i$, and for any $j \in Z, j \leq i, x^\lambda = x^0 + j$. Equivalently, we may define

$$x^\lambda = \lambda x^0 + (1 - \lambda)x^1,$$
$$\lambda = (i - j)/i.$$

In the following, we will use x as an argument for Ψ as if $Tx = Ix = \chi$ without losing generality. We make T explicit again when we speak of a general problem and not just the second stage.

Proposition 7. *Let $x^0 \in \Re, i, j \in Z$ with $j \le i, x^1 = x^0 + i, x^\lambda = x^0 + j$. Then*

$$\Psi(x^\lambda) \le \lambda \Psi(x^0) + (1 - \lambda)\Psi(x^1) \qquad (2.5)$$

with $\lambda = (i - j)/i$.

Proof: We prove that $\Psi(x+1) - \Psi(x)$ is a nondecreasing function of x. We leave it as an exercise to infer that this is a sufficient condition for (2.5) to hold. Using (3.3.16) and (3.3.17), we respectively obtain $u(x + 1) - u(x) = -(1 - F(x))$ and $v(x+1) - v(x) = \hat{F}(x+1)$, where F is again the cumulative distribution function of ξ and $\hat{F}$ is defined as in Section 3.3. With this,

$$\Psi(x + 1) - \Psi(x) = q^- \hat{F}(x + 1) - q^+(1 - F(x)).$$

The result follows as $q^+ \ge 0, q^- \ge 0$ and $\hat{F}$ and F are nondecreasing. $\square$

This means that we can draw a piecewise linear convex function through points that are integer length apart. Such a convex function is called a ρ-*approximation* rooted at x if it is drawn at points $x \pm \kappa, \kappa$ integer. In Figures 1 and 2, we provide the ρ-approximations rooted at $x = 0$ and $x = 0.5$, respectively, for the case in Example 3.1.

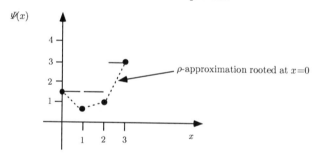

FIGURE 1. The ρ-approximation rooted at $x = 0$.

If we now turn to discrete random variables, we are interested in the different possible fractional values associated with a random variable. As an example, if ξ can take on the values 0.0, 1.0, 1.2, 1.6, 2.0, 2.2, 2.6, and 3.2 with some given probability, then the only possible fractional values are 0.0, 0.2, and 0.6. Let $s_1 < s_2 < \cdots < s_S$ denote the S ordered possible fractional values of ξ. Define $s_{S+1} = 1$. Let the extended list of fractionals be all points of the form $j + s_l, j \in Z, 1 \le l \le S$. This extended list is a countable list that contains many more elements than the possible values of ξ. In the example, $0.2, 0.6, 3.0, 3.6, 4.0, 4.2, \cdots$ are in the extended list of fractionals but are not possible values of ξ.

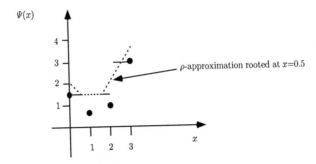

FIGURE 2. The ρ-approximation rooted at $x = 0.5$.

Lemma 8. *Let $\boldsymbol{\xi}$ be a discrete random variable. Assume that S is finite. Let $a \in \Re$. Define* **frac** *by $a = \lfloor a \rfloor + $* **frac** *and define j by $s_j \leq$* **frac** $< s_{j+1}$. *Then*

$\Psi(x)$ is constant within the open interval $\left(\lfloor a \rfloor + s_j, \lfloor a \rfloor + s_{j+1}\right)$,

$$\Psi(x) \geq \max\left\{\Psi(\lfloor a \rfloor + s_j), \Psi(\lfloor a \rfloor + s_{j+1})\right\},$$

$$\forall\, x \in \left(\lfloor a \rfloor + s_j, \lfloor a \rfloor + s_{j+1}\right).$$

Proof. The proof can be found in Louveaux and van der Vlerk [1993]. $\square$

The lemma states that $\Psi(x)$ is piecewise constant in the open interval between two consecutive elements of the extended list of fractionals and that the values in points between two consecutive elements of that list are always greater than or equal to the values of $\Psi(\cdot)$ at these two consecutive elements in the extended list. The reader can easily observe this property in the examples that have already been given.

Corollary 9. *Let $\boldsymbol{\xi}$ be a random variable with $S = 1$. Let $\rho(\cdot)$ be a ρ-approximation of $\Psi(\cdot)$ rooted at some point in the support of ξ. Then*

$$\rho(x) \leq \Psi(x), \ \forall\, x \in \Re.$$

Moreover, ρ is the convex hull of the function Ψ.

Proof. The function ρ is a lower bound of Ψ by Lemma 5. It is the convex hull of Ψ because it is convex, piecewise linear, and it coincides with Ψ in all points at integer distance from the root. $\square$

Among the cases where $S = 1$, the most natural one in the context of simple integer recourse is when $\boldsymbol{\xi}$ only takes on integer values. A well-known such case is the Poisson distribution. Then the ρ-approximation rooted at any integer point is the piecewise linear convex hull of Ψ that coincides with Ψ at all integer points.

We use Proposition 7 and Corollary 9 to derive finite algorithms for two classes of stochastic programs with simple integer recourse.

a. χs restricted to be integer

Integral χ is a natural assumption, because one would typically expect first-stage variables to be integer when second-stage variables are integer. It suffices then for T to have integer coefficients. By definition of a ρ-approximation rooted at an integer point, solving (2.1) is thus equivalent to solving

$$\min\{c^T x + \sum_{i=1}^{m_2} \rho_i(\chi_i)|Ax = b, \chi = Tx, x \in X\}, \qquad (2.6)$$

where T is such that $x \in X$ implies χ is integer, and ρ_i is a ρ-approximation of Ψ_i rooted at an integer point.

Because the objective in (2.6) is piecewise linear and convex, problem (2.6) can typically be solved by a dual decomposition method such as the L-shaped method. We recommend using the multicut version because we are especially concerned with generating individual cut information for each subproblem that may require many cuts. This amounts to solving a sequence of current problems of the form

$$\min_{x \in X, \theta \in \Re^{m_2}} \{c^T x + \sum_{i=1}^{m_2} \theta_i | Ax = b, \chi = Tx,$$

$$E_{il}\chi_i + \theta_i \geq e_{il}, i = 1, \cdots, m_2, l = 1, \cdots, s_i\}. \qquad (2.7)$$

In this problem, the last set of constraints consists of optimality cuts. They are used to define the epigraph of $\Psi_i, i = 1, \cdots, m_2$. Optimality cuts are generated only as needed. If χ_i^ν is a current iterate point with $\theta_i^\nu < \Psi_i(\chi_i^\nu)$, then an additional optimality cut is generated by defining

$$E_{ik} = \Psi_i(\chi_i^\nu) - \Psi_i(\chi_i^\nu + 1) \qquad (2.8)$$

and

$$e_{ik} = (\chi_i^\nu + 1)\Psi_i(\chi_i^\nu) - \chi_i^\nu \Psi_i(\chi_i^\nu + 1), \qquad (2.9)$$

which follows immediately by looking at a linear piece of the graph of Ψ_i. The algorithm iteratively solves the current problem (2.7) and generates optimality cuts until an iterate point (χ^ν, θ^ν) is found such that $\theta_i^\nu = \Psi_i(\chi_i^\nu), i = 1, \cdots, m_2$. It is important to observe that the algorithm is applicable for any type of random variable for which $\Psi_i s$ can be computed.

Example 3

Consider two products, $i = 1, 2$, which can be produced by two machines $j = 1, 2$. Demand for both goods follows a Poisson distribution with expectation 3. Production costs (in dollars) and times (in minutes) of the two products on the two machines are as follows:

	Machine	
	1	2
Product 1	3	2
2	4	5

Cost/Unit

	Machine		Finishing	
	1	2	1	2
Product 1	20	25	4	7
2	30	25	6	5

Time/Unit   Time/Unit

The total time for each machine is limited to 100 minutes. After machining, the products must be finished. Finishing time is a function of the machine used, with total available finishing time limited to 36 minutes. Production and demand correspond to an integer number of products. Product 1 sells at \$4 per unit. Product 2 sells at \$6 per unit. Unsold goods are lost.

Define x_{ij} = number of units of product i produced on machine j and $y_i(\xi)$ = amount of product i sold in state ξ. The problem reads as follows:

$$\min\ 3x_{11} + 2x_{12} + 4x_{21} + 5x_{22} + E_{\boldsymbol{\xi}}\{-4y_1(\xi) - 6y_2(\xi)\}$$
$$\text{s. t. } 20x_{11} + 30x_{21} \leq 100,\ y_1(\xi) \leq \boldsymbol{\xi}_1,$$
$$25x_{12} + 25x_{22} \leq 100,\ y_2(\xi) \leq \boldsymbol{\xi}_2,$$
$$4x_{11} + 7x_{12} + 6x_{21} + 5x_{22} \leq 36,\ y_1(\xi) \leq x_{11} + x_{12},$$
$$x_{ij} \geq 0 \text{ integer},\ y_2(\boldsymbol{\xi}) \leq x_{21} + x_{22},$$
$$y_1(\boldsymbol{\xi}), y_2(\boldsymbol{\xi}) \geq 0 \text{ integer}.$$

Letting $y_i^+(\xi) = \xi_i - y_i(\xi)$, one obtains an equivalent formulation,

$$\min\ 3x_{11} + 2x_{12} + 4x_{21} + 5x_{22} + E_{\boldsymbol{\xi}}\{4y_1^+(\boldsymbol{\xi}) + 6y_2^+(\boldsymbol{\xi})\} - 30$$
$$\text{s. t. } 20x_{11} + 30x_{21} \leq 100,\ y_1^+(\boldsymbol{\xi}) \geq \boldsymbol{\xi}_1 - \chi_1,$$
$$25x_{12} + 25x_{22} \leq 100,\ y_2^+(\boldsymbol{\xi}) \geq \boldsymbol{\xi}_2 - \chi_2,$$
$$4x_{11} + 7x_{12} + 6x_{21} + 5x_{22} \leq 36,\ y_1^+(\boldsymbol{\xi}), y_2^+(\boldsymbol{\xi}) \geq 0 \text{ and integer},$$
$$x_{11} + x_{12} = \chi_1,$$
$$x_{21} + x_{22} = \chi_2,$$
$$x_{ij} \geq 0 \text{ and integer}.$$

This representation puts the problem under the form of a simple recourse model with expected shortage only.

Let us start with the null solution, $x_{ij} = 0, \chi_i = 0, i, j = 1, 2$ with $\theta_i = -\infty, i = 1, 2$. We compute $u(0) = E[\boldsymbol{\xi}]^+ = \mu^+ = 3$; hence $\Psi_1(0) = 12, \Psi_2(0) = 18$, where we have dropped the constant, -30, from the objective for these computations. To construct the first optimality cuts, we also compute $u(1) = u(0) - 1 + F(0) = 2 + .05 = 2.05$. Thus, $E_{11} = 4(3 - 2.05) = 3.8, e_{11} = 4(1 * 3 - 0 * 2.05) = 12$, defining the optimality cut $\theta_1 + 3.8\chi_1 \geq 12$. As $\chi_2 = \chi_1, E_{21}$ and e_{21} are just 1.5 times E_{11} and e_1, respectively, yielding the optimality cut $\theta_2 + 5.7\chi_2 \geq 18$.

The current problem becomes

$$\min 3x_{11} + 2x_{12} + 4x_{21} + 5x_{22} - 30 + \theta_1 + \theta_2$$

$$\begin{aligned}
\text{s. t.} \quad & 20x_{11} + 30x_{21} \leq 100, \\
& 25x_{12} + 25x_{22} \leq 100, \\
& 4x_{11} + 7x_{12} + 6x_{21} + 5x_{22} \leq 36, \\
& x_{11} + x_{12} = \chi_1, \\
& x_{21} + x_{22} = \chi_2, \\
& \theta_1 + 3.8\chi_1 \geq 12, \\
& \theta_2 + 5.7\chi_2 \geq 18, \\
& x_{ij} \geq 0, \text{ integer.}
\end{aligned}$$

We obtain the solution $x_{11} = 0, x_{12} = 4, x_{21} = 1, x_{22} = 0, \theta_1 = -3.2, \theta_2 = 12.3$. We compute $u(4) = u(0) + \sum_{l=0}^{3}(F(l) - 1) = 0.31936$ and $\Psi_1(4) = 1.277 > \theta_1$. A new optimality cut is needed for $\Psi_1(\cdot)$. Because $\Psi(5) = 0.5385$, the cut is $0.739\chi_1 + \theta_1 \geq 4.233$. We also have $u(1) = 2.05$, hence $\Psi_2(1) = 12.3 = \theta_2$, so no new cut is generated for $\Psi_2(\cdot)$.

At the next iteration, with the extra optimality cut on θ_1, we obtain a new solution of the current problem as $x_{11} = 0$, $x_{12} = 2$, $x_{21} = 3$, $x_{22} = 0$, $\theta_1 = 4.4$, $\theta_2 = 0.9$. Here, two new optimality cuts are needed:

$$2.312\chi_1 + \theta_1 \geq 9.623$$

and

$$2.117\chi_2 + \theta_2 \geq 10.383.$$

The next iteration gives $x_{11} = 0$, $x_{12} = 3$, $x_{21} = 2$, $x_{22} = 0$, $\theta_1 = 2.688$, $\theta_2 = 6.6$ as a solution of the current problem. Because $\Psi_2(2) = 7.5 > \theta_2$, a new cut is generated, i.e., $3.467\chi_2 + \theta_2 \geq 14.435$. The next iteration point is $x_{11} = 0$, $x_{12} = 3$, $x_{21} = 2$, $x_{22} = 0$, $\theta_1 = 2.688$, $\theta_2 = 7.5$, which is the optimal solution with total objective value -5.812.

b. The case where $S = 1$, χ not integral

Details can again be found in Louveaux and van der Vlerk [1993]; we illustrate the results with an example. Consider Example 3 but with the x_{ij}s continuous. Because we still assume the random variables follow a Poisson distribution, the example indeed falls into the category $S = 1$; only integer realizations are possible.

For a given component i, the ρ_i-approximation rooted at an integer defines the convex hull of the function $\Psi_i(.)$. All optimality cuts defined at integer points are thus valid inequalities. If we take Example 3 again and impose all optimality cuts at integer points, the continuous solution is $x_{11} = 0, x_{12} = 3, x_{21} = 2, x_{22} = 0$, and no extra cuts are needed here. Now assume the objective coefficients of x_{12} and x_{21} are 1 and 4.5 (instead of 2 and 4). The solution of the stochastic program with continuous first-stage decisions and all optimality cuts imposed at integer points becomes $x_{11} = 0, x_{12} = 4, x_{21} = 1.33, x_{22} = 0$, and thus, $\chi_1 = 4, \chi_2 = 1.33$.

We now illustrate how to deal with a noninteger value of some χ_i. Now, $u(1.33) = 3 - 1 + F(0) = 2.05$ and therefore $\Psi_2(1.33) = 12.3 > \theta_2$. This requires imposing a new optimality cut. By Lemma 8, we know $\Psi_2(.)$ is constant within $(1, 2)$ with value 12.3.

$$\text{Let} \quad \delta_a = 1 \text{ if } \chi_2 > 1 \text{ and } 0 \text{ otherwise,}$$
$$\delta_b = 1 \text{ if } \chi_2 < 2 \text{ and } 0 \text{ otherwise.}$$

The cut imposes that $\theta_2 \geq 12.3$ if $1 < \chi_2 < 2$, i.e., if $\delta_a = \delta_b = 1$. This is realized by the following constraints:

$$\chi_2 \leq 1 + 10\delta_a, \chi_2 \geq (1 + \epsilon)\delta_a,$$
$$\chi_2 \leq 10 - (8 + \epsilon)\delta_b, \chi_2 \geq 2 - 2\delta_b,$$
$$\theta_2 \geq 12.3 - 100(2 - \delta_a - \delta_b),$$

where 10 and 100 are sufficiently large numbers to serve as bounds on χ_2 and $-\theta_2$, respectively, and ϵ is a very small number. Thus, defining a function $\Psi_i(.)$ to be constant in some interval requires two extra binary variables and three extra constraints. It is thus reasonable to first consider optimality cuts that define the convex hull.

Continuing the example, we solve the current problem with the three additional constraints. The solution is $x_{11} = 0, x_{12} = 3.43, x_{21} = 2, x_{22} = 0$ with $\chi_1 = 3.43, \chi_2 = 2, \theta_1 = 2.08, \theta_2 = 7.5$. Thus, one more set of cuts is needed to define Ψ_1 in the interval $(3, 4)$. The final solution is $x_{11} = 0, x_{12} = 3, x_{21} = 1, x_{22} = 0$, $\theta_1 = 2.689, \theta_2 = 12.3$, and $z = -7.51$.

8.3   Binary First-Stage Variables

In this section, we present optimality cuts derived by Laporte and Louveaux [1993] for problems with a binary first stage. These cuts are weaker than those defined by the integer L-shaped method. On the other hand, they apply even if the random variables take on many realizations, a situation where the integer L-shaped method would be inefficient. Observe that at this level no restriction is imposed on the second-stage problem. However, in addition to Assumption 3, we need the existence of a lower bound.

Assumption 10. *There exists a finite lower bound L satisfying*

$$L \leq \min_x \{\mathcal{Q}(x)|Ax = b, x \in X\}.$$

Assumption 3 seems like a minimal assumption one can make. In practice, however, it is far from being satisfied in all circumstances. As an example, if the second stage consists of a mixed integer program and the random variables are continuous, no general method is yet available to obtain $Q(x)$. On the other hand, many examples that satisfy Assumption 3 are available, in particular for location and routing problems. In Assumption 10, no requirement is made that the bound L should be tight, although it is desirable to have L as large as possible. Examples of how to find L will be given later.

Proposition 11. *Let $x_i = 1, i \in S$, and $x_i = 0, i \notin S$ be some first-stage feasible solution. Let $q_S = Q(x)$ be the corresponding recourse function value. Define the optimality cut as*

$$\theta \geq (q_S - L)(\sum_{i \in S} x_i - \sum_{i \notin S} x_i) - (q_S - L)(|S| - 1) + L. \qquad (3.1)$$

Then the set of optimality cuts (3.1) defined for all first-stage feasible solutions is a valid set of optimality cuts.

Proof: Define

$$\delta(x, S) = \sum_{i \in S} x_i - \sum_{i \notin S} x_i. \qquad (3.2)$$

Now, $\delta(x, S)$ is always less than or equal to $|S|$. It is equal to $|S|$ only if $x_i = 1, i \in S$, and $x_i = 0, i \notin S$. In that case, the right-hand side of (3.1) takes the value q_S and the constraint $\theta \geq q_S$ is valid as $Q(x) = q_S$. In all other cases, $\delta(x, S)$ is smaller than or equal to $|S| - 1$, which implies that the right-hand side of (3.1) takes a value smaller than or equal to L, which by Assumption 2 is a valid lower bound on $Q(x)$ for all feasible x. The result follows from the fact that there is only a finite (although potentially very large) number of first-stage decisions.

Readers more familiar with geometrical representations may see (3.1) as a half-space, in the (δ, θ) space, situated above a line going through the two points $(|S|, q_S)$ and $(|S| - 1, L)$. $\square$

Example 4

Consider a two-stage program, where the second stage is given by

$$\min -2y_1 - 3y_2,$$
$$\text{s. t. } y_1 + 2y_2 \leq \xi_1 - x_1,$$
$$y_1 \leq \xi_2 - x_2$$
$$y \geq 0, \text{ integer.}$$

Assume $\xi = (2, 2)^T$ or $(4, 3)^T$ with equal probability $1/2$ each. Find a lower bound L on $Q(x)$ and derive a cut of type (3.1) if the current iterate point is $x = (0, 1)^T$.

1. The second stage is equivalent to $-\max 2y_1 + 3y_2$. Because the first-stage decisions are binary, largest values of ys are obtained with $x = (0,0)^T$. To obtain a lower bound L, we simply drop the requirement that y should be integer and solve

$$\min -2y_1 - 3y_2$$
$$\text{s.t. } y_1 + 2y_2 \leq \xi_1,$$
$$y_1 \leq \xi_2,$$
$$y_1, y_2 \geq 0.$$

For $\xi = (2,2)^T$, the solution is $y = (2,0)^T$ and $Q(x,\xi) = -4$, while for $\xi = (4,3)^T$, the solution is $y = (3,0.5)^T$ with $Q(x,\xi) = -7.5$. This results in $L = 0.5 * (-4) + 0.5 * (-7.5) = -5.75$. (Alternatively, in this simple example, we may have maintained the requirement that y is integer and obtained the better bound $L = -5.5$. In general, this approach seems more difficult to implement. We continue here with $L = -5.75$.)

2. Here, $\delta(x, S) = x_2 - x_1$ because $x_1 = 0$ and $x_2 = 1$. For $\xi = (2,2)^T$, the second stage becomes

$$\min -2y_1 - 3y_2$$
$$\text{s.t. } y_1 + 2y_2 \leq 2,$$
$$y_1 \leq 1,$$
$$y_1, y_2 \geq 0, \text{ integer}$$

with solution $y = (0,1)^T$ and $Q(x,\xi) = -3$. For $\xi = (4,3)^T$, the second stage becomes

$$\min -2y_1 - 3y_2$$
$$\text{s.t. } y_1 + 2y_2 \leq 4,$$
$$y_1 \leq 2,$$
$$y_1, y_2 \geq 0, \text{ integer}$$

with solution $y = (2,1)^T$ and $Q(x,\xi) = -7$. We conclude that $q_S = -5$ and that the optimality cut (3.1) reads

$$\theta \geq 0.75(x_2 - x_1) - 5.75.$$

From a theoretical and practical point of view, it is not satisfactory to have a method that could lead to considering all first-stage feasible decision vectors. Improvements can be sought in two directions. One way is to find more efficient optimality cuts. The second approach is to append lower bounding functionals.

a. Improved optimality cuts

Improvement on (3.1) can be obtained when more information is available on $\mathcal{Q}(x)$, such as other bounds.

Define the set $N(s, S)$ of so-called *s-neighbors* of S as the set of solutions $\{x | Ax = b, x \in X, \delta(x, S) = |S| - s\}$, where $\delta(x, S)$ is as in (3.2). Let $\lambda(s, S) \leq \min_{x \in N(s, S)} \mathcal{Q}(x), s = 0, \cdots, |S|$ with $\lambda(0, S) = q_S$.

Proposition 12. *Let* $x_i = 1, i \in S, x_i = 0, i \notin S$ *be some solution with* $q_S = \mathcal{Q}(x)$. *Define* $a = \max\{q_S - \lambda(1, S), (q_S - L)/2\}$. *Then*

$$\theta \geq a(\sum_{i \in S} x_i - \sum_{i \notin S} x_i) + q_S - a|S| \qquad (3.3)$$

is a valid optimality cut.

Proof. For an s-neighbor, the right-hand side of (3.3) is equal to $q_S - as$. This is a valid lower bound on $\mathcal{Q}(x)$. This is obvious for $s = 0$. When $s = 1$, $q_S - a$ is, by construction, bounded above by $q_S - (q_S - \lambda(1, S)) = \lambda(1, S)$, which by definition is a lower bound on 1-neighbors. When $s = 2, q_S - 2a \leq q_S - 2(q_S - L)/2 = L$. Finally, for $s \geq 3, q_S - as \leq q_S - 2a$, because $a \geq 0$. Hence, $q_S - as \leq L$. Convergence is again guaranteed by $\theta \geq q_S$ when $\delta(x, S) = |S|$ and (3.3) improves on (3.1) for all 1-neighbors. The reader more familiar with geometrical representations may now see (3.3) as a halfspace in the (δ, θ) space, situated above a line going through the two points $(|S|, q_S)$ and $(|S| - 1, \lambda(1, S))$ when $a = q_S - \lambda(1, S)$, or the two points $(|S|, q_S)$ and $(|S| - 2, L)$ when $a = (q_S - L)/2$. $\square$

A further improvement for s-neighbors is sometimes possible.

Proposition 13. *Let* $x_i = 1, i \in S, x_i = 0, i \notin S$ *be some solution with* $q_S = \mathcal{Q}(x)$. *Let* $1 \leq t \leq |S|$ *be some integer. Then (3.3) holds with*

$$a = \max\{\max_{s \leq t}(q_S - \lambda(s, S))/s; (q_S - L)/(t + 1)\}. \qquad (3.4)$$

Proof. As before, for an s-neighbor, the right-hand side of (3.3) is $q_S - as$. By (3.4), $as \geq q_S - \lambda(s, S)$, for all $s \leq t$. Thus, $q_S - as \leq \lambda(s, S)$, which, by definition, is a lower bound on $\mathcal{Q}(x)$ for all s-neighbors. When $s \geq t + 1, q_S - as \leq L$, and (3.3) remains valid. $\square$

As computing $\lambda(s, S)$ for $s \leq t$ with t large may prove difficult, the following proposition is sometimes useful.

Proposition 14. *Define* $\lambda(0, S) = q_S$. *Assume* $q_S > \lambda(1, S)$. *Then, if* $\lambda(s - 1, S) - \lambda(s, S)$ *is nonincreasing in s for all* $1 \leq s \leq \lfloor(q_S - L)/(q_S - \lambda(1, S))\rfloor$*, (3.3) holds with* $a = q_S - \lambda(1, S)$.

Proof. We have to show that in applying Proposition 13, the maximum in (3.4) is obtained for $q_S - \lambda(1, S)$. Let $t = \lfloor(q_S - L)/(q_S - \lambda(1, S))\rfloor$. For

$s \leq t, q_S - \lambda(s,S) = \sum_{i=1}^{s} (\lambda(i-1,S) - \lambda(i,S))$. By assumption, each term of the sum is smaller than the first term of the sum, i.e., $\lambda(0,S) - \lambda(1,S) = q_S - \lambda(1,S)$ so the total is less than s times this first term. By definition of t, we have $t+1 \geq (q_S - L)/(q_S - \lambda(1,S))$, or $q_S - \lambda(1,S) \geq (q_S - L)/(t+1)$. $\square$

Clearly, much of the implementation is problem-dependent. We illustrate here the use of these propositions in one example.

Example 5

Let $i = 1, \cdots, n$ denote n inputs and $j = 1, \cdots, m$ denote m outputs. Each input can be used to produce various outputs. First-stage decisions are represented by binary variables x_{ij} with costs c_{ij} and are equal to 1 if i is used to produce j and equal to 0 otherwise. If input i is used for at least one output, some fixed cost f_i is paid. To this end, the auxiliary variable z_i is defined equal to 1 if input i is used and 0 otherwise. The level of output j obtained when $x_{ij} = 1$ is a non-negative random variable ξ_{ij}. A penalty r_j is incurred whenever the level of output j falls below a required threshold d_j. This is represented by the second-stage variable y_j^ξ taking the value 1. The problem can be defined as:

$$\min \sum_{i=1}^{n} f_i z_i + \sum_{i=1}^{n} \sum_{j=1}^{m} c_{ij} x_{ij} + E_\xi (\sum_{j=1}^{m} r_j y_j^\xi) \qquad (3.5)$$

$$\text{s.t. } x_{ij} \leq z_i, \quad i = 1, \cdots, n, \quad j = 1, \cdots m, \qquad (3.6)$$

$$\sum_{i=1}^{n} \xi_{ij} x_{ij} + d_j y_j^\xi \geq d_j, \quad j = 1, \cdots, m, \qquad (3.7)$$

$$x_{ij}, z_i, y_j^\xi \in \{0,1\}, \quad i = 1, \cdots, n, \quad j = 1, \cdots, m, \qquad (3.8)$$

where, in practice, the x_{ij} variables are only defined for the possible combinations of inputs and outputs. In this problem, the second-stage recourse function only depends on the x decisions so that the z variables may be left over in our analysis of optimality cuts. Moreover, the second stage is easily computed as

$$Q(x) = \sum_{j=1}^{m} r_j P(\sum_{i \in S(j)} \xi_{ij} < d_j), \qquad (3.9)$$

where

$$S(j) = \{i | x_{ij} = 1\}.$$

Let $S = \cup_{j=1}^{m} \{(i,j) | i \in S(j)\}$. To apply the propositions, we search for lower bounds, $\lambda(s,S)$, on the recourse function for all s-neighbors. To bound $q_S - \lambda(1,S)$, observe that 1-neighbors can be obtained in two distinct ways. The first way is to have one x_{ij}, with $(i,j) \in S$, going from one to zero and all other x_{ij}s being unchanged. This implies for that particular j that, in (3.9), $P(\sum_{i \in S(j)} \xi_{ij} < d_j)$ increases in the neighboring solution, as $S(j)$ would contain one fewer term. Thus, for this type of 1-neighbor, $Q(x)$ is increased.

The second way is to have one x_{ij}, with (i,j) not in S, going from zero to one, all other x_{ij}s being unchanged. For that particular j, $P(\sum_{i \in S(j)} \xi_{ij} < d_j)$ decreases in the neighboring solution. To bound the decrease of $Q(x)$, we simply assume $P(\sum_{i \in S(j)} \xi_{ij} < d_j)$ vanishes so that

$$q_S - \lambda(1, S) \leq \max_j \{ r_j P(\sum_{i \in S(j)} \xi_{ij} < d_j) \}. \tag{3.10}$$

Also observe that in this example, Proposition 14 applies. Indeed, $q_S - \lambda(s, S)$ can be taken as the sum of the s largest values of $\{ r_j P(\sum_{i \in S(j)} \xi_{ij} < d_j) \}$. It follows that $\lambda(s - 1, S) - \lambda(s, S)$ is nonincreasing in s.

Moreover, in this example, we can also find lower bounding functionals. By looking at (3.7), the optimal solution of the continuous relaxation of the second stage is easily seen to be

$$y_j^\xi = r_j (d_j - \sum_{i=1}^{n} \xi_{ij} x_{ij})^+ / d_j, j = 1, \dots, m,$$

and therefore,

$$C(x) = E_\xi [\sum_j r_j (d_j - \sum_{i=1}^{n} \xi_{ij} x_{ij})^+ / d_j]. \tag{3.11}$$

In fact, we just need to compute

$$C(x) = E_\xi \sum_j r_j (d_j - \sum_{i \in S(j)} \xi_{ij})^+ / d_j. \tag{3.12}$$

From (3.11), we may immediately apply Corollary 6 as

$$\theta \geq q_S + \sum_{ij \in S} a_{ij}(x_{ij} - 1) + \sum_{ij \notin S} a_{ij} x_{ij} \tag{3.13}$$

with

$$a_{ij} = -r_j / d_j E_\xi [\xi_{ij} P(\sum_{\substack{l \in S(j) \\ l \neq i}} \xi_{lj} \leq d_j - \xi_{ij})], \; i \in S(j),$$

$$a_{ij} = -r_j / d_j E_\xi [\xi_{ij} P(\sum_{l \in S(j)} \xi_{lj} < d_j)], \; i \notin S(j),$$

and

$$q_S = C(x) \text{ as in (3.12)}.$$

Example 6

We take Example 5 and consider the following numerical data. Let $n = 4, m = 6, f_i = 10$, for all $i, r_j = 40$ for all j. Let the c_{ij} coefficients take values between 5 and 15 as follows:

$$
\begin{array}{ccccccc}
j = & 1 & 2 & 3 & 4 & 5 & 6 \\
i = \quad 1 & 10 & 12 & 8 & 6 & 5 & 14 \\
2 & 8 & 5 & 10 & 15 & 9 & 12 \\
3 & 7 & 14 & 4 & 11 & 15 & 8 \\
4 & 5 & 8 & 12 & 10 & 10 & 10.
\end{array}
$$

Assume the ξ_{ij} are independent Poisson random variables with parameters

$$
\begin{array}{ccccccc}
j = & 1 & 2 & 3 & 4 & 5 & 6 \\
i = \quad 1 & 4 & 4 & 5 & 3 & 3 & 8 \\
2 & 5 & 2 & 4 & 8 & 5 & 6 \\
3 & 2 & 8 & 3 & 4 & 7 & 5 \\
4 & 3 & 5 & 6 & 4 & 6 & 5
\end{array}
$$

and, finally, let the demands d_j be given by

$$
\begin{array}{ccccccc}
j = & 1 & 2 & 3 & 4 & 5 & 6 \\
d_j & 8 & 4 & 6 & 3 & 5 & 8.
\end{array}
$$

As already said, we may apply Proposition 14 to this example. A second possibility is to use the separability of $\mathcal{Q}(x)$ as

$$\mathcal{Q}(x) = \sum_{j=1}^{m} \mathcal{Q}_j(x) \tag{3.14}$$

with

$$\mathcal{Q}_j(x) = r_j P\left(\sum_{i \in S(j)} \xi_{ij} < d_j \right). \tag{3.15}$$

Bounding each $Q_j(x)$ separately, we define

$$\theta = \sum_{j=1}^{m} \theta_j \tag{3.16}$$

and use Propositions 13 or 14 to define a valid set of cuts for each θ_j separately. Indeed, for one particular j, we have

$$\theta_j = r_j P\left(\sum_{i \in S(j)} \xi_{ij} < d_j \right) \tag{3.17}$$

and

$$\lambda_j(1, S) = r_j \min_{t \notin S(j)} P\left(\sum_{i \in S(j)} \xi_{ij} + \xi_{tj} < d_j \right), \tag{3.18}$$

where $\lambda_j(1, S)$ denotes a lower bound on $\mathcal{Q}_j(x)$ for 1-neighbors of the current solution obtained by changing x_{ij}s for that particular j only. Note that in practice finding t is rather easy. Indeed, because all random variables are independent Poisson, t is simply given by the random variable $\xi_{tj}, t \notin S(j)$, with the largest parameter value.

We illustrate the generation of cuts for $j = 1$. First, a lower bound is obtained by letting $x_{i1} = 1$, for all i. This gives $L_1 = 1.265$.

Assume a starting solution $x_{ij} = 0$, all i, j. For $j = 1$, the probability in the right-hand side of (3.15) is 1. Thus, $\mathcal{Q}_1(x) = r_1 = 40$. Cut (3.3) becomes $\theta_1 \geq 40 - 19.368(x_{11} + x_{21} + x_{31} + x_{41})$ with the coefficient $a = 19.368$ obtained from $(q_{S,1} - L_1)/2$, where $q_{S,1}$ is the notation for the value of $\mathcal{Q}_1(x)$. The continuous cut (3.13) is

$$\theta_1 \geq 40 - 20x_{11} - 25x_{21} - 10x_{31} - 15x_{41}.$$

The next iterate point is, e.g., $x_{11} = 1, x_{21} = 0, x_{31} = 0, x_{41} = 1$. Cut (3.3) becomes $\theta_1 \geq -16.788 + 20.368(x_{11} - x_{21} - x_{31} + x_{41})$ with the coefficient $a = 20.368$ now obtained from $(q_{S,1} - \lambda_1(1, S))$ while the continuous cut (3.13) is

$$\theta_1 \geq 29.164 - 11.974x_{11} - 14.968x_{21} - 5.987x_{31} - 8.981x_{41}.$$

Cut (3.3) is stronger than (3.13) at the current iterate point with value 23.948 instead of 8.309. Also, as the coefficient a comes from $(q_{S,1} - \lambda_1(1, S))$ and $\lambda_1(1, S)$ is obtained when x_{21} becomes 1, (3.3) gives an exact bound on the solution $x_{11} = 1, x_{21} = 1, x_{31} = 0, x_{41} = 1$. It provides a nontrivial but nonbinding bound for other cases, such as $x_{11} = 0, x_{21} = x_{31} = x_{41} = 1$. On the other hand, (3.13) provides a nontrivial (but nonbinding) bound for some cases such as $x_{11} = 0, x_{21} = 1, x_{31} = 1, x_{41} = 0$, where (3.3) does not.

The algorithm for the full example with six outputs was simulated by adding cuts each time a new iterate point was found, then restarting the branch and bound. Cuts (3.3) and (3.13) were added each time the amount of violation exceeded 0.1. The number of iterate points is dependent on the strategies used in the branch and bound. For this example, the largest number of iterate points was 21. In that case, the mean number of cuts per output was 6.833 cuts of type (3.13) and 2.5 cuts (3.3). As extreme cases, 10 improved optimality cuts were imposed for Output 1 and only 4 for Output 2, while 4 continuous cuts were imposed for Output 3 and only 1 for Output 5.

The optimal solution is $x_{11} = x_{13} = x_{15} = x_{16} = x_{21} = x_{22} = x_{24} = x_{41} = x_{42} = x_{43} = x_{45} = x_{46} = 1$; all other x_{ij}s are zero with first-stage cost 140 and penalty 13.26, for a total of 153.26. It strongly differs from the solution of the deterministic problem where outputs equal expected values: $x_{11} = x_{12} = x_{13} = x_{14} = x_{16} = x_{21} = x_{23} = x_{25} = 1$ with first-stage cost 97. The reason is that in the stochastic case, even if the expected output exceeds demand, the probability that the demand is not met is nonzero. In fact, the solution of the deterministic problem has a penalty of 87.59 for a total cost of 184.59 and a VSS of 31.33.

Exercises

1. Construct the cuts from the integer L-shaped method for Example 4, associated with the point $(0, 1)^T$.

   Compare the results by checking the bound on $\theta_1 + \theta_2$ by the integer L-shaped method and the bound in Example 4 on θ by (3.1) for the four possible points, $(0, 0), (0, 1), (1, 0), (1, 1)$ and, for some continuous points, $(1/2, 1/2), (1.2, 0), (0, 1.2)$, for example.

2. Extending (3.18), we obtain

$$\lambda_j(s, S) = r_j P\Big(\sum_{i \in S(j)} \xi_{ij} + \sum_{t \in J} \xi_{tj} < d_j \Big), \qquad (3.19)$$

   where J contains the indices of the s pairs $ij, i \notin S(j)$, with largest parameter values. Show that the assumptions of Proposition 13 hold.

3. Indicate why the wait-and-see solution cannot be reasonably computed in Example 5.

8.4 Other Approaches

a. Extensive forms and decomposition

Because structural properties of problems with mixed integer second-stage programs are unknown, problems of this type have been solved in practice by decomposing the second-stage variables into their discrete parts and continuous parts. Assuming a mixed second stage with binary variables, one can divide $y(\omega)^T = (y_B(\omega)^T, y_C(\omega)^T)$ where $y_B(\omega)$ is the vector of binary variables and $y_C(\omega)$ the vector of continuous variables. Partitioning q and W in a similar fashion, the classical two-stage program becomes

$$\min z = c^T x + E_{\boldsymbol{\xi}} q_B^T(\omega) y_B(\omega) + E_{\boldsymbol{\xi}} Q(x, y_B(\omega), \omega)$$
$$\text{s.t. } Ax = b,$$
$$x \in X, \ y_B(\omega) \in Y_B(\omega),$$

where

$$Q(x, y_B(\omega), \omega) = \min\{ q_C^T(\omega) y_C(\omega)$$
$$\text{s. t. } W_C y_C(\omega) \le h(\omega) - T(\omega)x - W_B y_B(\omega), y_C(\omega) \in Y_C(\omega) \}.$$

When ξ is a discrete random variable, this amounts to writing down the extensive form for the second-stage binary variables. When the number of realizations of ξ remains low, such a program is still solvable by the ordinary L-shaped method. An extension of this idea to a three-stage problem in the case of acquisition of resources can be found in Bienstock and Shapiro [1988].

The same idea applies for multistage stochastic programs having the block separable property defined in Section 3.5, provided the discrete variables correspond to the aggregate level decisions and the continuous variables correspond to the detailed level decisions. Then the multistage program is equivalent to a two-stage stochastic program, where the first stage is the extensive form of the aggregate level problems and the value function of the second stage for one realization of the random vector is the sum, weighted by the appropriate probabilities of the detailed level recourse functions for that realization and all its successors. This result is detailed in Louveaux [1986], where examples are provided.

Example 7

As an illustration, consider the warehouse location problem similar to those studied in Section 2.4. As usual, let

$$x_j = \begin{cases} 1 & \text{if plant } j \text{ is open,} \\ 0 & \text{otherwise,} \end{cases}$$

with fixed-cost c_j, and v_j, the size of plant j, with unit investment cost g_j, be the first-stage decision variables. Assume $k = 1, \cdots, K$ realizations of the demands d_i^k in the second stage. Let y_{ij}^k be the fraction of demand d_i^k served from j, with unit revenue q_{ij} (see Section 2.4c). Now, assume the possibility exists in the second stage to extend open plants by an extra capacity (size) of fixed value e_j at cost r_j. For simplicity, assume this extension can be made immediately available (zero construction delay).

To this end, let

$$w_j^k = \begin{cases} 1 & \text{if extra capacity is added to } j \\ & \text{when the second-stage realization is } k, \\ 0 & \text{otherwise.} \end{cases}$$

The two-stage stochastic program would normally read as

$$\max - \sum_{j=1}^n c_j x_j - \sum_{j=1}^n g_j v_j + \sum_{k=1}^K p_k \left(\max \sum_{i=1}^m \sum_{j=1}^n q_{ij} y_{ij}^k - \sum_{j=1}^n r_j w_j^k \right)$$

$$\text{s. t.} \sum_{j=1}^{n} y_{ij}^k \leq 1, k = 1, \cdots, K, i = 1, \ldots, m,$$

$$x_j \in \{0,1\}, v_j \geq 0, j = 1, \cdots, n,$$

$$\sum_{i=1}^{m} d_i^k y_{ij}^k - e_j w_j^k \leq v_j, \; k = 1, \cdots, K, j = 1, \cdots, n,$$

$$0 \leq y_{ij}^k \leq x_j, i = 1 \cdots, m, j = 1 \cdots, n,$$

$$k = 1 \ldots, K,$$

$$w_j^k \leq x_j, \; j = 1, \cdots, n, k = 1, \cdots, K$$

$$w_j^k \in \{0,1\}, \; j = 1, \cdots, n, k = 1 \ldots, K.$$

Using the extensive form for the binary variables, w_j^ks transforms it into

$$\max - \sum_{j=1}^{n} c_j x_j - \sum_{j=1}^{n} g_j v_j - \sum_{j=1}^{n} \sum_{k=1}^{K} p_k r_j w_j^k + \sum_{k=1}^{K} p_k \max \sum_{i=1}^{n} \sum_{j=1}^{n} q_{ij} y_{ij}^k$$

$$\text{s. t. } x_j \in \{0,1\}, v_j \geq 0, \; j = 1, \ldots, n,$$

$$\sum_{j=1}^{n} y_{ij}^k \leq 1, i = 1, \ldots, m, k = 1, \ldots, K,$$

$$w_j^k \leq x_j, \sum_{i=1}^{m} d_i^k y_{ij}^k \leq v_j + e_j w_j^k, j = 1, \ldots, n,$$

$$k = 1, \ldots, K,$$

$$w_j^k \in \{0,1\}, 0 \leq y_{ij}^k \leq x_j, i = 1, \ldots, m, j = 1, \ldots, n,$$

$$k = 1, \ldots, K.$$

Thus, at the price of expanding the first-stage program, one obtains a second stage that enjoys the good properties of continuous programs.

b. Asymptotic analysis

Approximation methods are commonly used in deterministic combinatorial optimization. This is justified by the fact that a large class of problems is known to be NP-hard. Loosely stated, for this class of problems, it is very unlikely that an algorithm will be found that would solve the problem in a number of operations polynomial in the problem data. This means that an exact solution of these problems can only be found for instances of small size. If the second stage of a stochastic problem corresponds to an NP-hard problem, it is pointless to design an exact method that would require the solution of the second stage for each realization of the random variable.

Approximation methods are typically based on the probabilistic error of some heuristic. We now illustrate one example, taken from Stougie [1987]. It is a hierarchical scheduling problem.

The first stage consists of determining the number x of identical machines at a unit cost of c. In the second stage, when the machines are available, they are used in parallel to process n jobs. The second stage consists of determining the optimal scheduling of the jobs. It is assumed that the processing times ξ_j are identically and independently distributed random variables with finite expectation μ and finite second moment. The second-stage objective consists of minimizing the completion time of the job completed last, also known as the *makespan*.

If we denote by $Q(x, \xi)$ the optimal makespan when x machines are available and the processing times are given by ξ (a vector of n components), the hierarchical scheduling problem takes the usual format

$$\min_x \{cx + \mathcal{Q}(x) | x \in Z_+\}. \tag{4.1}$$

Clearly, the difficulty is in finding $\mathcal{Q}(x)$. The idea is then to base the choice of x on an approximation of $\mathcal{Q}(x)$, obtained as follows. The total effective processing time is always equal to $\sum_{j=1}^n \xi_j$, whatever schedule is chosen. If the workload can be equally divided between the machines, the makespan is equal to $\frac{\sum_{j=1}^n \xi_j}{x}$. The approximation consists of taking this (idealistic) situation for $Q(x, \xi)$. Because processing times are independent and identically distributed with expectation μ, this corresponds to approximating $\mathcal{Q}(x)$ by $n\mu/x$. The original problem (4.1) becomes

$$\min_x \{cx + \frac{n\mu}{x} | x \in Z_+\}. \tag{4.2}$$

The optimal solution to (4.2), denoted by x^a for an approximate solution, is easily seen to be either $\lfloor \sqrt{\frac{n\mu}{c}} \rfloor$ or $\lceil \sqrt{\frac{n\mu}{c}} \rceil$, whichever gives the smallest value of $cx + n\mu/x$. There is no doubt that x^a may not always coincide with x^*, a solution to (4.1) and that $\frac{n\mu}{x}$ may strongly differ from $\mathcal{Q}(x)$. The reader may solve Exercise 2 to see this in a simple example.

On the other hand, this approximate solution has interesting asymptotic properties obtained by bounding $\mathcal{Q}(x)$ and letting n tend to infinity. We have already observed that $\frac{n\mu}{x}$ is a lower bound on $\mathcal{Q}(x)$. To obtain an upper bound, consider using the longest processing time first (LPT) rule in the second stage. This means that for a given vector of processing times, the jobs are ordered in decreasing processing times, then at each step assigned to the earliest available machine. If the workload is unevenly distributed, all jobs are started at last at $\frac{\sum_{j=1}^n \xi_j}{x}$. The job that is completed last has a processing time bounded by $\xi_{\max} = \max_{j=1,\cdots,n} \xi_j$.

It follows that

$$\frac{\sum_{j=1}^n \xi_j}{x} \leq Q(x, \xi) \leq Q^H(x, \xi) \leq \frac{\sum_{j=1}^n \xi_j}{x} + \xi_{\max}, \tag{4.3}$$

where $Q^H(x,\xi)$ denotes the makespan obtained when using the LPT heuristic.

Each term in (4.3) is divided by n and the limit when n tends to infinity is taken:

$$\lim_{n\to\infty}\frac{1}{n}\frac{\sum_{j=1}^{n}\xi_j}{x}\leq\lim_{n\to\infty}\frac{Q(x,\xi)}{n}$$

$$\leq\lim_{n\to\infty}\frac{Q^H(x,\xi)}{n}$$

$$\leq\lim_{n\to\infty}\left(\frac{1}{n}\frac{\sum_{j=1}^{n}\xi_j}{x}+\frac{1}{n}\xi_{\max}\right). \qquad (4.4)$$

By the strong law of large numbers (see, e.g., Chung [1974]), the first term equals $\frac{\mu}{x}$ with probability one. Convergence theorems on order statistics (see, e.g., Galambos [1978]) show that $\lim_{n\to\infty}\frac{\xi_{\max}}{n}=0$ with probability one. Hence, $\lim_{n\to\infty}\frac{Q(x)}{n}=\frac{\mu}{x}$, with probability one. It follows that indeed $\frac{\mu}{x}$ asymptotically coincides with $\frac{Q(x)}{n}$.

Now define $z^H=cx^a+Q^H(x^a)$ and $z^*=cx^*+Q(x^*)$. Adding cx to each term in (4.3), then taking expectations, it follows that

$$cx+\frac{n\mu}{x}\leq cx+Q(x)\leq cx+Q^H(x)\leq cx+\frac{n\mu}{x}+E(\xi_{\max}),$$

from which it turns out that

$$2\sqrt{n\mu c}\leq z^*\leq z^H\leq 2\sqrt{n\mu c}+E(\xi_{\max}),$$

and, finally,

$$\frac{z^H-z^*}{z^*}\leq\frac{E(\xi_{\max})}{\sqrt{2n\mu c}}. \qquad (4.5)$$

Under the assumption that the processing times have a finite second moment, the right-hand side tends to zero as $n\to\infty$, which proves the asymptotic optimality of x^a.

It can also be proven that z^H not only tends to z^* but also to WS, the wait-and-see value. This property has received the name of asymptotic clairvoyance (Lenstra, Rinnooy Kan, and Stougie [1984]). The asymptotic analysis of heuristics of other problems can be found in Rinnooy Kan and Stougie [1988]. A more recent study in the context of location routing is given by Simchi-Levi [1992]. One should be aware that in several cases, the results are truly asymptotic in the sense that they only apply for very large values of n. Psaraftis [1984] gives one such example.

c. Final comment

Many other approaches can be tried to solve stochastic integer programs. To cite just some, one may consider that Markov chains could be used for the analysis of multistage models with finite state space and finite action set. Also, dynamic programming may be a relevant tool to solve small or specially structured stochastic integer programs; see Lageweg et al. [1988] or Takriti [1994], as an example. One may also consider using Lagrangian decomposition techniques.

Exercises

1. In Example 7, assume a given construction delay for the warehouses in the second stage. Is it still possible to decompose the second stage?

2. Consider the hierarchical scheduling problem with $n = 3$ jobs. Assume the processing times are i.i.d. with a Bernoulli distribution with probability $1/2$. It follows that the number of jobs requiring one unit of time is $0, 1, 2$, or 3 with probability $1/8, 3/8, 3/8$, and $1/8$, respectively. (The other jobs require no processing time.)

   (a) Construct a table for $x = 0, 1, 2$, or 3 and ξ corresponding to 0, $1, 2$, or 3 processing times equal to 1. Obtain $\mathcal{Q}(x)$ and compare with $n\mu/x$. What is the largest relative error?

   (b) Obtain x^* and x^a for $c = 0.7, 0.4$, and 0.2, respectively. For each case, compare the decisions and, when relevant, the relative error in using x^a instead of x^* .

Part IV

Approximation and Sampling Methods

9

Evaluating and Approximating Expectations

The evaluation of the recourse function or the probability of satisfying a set of constraints can be quite complicated. This problem is basically one of numerical integration in high dimensions corresponding to the random variables. The general problem requires some form of approximation. General quadrature formulas do not fit the structure of stochastic programs because they typically apply to smooth functions in low dimensions that may not have any known convexity properties. In Section 1 of this chapter, we review some basic procedures, but in convex stochastic programs, we often do not have differentiability of the recourse function. Other forms of numerical integration are needed.

In the remaining sections of this chapter, we consider approximations that give lower and upper bounds on the expected recourse function value in two-stage problems. The intent of these procedures is to provide progressively tighter bounds until some a priori tolerance has been achieved. This chapter focuses on results for two-stage problems, while Chapter 11 discusses the multistage case. In Chapter 10, we will describe approximations built on Monte Carlo sampling.

Section 2 in this chapter discusses the most common type of approximations built on discretizations of the probability distribution. The lower bounds are extensions of midpoint approximations, while the upper bounds are extensions of trapezoidal approximations. The bounds are refined using partitions of the region. Other improvements are possible using more tightly constrained moment problem models of the approximation, as described in Section 5.

Section 3 discusses computational uses for bounds. The goal is to place the bounds effectively into computational methods. We present uses of the bounds in the L-shaped method, inner linearizations, and separable nonlinear programming procedures.

Section 4 discusses some basic bounding approaches for probabilistic constraints. General forms are presented briefly. These methods are based on fundamental inequalities from probability.

Section 5 presents a variety of extensions of the previous bounding approaches. It presents bounds based on approximations of the recourse function. The basic idea is to bound the objective function above and below by functions that are simply integrated, such as separable functions. We present the basic separable piecewise linear upper bounding function and various methods based on this approach. We also discuss results for particular moment problem solutions. We consider bounds based on second moment information and allowances for unbounded support regions.

Section 6 in this chapter gives basic results on convergence of approximations and bounding procedures. Most of the following results are based on these convergence ideas.

9.1   Direct Solutions with Multiple Integration

In this section, we again consider the basic stochastic program in the form

$$\min_{x}\{c^T x + \mathcal{Q}(x)|Ax = b, x \geq 0\}, \qquad (1.1)$$

where $\mathcal{Q}$ is the expected recourse function, $\int_{\Omega}[Q(x,\omega)]P(d\omega)$, where we use $P(d\omega)$ in place of $dF(\omega)$ to allow for general probability measure convergence. We again have $Q(x,\omega) =$

$$\min_{y(\omega)}\{q(\omega)^T y(\omega)|Wy(\omega) = h(\omega) - T(\omega)x, y(\omega) \geq 0\}, \qquad (1.2)$$

where we assume two stages and no probabilistic constraints for now.

As we mentioned previously, we can always treat (1.1) as a standard nonlinear program if we can evaluate $\mathcal{Q}(x)$ and perhaps its derivatives. The major difficulty of stochastic programming is, of course, just such an evaluation. These function evaluations all involve multiple integration with potentially large numbers (on the order of 1000 or more) of random variables. This section considers some of the basic techniques from numerical integration that have been attempted in the context of stochastic programming. Remaining sections consider various approximations that lead to computable problems.

Numerical integration procedures are generally built around formulas that apply only in small dimensions (see, e.g., Stroud [1971]). For some special functions defined over specific regions, efficient computations are possible, but these results do not generally carry over to the more general setting of the integrand, $Q(x, \omega)$. This function is piecewise linear in (1.2) as a function of ω and, hence, has many nondifferentiable points. The error analysis from standard smooth integrations (built on Peano's rule) cannot apply. In fact, quadrature formulas built on low-order polynomials may produce poor results when other simple calculations are exact (Exercise 1).

Generalizations of the basic trapezoid and midpoint approaches in numerical integration obtain bounds, however, when convexity properties of Q are exploited. Problem structure is in fact a key to obtaining computable approximations of the multiple integral.

The simple recourse example is the best case for exploitation of problem structure. In this case, $Q(x, \omega)$, becomes separable into functions of each component of $h(\omega)$, the right-hand side vector in (1.2). We obtain $Q(x) = \sum_{i=1}^{m_2} Q_i(x)$ as in (3.1.9), which only requires integration with respect to each h_i separately. As we described in Chapter 6, this allows the use of general nonlinear programming algorithms.

In general, the stochastic linear program recourse function can also be written in terms of bases in W. Suppose the set of bases in W is $\{B_i, i \in I\}$. Let $\pi_i(\omega)^T = q_{B_i}^T B_i^{-1}$. Then

$$Q(x, \omega) = \max_i \{\pi_i(\omega)^T (h(\omega) - T(\omega)x) | \pi_i(\omega)^T W \le q(\omega)\}, \qquad (1.3)$$

where, if $q(\omega)$ is constant (i.e., not random), the evaluation reduces to finding the maximum value of the inner product over the same feasible set for all ω. With $q(\omega)$ constant,

$$Q(x) = \sum_{i \in I} \int_{\Omega_i} \{\pi_i^T (h(\omega) - T(\omega)x)\} P(d\omega), \qquad (1.4)$$

where $\Omega_i = \{\omega | \pi_i^T(h(\omega) - T(\omega)x) \ge \pi_j^T(h(\omega) - T(\omega)x), j \ne i\}$. The integrand in (1.4) is linear, so we have

$$Q(x) = \sum_i \pi_i^T (\bar{h}_i - \bar{T}_i x), \qquad (1.5)$$

where $\bar{h}_i = \int_{\Omega_i} h_i P(d\omega)$ and $\bar{T}_i = \int_{\Omega_i} T_i P(d\omega)$. Thus, if each Ω_i can be found, then the numerical integration reduces to finding the expectations of the random parameters over the regions Ω_i, i.e., the conditional expectation on Ω_i.

The Ω_i are indeed polyhedral (Exercise 2). This yields direct procedures if these regions are simple enough to have explicit integration formulas. Unfortunately, this is not the case for the Ω_i regions that are common in stochastic programs with recourse. In problems with probabilistic constraints, however, there are possibilities for creating deterministic equivalents when the data are, for example, normal as in Theorem 3.18. In general, we must use some form of approximation.

In the following chapters, we explore several methods for approximating the value function and its subgradient in stochastic programming. The basic approaches are either approximations with known error bounds or approximations based on Monte Carlo procedures that may have associated confidence intervals. In the remainder of this chapter and Chapter 11, we will explore bounding approaches, while in Chapter 10 we also consider methods based on sampling.

Exercises

1. The principle of Gaussian quadrature is to find points and weights on those points that yield the correct integral over all polynomials of a certain degree. For example, we can solve for points, ξ_1, ξ_2, and weights, p_1, p_2, so that we have a probability $(p_1 + p_2 = 1)$ and distribution that matches the mean, $(p_1\xi_1 + p_2\xi_2 = \bar{\xi})$, the second moment, $(p_1\xi_1^2 + p_2\xi_2^2 = \bar{\xi}^{(2)})$, and the third moment, $(p_1\xi_1^3 + p_2\xi_2^3 = \bar{\xi}^{(3)})$. Solve this for a uniform distribution on $[0,1]$ to yield the two points, 0.211 and 0.788, each with probability 0.5.

   (a) Verify that this distribution matches the expectation of any polynomial up to degree three over $[0, 1]$.

   (b) Consider a piecewise linear function, f, with two linear pieces and $0 \leq f(\xi) \leq 1$ for $0 \leq \xi \leq 1$. How large a relative error can the Gaussian quadrature points give? Can you use two other points that are better?

2. Show that each Ω_i is polyhedral.

9.2   Discrete Bounding Approximations

The most common procedures in stochastic programming approximations are to find some relatively low cardinality discrete set of realizations that somehow represents a good approximation of the true underlying distribution or whatever is known about this distribution. The basic procedures are extensions of Jensen's inequality ([1906], generalization of the midpoint approximation) and an inequality due to Edmundson [1956] and Madansky [1959], the *Edmundson-Madansky inequality*, a generalization of the trapezoidal approximation. For convex functions in ξ, Jensen provides a lower bound while Edmundson-Madansky provides an upper bound. Significant refinements of these bounds appear in Huang, Ziemba, and Ben-Tal [1977], Kall and Stoyan [1982] and Frauendorfer [1988b].

We refer to a general integrand $g(x, \xi)$. Our goal is to bound $E(g(x)) = E_\xi[g(x,\xi)] = \int_\Xi g(x,\xi)P(d\xi)$. The basic ideas are to partition the support Ξ into a number of different regions (analogous to intervals in one-dimensional integration) and to apply bounds in each of those regions. We let the partition of Ξ be $\mathcal{S}^\nu = \{S^l, l = 1, \ldots, \nu\}$. Define $\xi^l = E[\boldsymbol{\xi}|S^l]$ and $p^l = P[\boldsymbol{\xi} \in S^l]$. The basic lower bounding result is the following.

Theorem 1. *Suppose that $g(x, \cdot)$ is convex for all $x \in D$. Then*

$$E(g(x)) \geq \sum_{l=1}^{\nu} p^l g(x, \xi^l). \tag{2.1}$$

Proof: Write $E(g(x))$ as

$$
\begin{aligned}
E(g(x)) &= \sum_{l=1}^{\nu} \int_{S^l} g(x, \boldsymbol{\xi}) P(d\boldsymbol{\xi}) \\
&= \sum_{l=1}^{\nu} p^l E[g(x, \boldsymbol{\xi})|S^l] \\
&\leq \sum_{l=1}^{\nu} p^l g(x, E[\boldsymbol{\xi}|S^l]),
\end{aligned}
\tag{2.2}
$$

where the last inequality follows from Jensen's inequality that the expectation of a convex function of some argument is always greater than or equal to the function evaluated at the expectation of its argument, i.e., $E(g(\boldsymbol{\xi})) \geq g(E(\boldsymbol{\xi}))$ (see Exercise 1). $\square$

This result applies directly to approximating $\mathcal{Q}(x)$ by $\mathcal{Q}^\nu(x) = \sum_{l=1}^{\nu} p^l Q(x, \xi^l)$. The approximating distribution P^ν is the discrete distribution with *atoms*, i.e., points ξ^l of probability $p^l > 0$ for $l = 1, \ldots, \nu$. By choosing $\mathcal{S}^{\nu+1}$ so that each $S^l \in \mathcal{S}^{\nu+1}$ is completely contained in some $S^{l'} \in \mathcal{S}^\nu$, the approximations actually improve, i.e.,

$$E(g(x)) \geq E^{\nu+1}(g(x)) \geq E^\nu(g(x)). \tag{2.3}$$

Various methods can achieve convergence in distribution of the P^ν to P. An example is given in Exercise 2.

In general, the goal of refining the partition from ν to $\nu + 1$ is to achieve as great an improvement as possible. We will describe the basic approaches; more details appear in Birge and Wets [1986], Frauendorfer and Kall [1988], and Birge and Wallace [1986].

Three basic decisions are to choose the cell, $S^{\nu^*} \in \mathcal{S}^\nu$, in which to make the partition, to choose the direction in which to split, S^{ν^*}, and to choose the point at which to make the split.

The reader should note that this section contains notation specific to bounding procedures. To keep the notation manageable, we reuse some from previous sections, including a and b for endpoints of rectangular regions and c for points within these intervals at which to subdivide the region. For ease of exposition, suppose that the sets S^l are all rectangular, defined by $[a_1^l, b_1^l] \times \cdots \times [a_N^l, b_N^l]$. The most basic refinement scheme for $l = \nu^*$ is to find i^* and $c_{i^*}^l$ so that $S^l(\nu)$ splits into $S^l(\nu+1) = [a_1^l, b_1^l] \times \cdots [a_{i^*}^l, c_{i^*}^l] \times [a_N^l, b_N^l]$ and $S^{\nu+1}(\nu+1) = [a_1^l, b_1^l] \times \cdots [c_{i^*}^l, b_{i^*}^l] \times [a_N^l, b_N^l]$.

If we also have an upper bound $UB(S^l) \geq E[g(x,\xi)|\xi \in S_l^\nu]$ for each cell S^l, then the most likely choice for S^{ν^*} is the cell that maximizes $p_l(UB(S^l) - g(x,\xi^l))$, which bounds the error attributable to the approximation on S^l. Reducing this greatest partition error appears to offer the most hope in reducing the error on the $\nu + 1$ approximation.

The direction choice is less clear. The general idea is to choose a direction in which the function g is "most nonlinear." The use of subgradient (dual price) information for this process was discussed in Birge and Wets [1986]. Frauendorfer and Kall [1988] improved on this and reported good results by considering all 2^{m+1} pairs, (α_j, β_j), of vertices of S^l, where $\alpha_j = (\gamma_1^l, \ldots, a_i^l, \ldots, \gamma_N^l)$ and $\beta_j = (\gamma_1^l, \ldots, b_i^l, \ldots, \gamma_N^l)$ with $\gamma_i^l = a_i^l$ or b_i^l. Given x, they assume a dual vector, π_{α_j}, at $Q(x, \alpha_j)$ and π_{β_j} at $Q(x, \beta_j)$. Because these represent subgradients of the recourse function $Q(x, \cdot)$, we have $Q(x, \beta_j) - (Q(x, \alpha^j) + \pi_{\alpha_j}^T(\beta_j - \alpha_j)) = \epsilon_j^1 \geq 0$ and $Q(x, \alpha_j) - (Q(x, \beta^j) + \pi_{\beta_j}^T(\alpha_j - \beta_j)) = \epsilon_j^2 \geq 0$. They then choose k^* that maximizes $\min\{\epsilon_k^1, \epsilon_k^2\}$ over k. They let i^* be i such that α_{k^*} and β_{k^*} differ in the ith coordinate. The position c_{i^*} is then chosen so that $Q(x, \beta^{k^*}) + \pi_{\beta_{k^*}}^T(c_{i^*} - b_{i^*}) = Q(x, \alpha^{k^*}) + \pi_{\alpha_{k^*}}^T(c_{i^*} - a_{i^*})$. (See Figure 1, where we use π for the subgradient at (a_1, b_2) and ρ for the subgradient at (a_1, a_2).) The general idea is then to choose the direction that yields the maximum of the minimum of linearization errors in each direction.

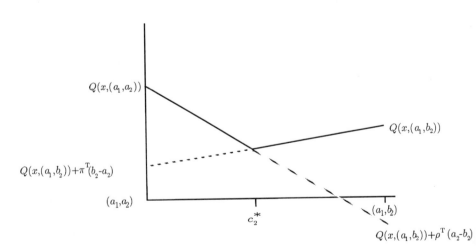

FIGURE 1. Choosing the direction according to the maximum of the minimum linearization errors.

Refinement schemes clearly depend on having upper bounds available. These bounds are generally based on convexity properties of g and the ability to obtain each ξ in terms of the extreme points. The fundamental result is the following theorem that also appears in Birge and Wets [1986]. In the following, we use P as the measure on Ω instead of Ξ because we wish to obtain a different measure derived from this. In context, this change should not cause confusion. We also let $ext\Xi$ be the set of extreme points of $co\Xi$ and $\mathcal{E}$ is *Borel field* of $ext\Xi$, in this case, the collections of all subsets of $ext\Xi$.

Theorem 2. *Suppose that $\xi \mapsto g(x,\xi)$ is convex and Ξ is compact. For all $\xi \in \Xi$, let $\phi(\xi, \cdot)$ be a probability measure on $(ext\Xi, \mathcal{E})$, such that*

$$\int_{e \in ext\Xi} e\phi(\xi, de) = \xi, \tag{2.4}$$

and $\omega \mapsto \phi(\xi(\omega), A)$ is measurable for all $A \in \mathcal{E}$. Then

$$E(g(x)) \leq \int_{e \in ext\Xi} g(x, e)\lambda(de), \tag{2.5}$$

where λ is the probability measure on $\mathcal{E}$ defined by

$$\lambda(A) = \int_{\Omega} \phi(\xi(\omega), A)P(d\omega). \tag{2.6}$$

Proof. Because g is convex in ξ, for ϕ,

$$g(x, \xi) \leq \int_{e \in ext\Xi} g(x, e)\phi(\xi, de). \tag{2.7}$$

Substituting $\xi(\omega)$ for ξ and integrating with respect to P the result in (2.5) is obtained. $\square$

This result states that if we can choose the appropriate ϕ and find λ, we can produce an upper bound. The key is to make the calculation of λ as simple as possible. Of course, the cardinality of $ext\Xi$ may also play a role in the computability of the bound.

One way to reduce the cardinality of the supporting extreme points is simply to choose the extreme point that has the highest value as an upper bound. Let this upper bound be $UB^{\max}(x) = \sup_{e \in ext\Xi} g(x, e) \geq \int_{e \in ext\Xi} g(x, e)\lambda(de) \geq E(g(x))$ from Theorem 2, regardless of the particular λ. While $UB^{\max}$ may only involve a single extreme point, it is often a poor bound (see the result from Exercise 3). Its calculation also often involves evaluating all the extreme points to maximize the convex function $g(x, \cdot)$.

In general, bounds built on the result in Theorem 2 construct the probability measure λ so that each extreme point e_j of Ξ has some weight, $p_j = \lambda(e_j)$. The following bounds, described in more detail in Birge and Wets [1986], find these weights in various cases. The first is general but involves some optimization. The second involves simplicial regions, and the third uses rectangular regions.

Because λ is constructed to be consistent with the distribution of $\boldsymbol{\xi}$, we must have that

$$
\begin{aligned}
\int_\Omega \xi(\omega) P(d\omega) &= \int_\Omega \int_{e \in ext\Xi} e\phi(\xi(\omega), de) P(d\omega) \\
&= \int_{e \in ext\Xi} e \int_\Omega \phi(\xi(\omega), de) P(d\omega) \qquad (2.8) \\
&= \int_{e \in ext\Xi} e\lambda(de).
\end{aligned}
$$

Hence, $\lambda \in \mathcal{P} = \{\mu | \mu$ is a probability measure on $\mathcal{E}$, and $E_\mu[e] = \bar{\xi}\}$. The next upper bound, originally suggested by Madansky [1960] and extended by Gassmann and Ziemba [1986], builds on this idea by finding an upper bound through a linear program to maximize the objective expectation over all probability measures in $\mathcal{P}$. We write this bound as UB^{mean}, where

$$
UB^{mean}(x) = \max_{p_1, \dots, p_K} \sum_{k=1}^K p_k g(x, e_k)
$$

$$
\text{s. t.} \sum_{k=1}^K p_k e_k = \bar{\xi},
$$

$$
\sum_{k=1}^K p_k = 1,
$$

$$
p_k \geq 0, k = 1, \dots, K. \qquad (2.9)
$$

As we shall see in Section 5, the probability measure that optimizes the linear program in (2.9) is the solution of a moment problem in which only the first moment is known. Another interpretation of this bound is that it represents the worst possible outcome if only the mean of the random variable is known. Optimizing with this bound, therefore, brings some form of risk avoidance if no other distribution information is available.

Assuming that the dimension of coΞ is N, Carathéodory's theorem states that $\bar{\xi}$ must be expressable as a convex combination of at most $N+1$ points in $ext\Xi$. Finding these $N + 1$ points may, however, again involve computations for the values at all extreme points. The number of extreme point representations may be much higher than $N+1$ if Ξ is, for example, rectangular, but lower if, for example, Ξ is a *simplex*, i.e., a convex combination of $N+1$ points, ξ^i, $i = 1, \dots, N+1$, such that $\xi^i - \xi^1$ are linearly independent for $i > 1$. The representation of interior points is, in fact, unique. Indeed, the p_j in this case are called the *barycentric coordinates* of $\bar{\xi}$.

Although Ξ may not be simplicial itself, it is often possible to extend $Q(x, \cdot)$ from Ξ to some simplex Σ including Ξ. The bound obtained with this approach is written UB^Σ. In this bound, the number of points used in the evaluation remains one more than the dimension of the affine hull of Ξ. Frauendorfer [1989, 1992] gives more details about this form of approximation and various methods for its refinement.

Often, Ξ is given as a rectangular region. In this case, the number of extreme points is 2^N. The number of simplices containing $\bar{\xi}$ may also be exponential in N. With relatively complete information about the correlations among random variables, however, bounds can be obtained that assign the same weight to each extreme point of Ξ (or a rectangular enclosing region), regardless of the value of x. This attribute is quite beneficial in algorithms where x may change frequently as an optimal solution is sought.

The basic bounds for rectangular regions follow Edmundson and Madansky, for which, the name *Edmundson-Madansky (E-M) bound* is used. They begin with the trapezoidal type of approximation on an interval. Here, if $\Xi = [a, b]$, we can easily construct $\phi(\xi, \cdot)$ in Theorem 2 as $\phi(\xi, a) = \pi(\xi)$ and $\phi(\xi, b) = 1 - \pi(\xi)$, where $\pi(\xi) = \frac{b-\xi}{b-a}$. Integrating over ω, we obtain

$$
\begin{aligned}
\lambda(a) &= \int_\Omega \phi(\xi(\omega), a) P(d\omega) \\
&= \int_\Omega \frac{b - \xi(\omega)}{b - a} P(d\omega) \\
&= \frac{b - \bar{\xi}}{b - a}.
\end{aligned}
\tag{2.10}
$$

We then also have $\lambda(b) = \frac{\bar{\xi}-a}{b-a}$. The bound obtained is $UB^{EM}(x) = \lambda(a)g(x, a) + \lambda(b)g(x, b) \geq E(g(x))$. Observe in Figure 2 that this bound represents approximating the integrand $g(x, \cdot)$ with the values formed as convex combinations of extreme point values. This is the same procedure as in trapezoidal approximation for numerical integration except that the endpoint weights may change for nonuniform probability distributions.

The E-M bound on an interval extends easily to multiple dimensions, where $\Xi = [a_1, b_1] \times \cdots \times [a_N, b_N]$ if either $g(x, \cdot)$ is separable in the components of ξ, in which case, the bound is applied in each component separately, or the components of ξ are stochastically independent. In this case, the bound is developed in each component $i = 1$ to N in order so that the full independent ξ_i bound contains the product of all combinations of each interval bound, i.e.,

$$
UB^{EM-I}(x) = \sum_{e \in ext\Xi} (\Pi_{i=1}^N \frac{|\bar{\xi}_i - e_i|}{b_i - a_i}) g(x, e),
\tag{2.11}
$$

where Ξ is again assumed polyhedral.

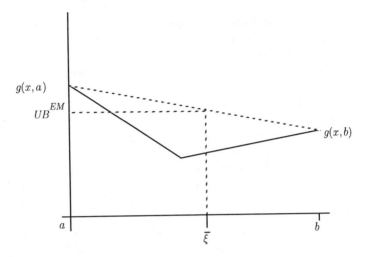

FIGURE 2. Example of the Edmundson-Madansky bound on an interval.

Example 1

We now consider an example to illustrate the bounds. Consider the following recourse problem with only **h** random:

$$Q(x, \boldsymbol{\xi}) = \begin{array}{ll} \min & \mathbf{y}_1^+ + \mathbf{y}_1^- \quad + \mathbf{y}_2^+ + \mathbf{y}_2^- \quad + \mathbf{y}_3 \\ \text{s. t.} & \mathbf{y}_1^+ - \mathbf{y}_1^- \qquad\qquad\qquad + \mathbf{y}_3 \; = \mathbf{h}_1 - x_1, \\ & \qquad\qquad \mathbf{y}_2^+ - \mathbf{y}_2^- \; + \mathbf{y}_3 \; = \mathbf{h}_2 - x_2, \\ & \mathbf{y}_1^+, \mathbf{y}_1^-, \quad \mathbf{y}_2^+, \mathbf{y}_2^-, \quad \mathbf{y}_3 \; \geq 0, \end{array}$$

where $\mathbf{h}_i$ is independently uniformly distributed on $[0,1]$ for $i = 1, 2$.

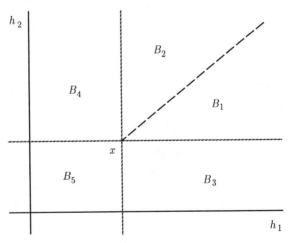

FIGURE 3. Optimal basis regions of Example 1.

The solution of this problem is illustrated in Figure 3. Here, the optimal bases are B_1 corresponding to (y_1^+, y_3), B_2 corresponding to (y_2^+, y_3), B_3

corresponding to (y_1^+, y_2^-), B_4 corresponding to (y_1^-, y_2^+), and B_5 corresponding to (y_1^-, y_2^-). Figure 3 shows the regions in which each of these bases is optimal.

Suppose $\bar{x} = (0.3, 0.3)^T$. Then integrating $Q(\bar{x}, \xi)$ yields $\mathcal{Q}(\bar{x}) = 0.466$. Our initial lower bound is $LB_1 = Q(\bar{x}, \xi = \bar{h} = (0.5, 0.5)^T = 0.2$.

The upper bounds can be found using the values at the extreme points of the support of $\mathbf{h}$. These values are $Q(\bar{x}, (0, 0)^T) = 0.6$, $Q(\bar{x}, (0, 1)^T) = 1.0$, $Q(\bar{x}, (1, 0)^T) = 1.0$, and $Q(\bar{x}, (1, 1)^T) = 0.7$. For $UB_1^{max}(\bar{x})$, we must take the highest of these values; hence $UB_1^{max}(\bar{x}) = 1.0$. For UB_1^{mean}, notice that $\bar{h} = (1/2)(1, 0)^T + (1/2)(0, 1)^T$, so $UB_1^{mean}(\bar{x}) = UB_1^{max}(\bar{x}) = 1.0$. For UB_1^{EM}, each extreme point is weighted equally, so $UB_1^{EM}(\bar{x}) = (1/4)(1 + 1 + .7 + .6) = 0.825$. For the simplicial approximation, let $\Sigma = co\{(0, 0), (2, 0), (0, 2)\}$, which includes the support of $\mathbf{h}$. In this case, the weights on the extreme points are $\lambda(0, 0) = 0.5$ and $\lambda(2, 0) = \lambda(0, 2) = 0.25$. The resulting upper bound is $UB^{\Sigma}(\bar{x}) = 0.5(.6) + 0.25(2)(2) = 1.3$.

To refine the bounds, we consider the dual multipliers at each extreme point. At $(0, 0)$, they are $(-1, -1)$. At $(1, 0)$, they are $(1, -1)$. At $(0, 1)$, they are $(-1, 1)$. At $(1, 1)$, both bases B_1 and B_2 are optimal, so the multipliers are $(0, 1)$, $(1, 0)$, or any convex combination. The linearization along the line segment from $(0, 0)$ to $(1, 0)$ is the minimum of $Q(\bar{x}, (1, 0)^T) - (Q(\bar{x}, (0, 0)^T) + (-1, -1)^T(1, 0) = 1 - (0.6 - 1) = 1.4$ and $Q(\bar{x}, (0, 0)^T) - (Q(\bar{x}, (1, 0)^T) + (1, -1)^T(-1, 0) = 0.6 - (1 - 1) = 0.6$. Hence, the minimum error on $(0, 0)$ to $(1, 0)$ is 0.6. Similarly, for $(0, 0)$ to $(0, 1)$, the error is 0.6. From $(1, 0)$ to $(1, 1)$, the minimum error is 0.3 if the $(0, 1)$ subgradient is used at $(1, 1)$; however, the minimum error on $(0, 1)$ to $(1, 1)$ is then $\min\{1 - (0.7 - 1), 0.7 - (1 - 1)\} = 0.7$. Thus, the maximum of these errors over each edge of the region is 0.7 for the edge $(0, 1)$ to $(1, 1)$.

To find the value of c_1^* to split the interval $[a_1 = 0, b_1 = 1]$, we need to find where $Q(\bar{x}, (0, 1)^T) - c_1^* = Q(\bar{x}, (1, 1)^T) + (c_1^* - 1)$ or where $1 - c_1^* = 0.7 - 1 + c_1^*$, i.e., where $c_1^* = 0.65$. We obtain two regions, $S_1 = [0, 0.65] \times [0, 1]$ and $S_2 = [0.65, 1] \times [0, 1]$, with $p_1 = 0.65$ and $p_2 = 0.35$.

The Jensen lower bound is now $LB_2 = 0.65(Q(\bar{x}, (0.325, 0.5)^T)) + (0.35)(Q(\bar{x}, (0.825, 0.5)^T)) = 0.65(0.2) + 0.35(0.525) = 0.31375$. The upper bounds are $UB_2^{max}(\bar{x}) = 0.65(1) + 0.35(1) = 1$, $UB_2^{mean}(\bar{x}) = 0.65(0.5)(1 + 0.65) + 0.35(0.5)(1 + 0.7) = 0.83375$, and $UB_2^{EM}(\bar{x}) = 0.65(0.25)(1 + 0.7 + 0.65 + 0.6) + 0.35(0.25)(0.7 + 0.7 + 1 + .65) = 0.74625$. (The simplicial bound is not given because we have split the region into rectangular parts.) Exercise 3 asks for these computations to continue until the lower and upper bounds are within 10% of each other.

Exercises

1. For Example 1, $\bar{x} = (0.1, 0.7)^T$, compute $\mathcal{Q}(\bar{x})$, the Jensen lower bound, and the upper bounds, UB^{mean}, UB^{max}, UB^{EM}, and UB^{Σ}.

2. Prove Jensen's inequality, $E(g(\boldsymbol{\xi})) \geq g(E(\boldsymbol{\xi}))$, by taking an expectation of the points on a supporting hyperplane to $g(\boldsymbol{\xi})$ at $g(E(\boldsymbol{\xi}))$.

3. Follow the splitting rules for Example 1 until the Edmundson-Madansky upper and Jensen lower bounds are within 10% of each other. Compare UB^{EM} to $UB^{\max}$ on each step.

9.3 Using Bounds in Algorithms

The bounds in Section 2 can be used in algorithms in a variety of ways. We describe three basic procedures in this section: (1) uses of lower bounds in the L-shaped method with stopping criteria provided by upper bounds; (2) uses of upper bounds in generalized programming with stopping rules given by lower bounds; and (3) uses of the dual formulation in the separable convex hull function. The first two approaches are described in Birge [1983] while the last is taken from Birge and Wets [1989].

The L-shaped method as described in Chapter 5 is based on iteratively providing a lower bound on the recourse objective, $\mathcal{Q}(x)$. If a lower bound, $\mathcal{Q}^L(x)$, is used in place of $\mathcal{Q}(x)$, then clearly for any supports, $E^L x + e^L$, if $\mathcal{Q}^L(x) \geq E^L x + e^L$, $\mathcal{Q}(x) \geq E^L x + e^L$. Thus, any cuts generated on a lower bounding approximation of $\mathcal{Q}(x)$ remain valid throughout a procedure that refines that lower bounding approximation. This observation leads to the following algorithm. We suppose that $\mathcal{Q}_j^L(x)$ and $\mathcal{Q}_j^U(x)$ are approximating lower and upper bounding approximations such that $\lim_{j\to\infty} \mathcal{Q}_j^L(x) = \mathcal{Q}(x)$ and $\lim_{j\to\infty} \mathcal{Q}_j^U(x) = \mathcal{Q}(x)$. We suppose that P_j^L is the jth lower bounding approximation measure so that $\mathcal{Q}_j^L(x) = \int_\Omega Q_j^L(x,\boldsymbol{\xi}) P_j^L(d\omega)$.

L-Shaped Method with Sequential Bounding Approximations

Step 0. Set $r = s = v = k = 0$.

Step 1. Set $\nu = \nu + 1$. Solve the linear program (3.1)–(3.3):

$$\min \ z = c^T x + \theta \tag{3.1}$$
$$\text{s.t.} \ Ax = b,$$
$$D_\ell \, x \geq d_\ell, \quad \ell = 1, \ldots, r, \tag{3.2}$$
$$E_\ell x + \theta \geq e_\ell, \quad \ell = 1, \ldots, s, \tag{3.3}$$
$$x \geq 0, \quad \theta \in \Re.$$

Let (x^ν, θ^ν) be an optimal solution. If no constraint (3.3) is present, θ is set equal to $-\infty$ and is ignored in the computation.

Step 2. Let the support of $\boldsymbol{\xi}$ for the current lower bounding approximation be Ξ^j. For any $\xi_j = (h_j, T_j, q_j) \in \Xi^j$, solve the linear program

$$\min \ w^1 = e^T v^+ + e^T v^- \tag{3.4}$$
$$\text{s.t.} \ \ Wy + Iv^+ - Iv^- = h_j - T_j x^\nu, \tag{3.5}$$
$$y \geq 0, \quad v^+ \geq 0, \quad v^- \geq 0,$$

where $e^T = (1, \ldots, 1)$, until, for some ξ_j, the optimal value $w^1 > 0$. Let σ^ν be the associated simplex multipliers and define

$$D_{r+1} = (\sigma^\nu)^T T_j \tag{3.6}$$

and

$$d_{r+1} = (\sigma^\nu)^T h_j \tag{3.7}$$

to generate a feasibility cut of type (3.2). Set $r = r + 1$ and return to Step 1. If for all ξ^j $w^1 = 0$, go to Step 3.

Step 3. Find $\mathcal{Q}_j^L(x^\nu) = \int_\Omega Q_j^L(x^\nu, \boldsymbol{\xi}) P_j^L(d\omega)$, the jth lower bounding approximation. Suppose $-(\pi^\nu(\boldsymbol{\xi}))^T \mathbf{T} \in \partial_x Q_j^L(x^\nu, \boldsymbol{\xi})$ (the simplex multipliers associated with the optimal solution of the recourse problem). Define

$$E_{s+1} = \int_\Omega (\pi^\nu(\boldsymbol{\xi}))^T \mathbf{T} P_j^L(d\omega) \tag{3.8}$$

and

$$e_{s+1} = \int_\Omega (\pi^\nu(\boldsymbol{\xi}))^T \mathbf{h} P_j^L(d\omega). \tag{3.9}$$

Let $w^\nu = e_{s+1} - E_{s+1} x^\nu = \mathcal{Q}_j^L(x^\nu)$. If $\theta^\nu \geq w^\nu$, x^ν is optimal, relative to the lower bound; go to Step 4. Otherwise, set $s = s + 1$ and return to Step 1.

Step 4. Find $\mathcal{Q}_j^U(x^\nu) = \int_\Omega Q_j^U(x^\nu, \boldsymbol{\xi}) P_j^U(\omega)$, the jth upper bounding approximation. If $\theta^\nu \geq \mathcal{Q}_j^U(x^\nu)$, stop; x^ν is optimal. Otherwise, refine the lower and upper bounding approximations from ν to $\nu + 1$. Let $\nu = \nu + 1$. Go to Step 3.

This form of the L-shaped method follows the same steps as the standard L-shaped method, except that we add an extra check with the upper bound to determine the stopping conditions. We also describe the calculation of $\mathcal{Q}_j^L$ somewhat generally to allow for more general types of approximating distributions and approximating recourse functions, $Q_j^L(x^\nu, \boldsymbol{\xi})$.

Example 2

Consider Example 1 from Chapter 5, where:

$$Q(x,\xi) = \begin{cases} \xi - x & \text{if } x \le \xi, \\ x - \xi & \text{if } x > \xi, \end{cases} \tag{3.10}$$

$c^T x = 0$, and $0 \le x \le 10$. Instead of a discrete distribution on ξ, however, assume that ξ is uniformly distributed on $[0, 5]$. For the bounding approximation, we use the Jensen lower bound and Edmundson-Madansky upper bound for Q^L and Q^U, respectively. We use the refinement procedure to split the cell that contributes most to the difference between Q^L and Q^U. We split this cell at the intersection of the supports from the two extreme points of this cell (here, interval).

The sequence of iterations is as follows.

Iteration 1:

Here, $x^1 = 0$. Find $Q_1^L(0) = Q(0, \bar{\xi} = 2.5) = 2.5$. $E_1 = -\partial_x Q_1^L(0, 2.5) = -(-1)$ and $e_1 = -\partial_x Q_1^L(0, 2.5)(h = 2.5) = -(-1)(2.5) = 2.5$. Add the cut:

$$\theta \ge 2.5 - x. \tag{3.11}$$

Iteration 2:

Here, $x^2 = 10$, $\theta = -7.5$, but $Q_1^L(10) = Q(10, \bar{\xi} = 2.5) = 7.5$. At this point, the subgradient of $Q_1^L(10)$ is 1. $E_2 = -\partial_x Q_1^L(10, 2.5) = -1$, and $e_1 = -\partial_x Q_1^L(0, 2.5)(h = 2.5) = -(1)(2.5) = -2.5$. Add the cut:

$$\theta \ge -2.5 + x. \tag{3.12}$$

Iteration 3:

Here, $x^3 = .25$, $\theta = 0$, $Q_1^L(2.5) = Q(.25, \bar{\xi} = 2.5) = 0$. Hence we meet the condition for optimality of the first lower bounding approximation. Now, go to Step 4 and consider the first upper bounding approximation with equal weights of 0.5 on $\xi = 0$ and $\xi = 5$. In this case, $Q_1^U(2.5) = 0.5 * (Q(2.5, 0) + Q(2.5, 5)) = 2.5$. Thus, we must refine the approximation. Using the subgradient of -1 at $\xi = 0$ and 1 at $\xi = 5$, split at $c^* = 2.5$.

The new lower bounding approximation has equal weights of 0.5 on $\xi = 1.25$ and $\xi = 3.75$. In this case, $Q_2^L(2.5) = 0.5 * (Q(2.5, 1.25) + Q(2.5, 3.75)) = 1.25$. Now, we add the cut $E_2 = 0.5(-\partial_x Q(2.5, 1.25) - \partial_x Q(2.5, 3.75)) = 0$ and $e_1 = 0.5(-\partial_x Q(2.5, 1.25)(1.25) - \partial_x Q(2.5, 3.75)(3.75)) = (0.5)(-1.25 + 3.75) = 1.25$. Thus, we add the cut:

$$\theta \ge 1.25. \tag{3.13}$$

Iteration 4:

Here, keep $x^4 = x^3 = 2.5$ (although other optima are possible) and $\theta = 1.25$. Again, $Q_2^L(2.5) = 1.25$, so proceed to Step 4.

Checking the upper bound, we find that the upper bound places equal weights on the endpoints of each interval, $[0, 2.5]$ and $[2.5, 5]$. Thus, $Q_2^U(2.5) = 0.5 * (Q(2.5, 2.5)) + (0.25) * (Q(2.5, 0) + Q(2.5, 5)) = 1.25$, and $\theta = Q_2^U(2.5)$. Stop with an optimal solution.

The steps are illustrated in Figure 4. We show the true $Q(x)$ as a solid line, with dashed lines representing the approximations (lower and upper). Note that the method may not have converged as quickly if we had chosen some point other than $x^4 = x^3 = 2.5$. The upper and lower bounds meet at this point, because we chose the division precisely at the link between the linear pieces of the recourse function $Q(x, \cdot)$.

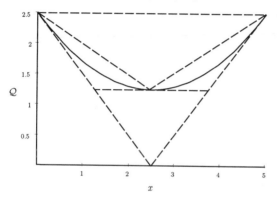

FIGURE 4. Example of L-Shaped method with sequential approximation.

Bounds with generalized programming

In generalized linear programming, the same types of procedures can be applied. The difference is that because the generalized programming method uses inner linearization instead of outer linearization, the bounds used should be upper bounds. We would thus substitute Ψ_j^U for Ψ in (5.5.6). The same steps are followed again with Ψ_j^U until optimality relative to Ψ_j^U is achieved. At this point, as in Step 4 of the L-shaped method with sequential bounding approximations, overall convergence is tested by solving (5.5.10) with a lower bounding Ψ_j^L in place of Ψ. If this value is again nonnegative, then the procedure stops. If not, refinement is made until a new upper bounding column is generated or no solution of (5.5.10) is negative for a lower bounding approximation.

As stated in Chapter 5, generalized programming is most useful if the recourse function, $\Psi(\chi)$, is separable in the components of χ. The separable upper bounding procedure is a natural use for this approach. A separable lower bound can be obtained by using a supporting hyperplane. This leads to the Jensen lower bound.

This generalized programming approach applies most directly when a single basis separable approximation is used. With the convex hull operation, we would still have the problem of evaluating this function. This difficulty is, however, overcome by dualizing the problem. In this case, we suppose that the original problem using a set $\mathcal{D}$ of bases is to find $x \in \Re^{n_1}, \chi \in \Re^{m_2}$ to

$$
\begin{array}{lll}
\min & c^T x & +co\{\Psi_D, D \in \mathcal{D}\}(\chi) \\
\text{s. t.} & Ax & = b, \\
& Tx & -\chi & = 0, \\
& x & & \geq 0.
\end{array}
\tag{3.14}
$$

The main result is the following theorem. Recall the conjugate function defined in Section 2.9.

Theorem 3. *A dual program to (3.14) is to find $\sigma \in \Re^{m_1}, \pi \in \Re^{m_2}$ to*

$$
\begin{array}{lll}
\max & \sigma^T b & - \sup\{\Psi_D^*, D \in \mathcal{D}\}(-\pi) \\
\text{s. t.} & \sigma^T A & +\pi^T T & \leq c^T,
\end{array}
\tag{3.15}
$$

where Ψ_D^* is the conjugate function and (3.14) and (3.15) have equal optimal values.

Proof. Let $\gamma(\chi) = co\{\Psi_D, D \in \mathcal{D}\}(\chi)$. Then a dual to (3.14) (see, e.g., Geoffrion [1971], Rockafellar [1974]) is

$$
\max_{\pi,\sigma}\{ \inf_{x\geq 0,\chi} [c^T x + \gamma(\chi) + \sigma^T(b - Ax) + \pi^T(\chi - Tx)]\},
$$

which is equivalently

$$
\max_{\pi,\sigma}\{ \inf_{x\geq 0,\chi} [(c^T - \sigma^T A - \pi^T T)x + \sigma^T(b) - (-\pi^T\chi - \gamma(\chi))]\}
$$
$$
= \max_{\sigma^T A+\pi^T T\leq c^T}\{\sigma^T b - \gamma^*(-\pi)\}.
\tag{3.16}
$$

Problem (3.16) immediately gives (3.15) because $(co\{\Psi_D, D \in \mathcal{D}\}(\chi))^*(-\pi) = \sup\{\Psi_D^*, D \in \mathcal{D}\}(-\pi)$ (Rockafellar [1969, Theorem 16.5]). $\square$

Problem (3.15) only involves finding the supremum of convex functions, which is again a convex function. The main difficulty is in finding expressions for the Ψ_D^*. These are, however, relatively straightforward to evaluate (Exercise 2). They can be used in a variety of optimization procedures, but the objective is nondifferentiable. In Birge and Wets [1989], this difficulty is overcome by making each Ψ_D^* a lower bound on some parameter that replaces $\sup\{\Psi_D^*, D \in \mathcal{D}\}$ in the objective.

The main refinement choice in the separable optimization procedure using (3.15) is to determine how to update the set $\mathcal{D}$. Choices of bases that are optimal for $\bar{\xi}$ and then $\bar{\xi}\pm\delta e_i\sigma_i$ for increasing values of δ appear to give a rich set $\mathcal{D}$ as in Birge and Wets [1989]. Any sense of *optimal* refinements or basis choice is, however, an open question.

Exercises

1. Consider Example 2 where we redefine Q as

$$Q(x,\boldsymbol{\xi}) = \begin{cases} 2(\boldsymbol{\xi} - x) & \text{if } x \leq \boldsymbol{\xi}, \\ x - \boldsymbol{\xi} & \text{if } x > \boldsymbol{\xi}, \end{cases}$$

   with $\boldsymbol{\xi}$ uniformly distributed on $[0,5]$, $c^T x = 0$, and $0 \leq x \leq 10$. Follow the L-shaped sequential approximation method until achieving a solution with two significant digits of accuracy.

2. Find $\Psi_D^*(-\pi)$ and $\partial\Psi_D^*(-\pi)$. A useful set may be $\gamma_{D^i}(p) = \{y | P_{D^i}(y)^- \leq p \leq P_{D^i}(y)\}$.

3. Use the dualization procedure to solve a stochastic linear program with $c^T x = x$, $0 \leq x \leq 1$, and the recourse function in Example 1.

9.4 Bounds in Chance-Constrained Problems

Our procedures have so far concentrated on methods for recourse problems as we have throughout this book. In many cases, of course, probabilistic constraints may also be in the formulation or may be the critical part of the model. The basic results are aimed at finding some inequalities $\tilde{A}x \geq \tilde{h}$ (or, perhaps, nonlinear inequalities) that imply that $P\{\mathbf{A}x \geq \mathbf{h}\} \leq \alpha$. In Section 3.2, we found some deterministic equivalents for specific forms of the distribution, but these are not always available. In these cases, it is useful to have upper and lower bounds on $P\{\mathbf{A}x \geq \mathbf{h}\}$ for any x such that $\tilde{A}x \leq \tilde{h}$.

The bounds in this case are generally of two types: bounds for a single inequality such as $P\{\mathbf{A}_i x \geq \mathbf{h}_i\}$ and bounds for the set of inequalities in terms of results in lower dimensions. In algorithms, (see Prékopa [1988]), it is often common to place the probabilistic constraint into the objective and to use a Lagrangian relaxation or parametric solution procedure.

For bounds with a single constraint, the basic results are extensions of Chebyshev's inequality. We refer to Hoeffding [1963] and Pintér [1989] for many of these results. The basic Chebyshev inequality is (see, e.g., Feller [1971, section V.7]) that if $\boldsymbol{\xi}$ has a finite second moment, then

$$P\{|\boldsymbol{\xi}| \geq a\} \leq \frac{E[\boldsymbol{\xi}^2]}{a^2}, \tag{4.1}$$

and for σ^2, the variance of $\boldsymbol{\xi}$,

$$P\{|\boldsymbol{\xi} - \bar{\xi}| \geq a\} \leq \frac{\sigma^2}{a^2}. \tag{4.2}$$

Another useful inequality is the one-sided inequality that

$$P\{\boldsymbol{\xi} - \bar{\xi} \geq a\} \leq \frac{\sigma^2}{\sigma^2 + a^2}. \tag{4.3}$$

To apply (4.2) and (4.3) in the context of stochastic programming, we suppose that we can represent $\mathbf{A}_i x \geq \mathbf{h}_i$ as $\boldsymbol{\xi}_0 + \boldsymbol{\xi}^T x \geq r_0 + r^T x$, where $\mathbf{A}_{ij} = \boldsymbol{\xi}_j - t_j$ and $\mathbf{h}_i = -\boldsymbol{\xi}_0 + r_0$, to distinguish random elements from those that are not random and to allow us to set $\bar{\xi}_j = 0$ for $j = 0, \ldots, n$. If $\boldsymbol{\xi}$ has covariance matrix, C, then the variance of $\boldsymbol{\xi}_0 + \boldsymbol{\xi}^T x$ is $\hat{x}^T C \hat{x}$, where $\hat{x} = \begin{pmatrix} 1 \\ x \end{pmatrix}$. In this case, substituting $\hat{x}^T C \hat{x}$ for σ^2 and $r_0 + r^T x = \hat{r}^T \hat{x}$ for r in (4.3) yields:

$$P\{\mathbf{A}_i x \geq \mathbf{h}_i\} \leq \frac{\hat{x}^T C \hat{x}}{\hat{x}^T C \hat{x} + (\hat{r}^T \hat{x})^2}, \tag{4.4}$$

which implies that if x satisfies

$$\hat{x}^T C \hat{x}(1 - \alpha) \leq \alpha(\hat{r}^T \hat{x})^2, \tag{4.5}$$

then

$$P\{\mathbf{A}_i x \geq \mathbf{h}_i\} \leq \alpha. \tag{4.6}$$

Alternatively, if

$$P\{\mathbf{A}_i x \geq \mathbf{h}_i\} \geq \alpha, \tag{4.7}$$

then

$$\hat{x}^T C \hat{x}(1 - \alpha) \geq \alpha(\hat{r}^T \hat{x})^2. \tag{4.8}$$

Thus, adding constraint (4.8) in place of (4.7) in a stochastic program allows a large feasible region and in a minimization problem, would produce a lower bound on the objective value with constraint (4.7). For an upper bound, we could note that $P\{\mathbf{A}_i x \geq \mathbf{h}_i\} \geq \alpha$ is equivalent to $P\{\mathbf{A}_i x \leq \mathbf{h}_i\} \leq 1 - \alpha$ or $P\{\mathbf{h}_i - \mathbf{A}_i x \geq 0\} \leq 1 - \alpha$. We just replace the previous $\boldsymbol{\xi}$ and t with $-\boldsymbol{\xi}$ and $-t$ and replace α with $(1 - \alpha)$ to obtain that if

$$\hat{x}^T C \hat{x}(\alpha) \leq (1 - \alpha)(\hat{r}^T \hat{x})^2, \tag{4.9}$$

then (4.7). Hence, replacing (4.7) with (4.9) yields a smaller region and an upper bound in a minimization problem.

Other information, such as ranges, can also be used to obtain sharper bounds. A particularly useful inequality (see, again, Feller [1971]) is that, for any function $u(\xi)$ such that $u(\xi) > \epsilon > 0$, for all $\xi \geq t$,

$$P\{\boldsymbol{\xi} \geq t\} \leq \frac{1}{\epsilon} E[u(\boldsymbol{\xi})]. \tag{4.10}$$

In fact, using, $u(\xi) = (\xi + \frac{\sigma^2}{a})^2$ yields (4.3) from (4.10). A difficulty in using bounds based on (4.3) is that the constraint in (4.8) or (4.9) may be quite difficult to include in an optimization problem. Various linearizations around certain values of x of this constraint can be used in place of (4.8) or (4.9). Other approaches, as in Pintér [1989], are based on the expectations of exponential functions of $\boldsymbol{\xi}_i$ that can in turn be bounded using the Jensen inequality.

Given these approaches or deterministic equivalents for a single inequality as in Section 3.2, we wish to find approximations for multiple inequalities, $P\{\mathbf{A}x \leq \mathbf{h}\}$. With relatively few inequalities and special distributions, such as the multivariate gamma described in Szántai [1986], deterministic equivalents can again be found. The general cases are, however, most often treated with approximations based on Boole-Bonferroni inequalities. A thorough description is found in Prékopa [1988].

We suppose that $\mathbf{A} \in \Re^{m \times n}$ and that $\mathbf{h} \in \Re^m$. The Boole-Bonferroni inequality bounds are based on evaluating $P\{\mathbf{A}_i x \leq \mathbf{h}_i\}$ and $P\{\mathbf{A}_i x \leq \mathbf{h}_i, \mathbf{A}_j x \leq \mathbf{h}_j\}$ for each i and j and using these values to bound the complete expression $P\{\mathbf{A}x \leq \mathbf{h}\}$. To distinguish among the rows of $\mathbf{A}$, we let $\mathbf{A}_{ij} = \xi_j^i - t_j^i$ and $\mathbf{h}_i = -\xi_0^i + t_0^i$. A main result is then the following.

Theorem 4. *Given these assumptions,*

$$P\{\mathbf{A}x \leq \mathbf{h}\} = 1 - (a - \frac{2b}{m}) + \lambda[\frac{(c-1)a}{c+1} - \frac{2(-m+c(c+1))b}{m(c(c+1))}], \quad (4.11)$$

$$a = \sum_{1 \leq i \leq m+1} P(\boldsymbol{\eta}_i > s_i(x)),$$

$$b = \sum_{1 \leq i < j \leq m+1} P(\boldsymbol{\eta}_i > s_i(x), \boldsymbol{\eta}_i > s_j(x)),$$

$$c = \lfloor \frac{2b}{a} \rfloor,$$

$$0 \leq \lambda \leq 1,$$

$$\boldsymbol{\eta}_i = (\boldsymbol{\xi}^i)^T \hat{x},$$

$$s_i(x) = (r^i)^T \hat{x}.$$

Proof.   Denote the event $\eta_i \leq s_i(x)$ by A_i. Then

$$P(\mathbf{A}x \leq \mathbf{h}) = P(A_1 \cdots A_m) = 1 - P(\hat{A}_1 + \ldots + \hat{A}_m), \quad (4.12)$$

where $\hat{S}$ for a set S indicates the complement of A, i.e., the set of points not in A.

By the inequality of Dawson and Sankoff [1967] ((7) of Prékopa [1988]),

$$P(\hat{A}_1 + \ldots + \hat{A}_m) \geq \frac{2}{c+1}a - \frac{2}{c(c+1)}b, \quad (4.13)$$

where

$$a = \sum_{1 \leq i \leq m} P(\hat{A}_i) = \sum_{1 \leq i \leq m} P(\boldsymbol{\eta}_i > s_i(x)),$$

$$b = \sum_{1 \leq i < j \leq m} P(\hat{A}_i \cdot \hat{A}_j)$$

$$= \sum_{1 \leq i < j \leq m} P(\boldsymbol{\eta}_i > s_i(x), \boldsymbol{\eta}_j > s_j(x)),$$

$$c = \lfloor \frac{2b}{a} \rfloor.$$

Similarly, by the inequality of Sathe, Pradhan, and Shah [1980] ((8) of Prékopa [1988]),

$$P(\hat{A}_1 + ... + \hat{A}_m) \leq a - \frac{2}{m}b. \qquad (4.14)$$

Combining (4.12)–(4.14), we obtain (4.11). □

We may use (4.11) to approximate $P\{\mathbf{A}x \leq \mathbf{h}\}$ by assigning a in $[0,1]$, e.g., 0.5. With the marginal distribution of $\boldsymbol{\eta}_i$ and the joint distribution of $\boldsymbol{\eta}_i$ and $\boldsymbol{\eta}_j$, we can again use bounds on the variances of these random variables to calculate bounds from (4.11). Of course, with normally distributed random variables, we may again obtain the η_i to be normally distributed or may obtain such limiting distributions (see, e.g., Salinetti [1983]). In this case, besides the exact results in Section 3.2, we should mention the specializations of Gassmann [1988] and Deák [1980]. They also combine these inequalities with Monte Carlo simulation schemes (see, e.g., Rubinstein [1981]). In general, the inequalities from (4.11) can reduce the variance of Monte Carlo schemes. For this approach and the bivariate gamma, we again refer to Szántai [1986].

Before closing this section, we should also mention that approximating probabilities is also quite useful in recourse problems because the gradient of the linear recourse function with fixed q and T is simply a weighted probability of given bases' optimality. From Theorem 3.11, if x is in the interior of K_2, then

$$\partial Q(x) = E_{\boldsymbol{\xi}}[-\pi(\boldsymbol{\xi})^T T] = \sum_{j=1}^{J} -\pi^j TP\{(\pi^j)^T(\mathbf{h} - Tx)$$
$$\geq \pi^T(\mathbf{h} - Tx), \forall \pi^T W \leq q\}, \qquad (4.15)$$

where $\{\pi^1, ..., \pi^J\}$ is the set of extreme values of $\{\pi | \pi^T W \leq q\}$. Because $(\pi^j)^T = (W^j)^{-1}q^T$ is optimal, if and only if $(W^j)^{-1}(\mathbf{h} - Tx) \geq 0$, the result reduces to finding the probability that $(W^j)^{-1}(\mathbf{h} - Tx) \geq 0$. This observation can be useful in guiding algorithms based on subgradient information. This idea is explored in Birge and Qi [1995].

Other model forms also lead to bounds of this type that can in some cases be stronger because of the structure of $\mathbf{A}$. A particular case is when $\mathbf{A}$ represents a network. In this case, bounds on project network completion times can be found in Maddox and Birge [1991]. These bounds, as well as those given earlier, can be derived from solutions of a generalized moment problem. That is one of the main topics of the generalizations in the next section.

Exercises

1. Derive (4.10).

2. Define u in (4.10) as $u(\xi) = c\sigma^2 - (\xi - \frac{u+t}{2})^2$, where it is known, however, that $\xi \leq U = \beta a$, a.s., for some finite β. For given β and a, can you find c such that (4.10) gives a better bound with this u than with the u used to obtain (3.3)?

3. Suppose $\xi_i, i = 1, 2, 3$, are jointly multivariate normally distributed with zero means and variance-covariance matrix

$$
C = \begin{pmatrix} 1 & .25 & -.25 \\ .25 & 1 & -.5 \\ -.25 & -.5 & 1 \end{pmatrix}.
$$

Use Theorem 4 to bound $P\{\xi \leq 1, i = 1, 2, 3\}$. What is the exact result? (Hint:Try a transformation to independent normal random variables.)

9.5 Generalized Bounds

a. Extensions of basic bounds

When the components of ξ are correlated, a bound is still tractable (see Frauendorfer [1988b]), although somewhat more difficult to evaluate. In this subsection, we give the necessary generalizations. The notation here is particularly cumbersome, although the results are straightforward.
    For the general results, we define:

$$
\eta(e, \xi_i) = \begin{cases} (\xi_i - a_i) & \text{if } e_i = a_i, \\ (b_i - \xi_i) & \text{if } e_i = b_i. \end{cases} \tag{5.1}
$$

Then we have (Exercise 1) that

$$
\phi(\xi, e) = \Pi_{i=1}^{N} \frac{\eta(e, \xi_i)}{(b_i - a_i)}. \tag{5.2}
$$

The $\lambda(e)$ values can be found by integrating over ω. This may involve all products of the ξ_i components. Defining $\mathcal{M} = \{M | M \subset \{1, \ldots, N\}\}$, and $\rho_M = E[\Pi_{i \in M} \xi_i] - \Pi_{i \in M} \bar{\xi}_i$, we obtain the general E-M extension:

$$UB^{EM-D}(x) = UB^{EM-I}(x) + \sum_{e \in ext\Xi} \frac{1}{\Pi_{i=1}^N (b_i - a_i)}$$
$$\{ \sum_{M \in \mathcal{M}} [\Pi_{i \notin M} (-1)^{\frac{e_i - a_i}{b_i - a_i}} (a_i(\frac{e_i - a_i}{b_i - a_i}) + b_i(\frac{b_i - e_i}{b_i - a_i}))$$
$$\times \Pi_{i \in M} (-1)^{1 - \frac{e_i - a_i}{b_i - a_i}}] \rho_M \} g(x, e). \tag{5.3}$$

Notice, in (5.3), that if the components of ξ are independent, then $\rho_M = 0$ for all M and $UB^{EM-D}(x) = UB^{EM-I}(x)$, as expected.

Each of these upper bounds is a solution of a corresponding moment problem in which the highest expected function value is found over all probability distributions with the given moment information. The upper bounds derived so far all used first moment information plus some information about correlations. In Subsection c, we will explore the possibilities for higher moments and methods for constructing bounds with this additional information.

For different support regions, Ξ, we can combine the bounds or use enclosing regions as we mentioned in terms of simplicial approximation. To apply the bounds in a convergent method, the partitioning scheme in Theorem 1 is again applied. Instead of applying the bounds on Ξ in its entirety, they are applied on each S^l. The dimension of these cells may, however, make computations quite cumbersome, especially if the S^l have exponential numbers of extreme points. For this reason, algorithms primarily concentrate on a lower bounding approximation for most computations and only use the upper bound to check optimality and stopping conditions.

So far, we have only considered convex $g(x, \cdot)$. In the recourse problem, $Q(x, \xi(\omega))$ is generally convex in $h(\omega)$ and $T(\omega)$ but concave in $q(\omega)$. In this general case, the Jensen-type bounds provide an upper bound on $\mathcal{Q}$ in terms of q while the extreme point bounds provide lower bounds in q. We can combine these results with the convex function results to obtain overall bounds by, for example, determining $UB(x) = \int_\Omega UB(x, \mathbf{q}) P(d\omega)$ where $UB(x, \mathbf{q}) = UB_{\mathbf{h}, \mathbf{T}}(Q(x, \xi))$, where the last upper bound is taken with respect to the $\mathbf{h}$ and $\mathbf{T}$ with q fixed. The difficulty of evaluating $\int_\Omega UB(x, \mathbf{q}) P(d\omega)$ may determine the success of this effort. In the case of $\mathbf{q}$ independent of $\mathbf{h}$ and $\mathbf{T}$, it is simple. In other cases, linear upper bounding hulls may be constructed to allow relatively straightforward computation (Frauendorfer [1988a]) or extensions of the approach in UB^{mean} may be used (Edirisinghe [1991]).

For the procedure in Frauendorfer [1988a], assume that Ξ is compact and rectangular with $q \in \Xi_1 = [c_1, d_1] \times \cdots \times [c_{n_2}, d_{n_2}]$ and $(h, T)^T \in \Xi_2 = [a_1, b_1] \times \cdots \times [a_{N-n_2}, b_{N-n_2}]$. For convenience here, we consider T

as a single vector of all components in order, $T_1., \ldots, T_{m_2}.$. We also delete transposes on vectors when they are used as function arguments.

Let the extreme points of the support of $\mathbf{q}$ be $e_l, 1, \ldots, L$, and the extreme points of the support of $(\mathbf{h}, \mathbf{T})$ be $e_j, k = 1, \ldots, K$. In this case, because $Q(x, \cdot)$ is convex in (h, T), for any e_l, we can take any support $\pi(e_l)$ such that $\pi(e_l)^T W \leq e_l$ and obtain a lower bound on $Q(x, (e_l, h, T))$ as

$$\pi(e_l)^T (h - Tx) \leq Q(x, (e_l, h, T))). \tag{5.4}$$

We can also let $\phi(q, e_l) = \Pi_{i=1}^{n_2} \frac{\eta(e_l, q_i)}{(d_i - c_i)}$, where η is as defined earlier with c replacing a and d replacing b. Because for any (h, T), $Q(x, (q, h, T))$ is concave in q, we have that

$$Q(x, (q, h, T)) \geq \sum_{l=1}^{L} \phi(q, e_l) Q(x, (e_l, h, T)) \geq \sum_{l=1}^{L} \phi(q, e_l) \pi(e_l)^T (h - Tx), \tag{5.5}$$

where we note that $\pi(e_l)$ need not depend on (h, T). A bound is obtained by integrating over $(h(\omega), T(\omega))$ in (5.5), so that

$$\mathcal{Q}(x) \geq \sum_{l=1}^{L} \int_{\Omega} \Pi_{i=1}^{n_2} \frac{\eta(e_l, \mathbf{q}_i)}{(d_i - c_i)} \pi(e_l)^T (\mathbf{h} - \mathbf{T}x) P(d\omega). \tag{5.6}$$

Note the terms in (5.6) just involve products of the components of $\mathbf{q}$ and each component of $\mathbf{h}$ or $\mathbf{T}x$ singly. Following Frauendorfer [1988a], we let $\mathcal{L} = \{\Lambda | \Lambda \subset \{1, \ldots, n_2\}\}$ and define

$$c_\Lambda(e_l) = \frac{1}{\Pi_{i=1}^{n_2}(d_i - c_i)} [\Pi_{i \notin \Lambda}(-1)^{\frac{e_{l,i} - c_i}{d_i - c_i}} (c_i \frac{e_{l,i} - c_i}{d_i - c_i} + d_i \frac{d_i - e_{l,i}}{d_i - c_i})]$$
$$\times [\Pi_{i \in \Lambda}(-1)^{1 - \frac{e_{l,i} - c_i}{d_i - c_i}}], \tag{5.7}$$

$$m_\Lambda = \int_{\Omega} \Pi_{i \in \Lambda} \mathbf{q}_i P(d\omega), \tag{5.8}$$

and

$$m_{j,\Lambda} = \int_{\Omega} \mathbf{h}_j \Pi_{i \in \Lambda} \mathbf{q}_i P(d\omega), \tag{5.9}$$

where $j = 1, \ldots, m_2$. We may also include stochastic components of $\mathbf{T}$ in place of $\mathbf{h}_j$ in (5.9). For simplicity, however, we only consider $\mathbf{h}$ stochastic next.

Assuming that $\sum_{\Lambda \in \mathcal{L}} c_\Lambda(e_l) m_\Lambda > 0$ for all $l = 1, \ldots, L$, the integration in (5.6) yields a lower bound. With the definitions in (5.7)–(5-9), we can define a general dependent lower bound, $LB^{q,h}(x)$, as

$$
LB^{q,h}(x) = \sum_{l=1}^{L} (\sum_{\Lambda \in \mathcal{L}} c_\Lambda(e_l) m_\lambda) [\sum_{j=1}^{m_2} \pi(e_l, j) (\frac{\sum_{\Lambda \in \mathcal{L}} c_\Lambda(e_l) m_{j,\lambda}}{\sum_{\Lambda \in \mathcal{L}} c_\Lambda(e_l) m_\lambda} - (Tx)_j)]
$$

$$
= \sum_{l=1}^{L} (\sum_{\Lambda \in \mathcal{L}} c_\Lambda(e_l) m_\lambda) Q(x, e_l, \frac{\sum_{\Lambda \in \mathcal{L}} c_\Lambda(e_l) m_{j,\lambda}}{\sum_{\Lambda \in \mathcal{L}} c_\Lambda(e_l) m_\lambda})
$$

$$
\leq Q(x),
$$

$$(5.10)$$

where $\pi(e_l)$ is chosen so that

$$
Q(x, e_l, \frac{\sum_{\Lambda \in \mathcal{L}} c_\Lambda(e_l) m_{j,\lambda}}{\sum_{\Lambda \in \mathcal{L}} c_\Lambda(e_l) m_\lambda}) = [\sum_{j=1}^{m_2} \pi(e_l, j) (\frac{\sum_{\Lambda \in \mathcal{L}} c_\Lambda(e_l) m_{j,\lambda}}{\sum_{\Lambda \in \mathcal{L}} c_\Lambda(e_l) m_\lambda} - (Tx)_j)].
$$

When $\sum_{\Lambda \in \mathcal{L}} c_\Lambda(e_l) m_\lambda = 0$, we also have $\sum_{\Lambda \in \mathcal{L}} c_\Lambda(e_l) m_{j,\lambda} = 0$ (Exercise 5) making the lth component of the bound zero in that case. A completely analogous upper bound is also available then.

Dependency can be removed if the random variables, $\mathbf{h}$, can be written as linear transformations of independent random variables. Here, the independent case needs only to be slightly altered. A discussion appears in Birge and Wallace [1986].

The difficulty with the upper bounds for convex $g(x, \cdot)$ and the other bounds with concave components is that they minimally require function evaluations at the support of the random vectors. They also may require joint moment information that is not available. These factors make bounds based on extreme points unattractive for practical computation with more than a small number of random elements. As we saw earlier, in the case of simplicial support, we can reduce the effort to only being linear in the dimension of the support, but the bounds generally become imprecise.

Another problem with the upper bounds described so far in this chapter is that they require bounded support. In Subsection c, we will describe generalizations to eliminate this requirement for Edmundson-Madansky types of bounds. In the next subsection, we consider other bounds that do not have this limitation. They are based on exploiting separable structure in the problem. The goal in this case is to avoid exponential growth in effort as the number of random variables increases. The bounds of Section 3 are still quite useful for low dimensions.

b. Bounds based on separable functions

As we observed earlier, simple recourse problems are especially attractive because they only require simple integrals to evaluate. The basic idea in this section is to construct approximating functions that are separable and, therefore, easy to integrate. This idea can be extended to separate low-dimension approximations, which can then be combined with the bounds in Section 3.

In the simple recourse problem (Section 3.1d), we noticed that $\Psi(\chi)$ can be written as

$$\Psi(\chi) = \sum_{i=1}^{m_2} \Psi_i(\chi_i), \tag{5.11}$$

in the case when only $\mathbf{h}$ is random in the recourse problem. We again consider this case and build approximations on it. These results appear in Birge and Wets [1986, 1989], Birge and Wallace [1988], and, for network problems, Wallace [1987].

The basic simple recourse approximation is to consider an optimal response to changes in each component of $\mathbf{h}$ separately and to combine those responses into an approximating function. For the ith component of $\mathbf{h}$, this response is the pair of optimal solutions, $y^{i,+}, y^{i,-}$, to:

$$\begin{aligned} \min \, & q^T y \\ \text{s. t. } & Wy = \pm e_i, \\ & y \geq 0, \end{aligned} \tag{5.12}$$

where e_i is the ith coordinate direction, $y^{i,+}$ corresponds to a right-hand side of e_i, and $y^{i,-}$ corresponds to a right-hand side of $-e_i$. Thus, for any value h_i of $\mathbf{h}_i$, the approximating response of $y^{i,+}(h_i - \chi_i)$ if $h_i \geq \chi_i$ and $y^{i,-}(\chi_i - h_i)$ if $h_i < \chi_i$. We have thus used the positive homogeneity of $\psi(\chi, h + \chi)$.

Using $y^{i,+}$ and $y^{i,-}$, we then obtain the approximate simple recourse functions:

$$\psi_{I(i)}(\chi_i, h_i) = \begin{cases} q^T y^{i,+}(h_i - \chi_i) & \text{if } h_i \geq \chi_i, \\ q^T y^{i,-}(\chi_i - h_i) & \text{if } h_i < \chi_i, \end{cases} \tag{5.13}$$

which are integrated to form

$$\Psi_{I(i)}(\chi_i) = \int_{\mathbf{h}_i} \psi_{I(i)}(\chi_i, h_i) P_i(d\mathbf{h}_i), \tag{5.14}$$

where we let P_i be the marginal probability measure of $\mathbf{h}_i$. Note that the calculation in (5.14) only requires the conditional expectation of $\mathbf{h}_i$ on each interval $(-\infty, \chi_i]$ and (χ_i, ∞) and the expectation of these intervals.

The $\Psi_{I(i)}$ functions combine to form

$$\Psi_I(\chi) = \sum_{i=1}^{m_2} \Psi_{I(i)}(\chi_i), \tag{5.15}$$

which is a simple recourse function. The next theorem states the main result of this section.

Theorem 5. *The function $\Psi_I(\chi)$ constructed in (5.13)–(5.15) represents an upper bound on the recourse function $\Psi(\chi)$, i.e.,*

$$\Psi(\chi) \le \Psi_I(\chi), \tag{5.16}$$

for all χ.

Proof: Consider the solution $y_I = \sum_{i=1}^{m_2}[y^{i,+}(h_i - \chi_i)^+ + y^{i,-}(-)(h_i - \chi_i)^-]$. Note that y_I is feasible in the recourse problem for h. Thus

$$\Psi(\chi) = \int_\Omega \psi(\chi, h)P(d\omega)$$
$$\le \int_\Omega q^T y_I P(d\omega) = \sum_{i=1}^{m_2} \Psi_I(\chi_i) = \Psi_I(\chi). \tag{5.17}$$

$\square$

The result in Theorem 5 is straightforward but useful. In particular, we can construct other approximations that use different representations of a solution to the recourse problem with right-hand side $h - \chi$. A particularly useful type of this approximation is to consider a set of vectors, $V = \{v_1, \ldots, v_\nu\}$, such that any vector in $\Re^{m_2}$ can be written as a non-negative linear combination of the vectors in V. This defines V as a *positive linear basis* of $\Re^{m_2}$. For such V, we suppose that $y^{V,i}$ solves:

$$\begin{aligned} \min\; & q^T y \\ \text{s. t. } & Wy = v_i, \\ & y \ge 0. \end{aligned} \tag{5.18}$$

We can then represent any $h - \chi$ in terms of non-negative combinations of the v_i or W times the corresponding non-negative combination of the $y^{V,i}$. Thus, we construct a feasible solution that responds separately to the components of V.

If V is a simplex, the construction of $h - \chi$ from V corresponds to a barycentric coordinate system. Bounds based on this idea are explored in Dulá [1991]. Another option is to let V be the set of positive and negative components of a basis $D = [d_1| \cdots |d_{m_2}]$ of $\Re^{m_2}$, or, $V = \{d_1, \ldots, d_{m_2}, -d_1, \ldots, -d_{m_2}\}$. This yields solutions, $y^{D,i,+}$, to (5.18) when $v_i = d_i$ and $y^{D,i,-}$ when $v_i = -d_i$. To use these in approximating a recourse problem with right-hand side, $h - \chi$, we want the values of ζ such that $D\zeta = h - \chi$ or $\zeta = D^{-1}(h - \chi)$. Then the weight on d_i is ζ_i if $\zeta_i \ge 0$ and the weight on $-d_i$ is $-\zeta_i$ if $\zeta_i < 0$. We thus construct simple recourse-type functions,

$$\psi_{D^i}(\zeta_i) = \begin{cases} q^T y^{D,i,+}(\zeta_i) & \text{if } \zeta_i \ge 0, \\ q^T y^{D,i,-}(-\zeta_i) & \text{if } \zeta_i < 0, \end{cases} \tag{5.19}$$

which are integrated to form

$$\Psi_{D^i}(\chi) = \int_{\zeta_i} \psi_{D^i}(\zeta_i)P_{D^i}(d\zeta_i), \tag{5.20}$$

where P_{D^i} is the marginal probability measure of ζ_i. Again, these are added to create a new upper bound,

$$\Psi_D(\chi) = \sum_{i=1}^{m_2} \Psi_{D^i}(\chi) \geq \Psi(\chi). \tag{5.21}$$

Now, computation of Ψ_D relies on the ability to find the distribution of ζ_i. In special cases, such as when $\mathbf{h}$ is normally distributed, then ζ, the affine transformation of a normal vector is also normally distributed so that the marginal ζ_i can be easily calculated. In other cases, full distributional information of $\mathbf{h}$ may not be known. In this case, first or higher moments of ζ_i can be calculated and bounds such as those in Section 2 or those based on the moment problem in Subsection c, can be used. In either case, the calculation of Ψ_D reduces to evaluating or bounding the expectation of a function of a single random variable.

Of course, if a set of bases, $\mathcal{D}$, is available, then the best bound within this set can be used. In fact, the convex hull of all approximations, Ψ_D, for $D \in \mathcal{D}$, is also a bound. We write this function as:

$$co\{\Psi_D, D \in \mathcal{D}\}(\chi) = \inf\{\sum_{i=1}^{K} \lambda^i \Psi_{D^i}(\chi^i) | \sum_{i=1}^{K} \lambda^i \chi^i = \chi,$$

$$\sum_{i=1}^{j} \lambda^i = 1, \lambda^i \geq 0, i = 1, \ldots, K\}, \tag{5.22}$$

where $\mathcal{D} = \{D^1, \ldots, D^j\}$. This definition yields the following.

Theorem 6. *For any set $\mathcal{D}$ of linear bases of $\Re^{m_2}$,*

$$\Psi(\chi) \leq co\{\Psi_D, D \in \mathcal{D}\}(\chi). \tag{5.23}$$

Proof. From earlier,

$$\Psi(\chi^i) \leq \Psi_{D^i}(\chi^i) \tag{5.24}$$

for each $i = 1, \ldots, K$ and choice of χ^i. By convexity of Ψ, $\Psi(\chi) \leq \sum_{i=1}^{j} \lambda^i \Psi(\chi^i)$ where

$$\sum_{i=1}^{K} \lambda^i \chi^i = \chi, \sum_{i=1}^{j} \lambda^i = 1, \lambda^i \geq 0, i = 1, \ldots, K. \tag{5.25}$$

Combining (5.24) and (5.25) with the definition in (5.22) yields (5.23). $\square$

From Theorem 6, we continue to add bases D^i to $\mathcal{D}$ to improve the bound on $\Psi(\chi)$. Even if $\mathcal{D}(W)$, the set of all bases in W are included; however, the bound is not exact. In this case, $co\{\psi_D(D^{-1}(\mathbf{h} - \chi)) | D \in \mathcal{D}(W)\} =$

$\psi(\chi, \mathbf{h})$ because $\psi(\chi, \mathbf{h}) = q^T \mathbf{y}^* = q^T (D^*)^{-1} (\mathbf{h} - \chi)$ for some $D^* \in \mathcal{D}(W)$. However,

$$
\begin{aligned}
\Psi(\chi) &= \int co\{\psi_D(D^{-1}(\mathbf{h} - \chi)) | D \in \mathcal{D}(W)\} P(d\mathbf{h}) \\
&\leq co\{\int \psi_D(D^{-1}(\mathbf{h} - \chi)) P(d\mathbf{h}) | D \in \mathcal{D}(W)\} \qquad (5.26) \\
&= co\{\Psi_D, D \in \mathcal{D}\}(\chi),
\end{aligned}
$$

where the inequality is generally strict except for unusual cases (such as Ψ linear in χ).

As we shall see in an example later, the main intention of this approximation is to provide a means to find the optimal x value. Thus, the most important consideration is whether the subgradients of $co\{\Psi_D, D \in \mathcal{D}\}(\chi)$ are approximately the same as those for $\Psi(\chi)$. In this case, the approximation appears to perform quite well (see Birge and Wets [1989]).

Example 1 (continued)

Let us consider Example 1 again, as in Section 2. The optimal bases and their regions of optimality were given there. In this case, we let $D^1 = B^1$, $D^2 = B^2$, and $D^3 = B^3$. Note that this last approximation is derived for B^4 and B^5 because they correspond to the same positive linear basis as $[B_3, -B_3]$. At $\chi = (0.3, 0.3)^T$, we can evaluate each of the bounds, Ψ_{D^i}. For $i = 1$, we have $(D^1)^{-1} = \begin{pmatrix} 1 & -1 \\ 0 & 1 \end{pmatrix}$, so that $\zeta_1^1 = \mathbf{h}_1 - \mathbf{h}_2$ and $\zeta_2^1 = \mathbf{h}_2 - \chi_2 = \mathbf{h}_2 - 0.3$. In this case, $y^{D^1,1,+} = (y_1^+, y_1^-, y_2^+, y_2^-, y_3)^T = (1, 0, 0, 0, 0)^T$, $y^{D^1,1,-} = (0, 1, 0, 0, 0)^T$, $y^{D^1,2,+} = (0, 0, 0, 0, 1)^T$, and $y^{D^1,2,-} = (0, 1, 0, 1, 0)^T$. Integrating out each ζ_i^1, we obtain $\Psi_{D^1}(0.3, 0.3) = 0.668$. Symmetrically, $\Psi_{D^2}(0.3, 0.3) = 0.668$. For $\Psi_{D^3}(0.3, 0.3)$, we note that each component is simply the probability that $\mathbf{h}_i \leq 0.3$ times the conditional expectation of $\mathbf{h}_i - 0.3$ given $\mathbf{h}_i \leq 0.3$ plus the probability that $\mathbf{h}_i > 0.3$ times the conditional expectation of $\mathbf{h}_i - 0.3$ given $\mathbf{h}_i > 0.3$. Thus, $\Psi_{D^3}(0.3, 0.3) = 2[(0.3)(0.15) + (0.7)(0.35)] = 0.580$.

Comparing the best of these bounds with those in the previous chapters leads to a more accurate approximation. We should note, however, that this approach requires more distributional information.

Taking convex hulls can produce even better bounds. The convex hull operation is, however, a nonconvex optimization problem. The dual gives some computational advantage. To give an idea of the advantage of the convex hull, however, consider Figure 5, where the graphs of Ψ_{D^i} are displayed with that of Ψ as functions of χ_1 for $\chi_2 = 0.1$. Note how the convex hull of the graphs of the approximations appears to have similar subgradients to that of Ψ. This observation appears to hold quite generally, as indicated by the computational tests in Birge and Wets [1989].

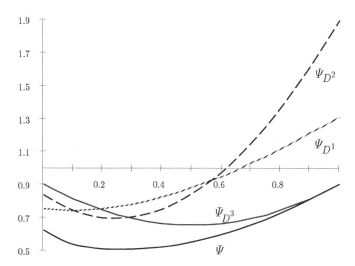

FIGURE 5. Graphs of Ψ (solid line) and the approximations, Ψ_{D^i} (dashed lines).

The separable bounds in Ψ_{D^i} can also be enhanced by, for example, including fixed values (due to known entries in h) into the right-hand sides of (5.18). Other possibilities are to combine the component approximations on an interval instead of assuming that they may apply for all positive multiples of the v_i. In this case, the solution for some interval of v_i multiples can serve as a constraint for determining solutions for the next v_{i+1}. This procedure is carried out in Birge and Wallace [1988]. It appears especially useful for problems with bounded random variables and networks (Wallace [1987]).

To improve on these bounds and obtain some form of convergence requires relaxation of complete separability. For example, pairs of random variables can be considered together. In this way, more precise bounds can be found. Determination of these terms is, however, problem-specific. In general, the structure of the problem must be used to obtain the most efficient improvements on the basic separable approximation bounds.

So far, we have presented bounds for the recourse function with a fixed χ value. In the next subsection, we consider how to combine these approximations into solution algorithms where x varies from iteration to iteration. In the case of the separable bounds, this implementation results from a dualization that turns the difficult convex hull operation into a simpler supremum operation.

c. Extensions with general moments

Many other bounds are possible in addition to those presented so far. A general form for many of these bounds is found through the solution of an abstract linear program, called a *generalized moment problem*. This problem provides the lowest or highest expected probabilities or objective values that are possible given certain distribution information that can be written as generalized moments. In this subsection, we present this basic framework, some results using second moments, and generalizations to nonlinear functions. Concepts from measure theory appear again in this development.

To obtain bounds that hold for all distributions with certain properties, we can find $p \in \mathcal{P}$ a set of probability measures on $(\Xi, \mathcal{B}^N)$ to extremize a moment problem. We let $\mathcal{B}^N$ be the Borel field of $\Re^N$ where $\Re^N \supset \Xi$. We use probability measures defined directly on $\mathcal{B}^N$ to simplify the following discussion. We wish to find

$$P \in \mathcal{P} \text{ a set of probability measures on } (\Xi, \mathcal{B}^N) \text{ s. t.}$$

$$
\begin{aligned}
\int_\Xi v_i(\xi) P(d\xi) &\le \alpha_i, i = 1, \dots, s, \\
\int_\Xi v_i(\xi) P(d\xi) &= \beta_i, i = s + 1, \dots, M,
\end{aligned}
\tag{5.27}
$$

$$\text{to maximize } \int_\Xi g(\xi) P(d\xi),$$

where M is finite and the v_i are bounded, continuous functions. A solution of (5.27) obtains an upper bound on the expectation of g with respect to any probability measure satisfying the conditions given earlier. We could equally well have posed this to find a lower bound.

Problem (5.27) is a *generalized moment problem* (Krein and Nudel'man [1977]). When the v_i are powers of ξ, the constraints restrict the moments of ξ with respect to P. In this context, (5.27) determines an upper bound when only limited moment information on a distribution is available.

Problem (5.27) can also be interpreted as an *abstract linear program*, i.e., a linear program defined over an abstract space, because the objective and constraints are linear functions of the probability measure. The solution is then an extreme point in the infinite-dimensional space of probability measures. The following theorem, proven in Karr [1983, Theorem 2.1], gives the explicit solution properties. We state it without proof because our main interests here are in the results and not the particular form of these solutions. Readers with statistics backgrounds may compare the result with the

Neyman-Pearson lemma and the proof of the optimality conditions as in Dantzig and Wald [1951]. For details on the weak* topology that appears in the theorem, we refer the reader to Royden [1968].

Theorem 7. *Suppose Ξ is compact. Then the set of feasible measures in (5.27), $\mathcal{P}$, is convex and compact (with respect to the weak* topology), and $\mathcal{P}$ is the closure of the convex hull of the extreme points of $\mathcal{P}$. If g is continuous relative to Ξ, then an optimum (maximum or minimum) of $\int_\Xi g(x,\xi)P(d\xi)$ is attained at an extreme point of $\mathcal{P}$. The extremal measures of $\mathcal{P}$ are those measures that have finite support, $\{\xi_1,\ldots,\xi_L\}$, with $L \leq M + 1$, such that the vectors*

$$\begin{pmatrix} v_1(\xi_1) \\ v_2(\xi_1) \\ \vdots \\ v_M(\xi_1) \\ 1 \end{pmatrix}, \ldots, \begin{pmatrix} v_1(\xi_L) \\ v_2(\xi_L) \\ \vdots \\ v_M(\xi_L) \\ 1 \end{pmatrix} \tag{5.28}$$

are linearly independent.

Kemperman [1968] showed that the supremum is attained under more general continuity assumptions and provides conditions for $\mathcal{P}$ to be nonempty. Dupačová (formerly Žáčková) [1976, 1977, 1966] pioneered the use of the moment problem as a bounding procedure for stochastic programs in her work on a minimax approach to stochastic programming. She showed that (5.27) attains the Edmundson-Madansky bound (and the Jensen bound if the objective is minimized) when the only constraint in (5.27) is $v_1 = \xi$, i.e., the constraints fix the first moment of the probability measure. She also provided some properties of the solution with an additional second moment constraint ($v_2(x) = \xi^2$) for a specific objective function g. Frauendorfer's [1988b] results can be viewed as solutions of (5.27) when the constraints satisfy all of the joint moment conditions.

To solve (5.27) generally, we consider a generalized linear programming procedure.

Generalized Linear Programming Procedure for the Generalized Moment Problem (GLP)

Step 0. Initialization. Identify a set of $L \leq M + 1$ linearly independent vectors as in (5.28) that satisfy the constraints in (5.27). (Note that a phase one–objective (Dantzig [1963]) may be used if such a starting solution is not immediately available. For $N = 1$, the Gaussian quadrature points may be used.) Let $r = L$, $\nu = 1$; go to 1.

Step 1. Master problem solution. Find $p_1 \geq 0, \ldots, p_r \geq 0$ such that

$$\sum_{l=1}^{r} p_l = 1,$$

$$\sum_{l=1}^{r} v_l(\xi_l)p_l \leq \beta_i, i = 1, \ldots, s,$$

$$\sum_{l=1}^{r} v_l(\xi_l)p_l = \beta_i, i = s+1, \ldots, M, \text{and} \tag{5.29}$$

$$z = \sum_{l=1}^{r} g(\xi_l)p_l \text{ is maximized.}$$

Let $\{p_1^j, \ldots, p_r^j\}$ attain the optimum in (5.29), and let $\{\sigma^j, \pi_1^j, \ldots, \pi_M^j\}$ be the associated *dual multipliers* such that

$$\sigma^j + \sum_{i=1}^{M} \pi_i^j v_i(\xi_l) = g(\xi_l), \text{ if } p^l > 0, l = 1, \ldots, r,$$

$$\sigma^j + \sum_{i=1}^{M} \pi_i^j v_i(\xi_l) \geq g(\xi_l), \text{ if } p^l = 0, l = 1, \ldots, r, \tag{5.30}$$

$$\pi_i^j \geq 0, i = 1, \ldots, s.$$

Step 2. Subproblem solution. Find ξ^{r+1} that maximizes

$$\gamma(\xi, \sigma^j, \pi^j) = g(\xi) - \sigma^j - \sum_{i=1}^{M} \pi_i^j v_i(\xi). \tag{5.31}$$

If $\gamma(\xi^{r+1}, \sigma^j, \pi^j) > 0$, let $r = r+1$, $\nu = \nu + 1$ and go to Step 1. Otherwise stop; $\{p_1^j, \ldots, p_r^j\}$ are the optimal probabilities associated with $\{\xi_1, \ldots, \xi_r\}$ in a solution to (5.27).

As we saw in Chapter 3, the generalized programming approach is useful in problems with a potentially large number of variables. This approach is used in Ermoliev, Gaivoronski, and Nedeva [1985] to solve a class of problems (5.27). The difficulty in GLP is in the solution of the subproblem (5.31), which generally involves a nonconvex function. Birge and Wets [1986] describe how to solve (5.31) with constrained first and second moments, if convexity properties of γ can be identified. Cipra [1985] describes other methods for this problem based on discretizations and random selections of candidate points, x_i. Dulá [1991] gives results when g is sublinear and has simplicial level sets. Kall [1991] gives the results for sublinear, polyhedral functions with known generators. Edirisinghe [1996] also finds bounds using second moment information that is somewhat looser than the generalized moment solution.

Kall's result is useful when the optimal recourse problem multipliers are known, so that

$$Q(x, \boldsymbol{\xi}) = \max_{i=1,\ldots,K} \pi_i^T (\mathbf{h} - Tx), \qquad (5.32)$$

where we again assume that $\boldsymbol{\xi} = \mathbf{h}$ or that T and q are known. Kall's result pertains to having known means for all $\mathbf{h}_i$ and a limit ρ on the *total second moment*, defined as

$$\rho = \int_\Xi \|\xi\|^2 P(d\xi). \qquad (5.33)$$

The moment problem becomes:

$$\sup_{P \in \mathcal{P}} \int_\Omega Q(x, \xi) P(d\xi)$$

$$\text{s. t. } \int_\Xi \xi P(d\xi) = \bar{h} \text{ and (5.33)}, \qquad (5.34)$$

where $\mathcal{P}$ is a set of probability measures with support, Ξ.

Kall shows that the solution of (5.34) with Q defined as in (5.32) is equivalent to the following finite-dimensional optimization problem:

$$\inf_{y \in \Re^m} \{ \max_{i=1,\ldots,K} \sqrt{\rho - 2(\bar{h})^T Tx + \|Tx\|^2} \|\pi_i - y\| + (\bar{h} - Tx)^T y \}. \qquad (5.35)$$

Dulá obtained similar results for strictly simplicial Q. Note that when $\bar{h} = Tx$, this reduces to a form of location problem to minimize the maximum weighted distance from π_i to y. The solution to (5.34) may involve calculations with each of these recourse problem solutions, but the resulting distribution P that solves (5.34) still has only $m_2 + 2$ points of support. These are found by solving for the Karush-Kuhn-Tucker conditions for problem (5.34), where the y values correspond to multipliers for the mean value constraints.

Other bounds are also possible for different types of objective functions. In particular, we consider functions built around separable properties. The use of the generalized programming formulation is limited in multiple dimensions because of the difficulty in solving subproblem (5.32). These computational disadvantages for large values of N suggest that a looser but more computationally efficient upper bound on the value of (5.27) may be more useful than solving (5.27) exactly for large N.

If a separable function, $\eta(x) = \sum_{i=1}^N \eta_i(x(i))$, is available, it offers an obvious advantage by only requiring single integrals, as we stated earlier. Here, we would also like to show that these bounds can be extended to nonlinear recourse functions. We suppose that the recourse function becomes some general $g(\xi(\omega))$, where

$$g(\xi) = \inf_y \{ q(y) | g(y) \le \xi \}. \qquad (5.36)$$

In this case, we would like to find $\eta(\xi) = \sum_{i=1}^{N} \eta_i(\xi(i)) \geq g(\xi)$ where each $\eta_i(\xi(i))$ is a convex function. Methods for constructing these functions to bound the optimal value of a linear program with random right-hand sides were discussed in Subsection b. We next give the results for the general problem in (5.36).

Lemma 8. *If g is defined as in (5.36), then g is a convex function of ξ.*

Proof: Let y_1 solve the optimization problem in (5.36) for ξ_1 and let y_2 solve the corresponding problem for ξ_2. Consider $\xi = \lambda\xi_1 + (1-\lambda)\xi_2$. In this case, $g(\lambda y_1 + (1-\lambda)y_2) \leq \lambda g(y_1) + (1-\lambda)g(y_2) \leq \lambda\xi_1 + (1-\lambda)\xi_2$. So $g(\lambda\xi_1 + (1-\lambda)\xi_2) \leq q(\lambda y_1 + (1-\lambda)y_2) \leq \lambda g(\xi_1) + (1-\lambda)g(\xi_2)$, giving the result. $\square$

Let

$$\eta_i(\xi(i)) \equiv \frac{1}{N}g(N\xi(i)e_i), \qquad (5.37)$$

which is the optimal value of a parametric mathematical program. The following theorem shows that these values supply the separable bound required. Related bounds are possible by defining η_i with other right scalar multiples, $g\lambda_i(\xi(i)e_i)$ (see Rockafellar [1969] for general properties), where $\sum_{i=1}^{N} \lambda_i = 1$. The following proof below is easily extended to these cases and to translations of the constraints and explicit variable bounds.

Theorem 9. *The function $\eta(\xi) = \sum_{i=1}^{N} \eta_i(\xi(i)) \geq g(\xi)$, where g is defined as in (5.36).*

Proof: In this case, let $y_i(\xi(i))$ solve (5.36), where $\xi(\omega) = N\xi(i)e_i$. Then, $g(\frac{\sum_{i=1}^{N} y_i(\xi(i))}{N}) \leq \sum_{i=1}^{N}(\frac{1}{N})[g(y_i(\xi(i)))] \leq \sum_{i=1}^{N}(\frac{1}{N})N\xi(i)e_i = \xi$. Next, let y^* solve (5.36) for ξ in the right-hand side of the constraints. By feasibility of $\sum_{i=1}^{N} \frac{y_i(\xi(i))}{N}$, $g(\xi) = q(y^*) \leq q(\sum_{i=1}^{N} \frac{y_i(\xi(i))}{N}) \leq \sum_{i=1}^{N}(\frac{1}{N})q(y_i(\xi_i)) = \sum_{i=1}^{N} \eta_i(\xi(i)) = \eta(\xi)$. $\square$

This result demonstrates that a parametric optimization of (5.36) in $i = 1,\ldots,N$ yields an upper bound on $g(\xi)$ for any ξ. The bound may be tight, as in some examples for stochastic linear programs as given in Subsection b.

Generalizations of the stochastic linear program bound as in Subsection b can also be given for the general bound in Theorem 9. For example, we may apply a linear transformation T to ξ to obtain $u = T\xi$. The constraints become $g(y) \leq G^{-1}(u)$. To use any bound of the general type in Theorem 9 to bound $\int_{\Re^N} g(\xi)dg(\xi)$ requires a bound on $\int_{\Re} \eta_i(\xi(i))\,dF_i(\xi_i)$ or $\int_{\Re} \mu_i(u(i))dF_{u_i}(u(i))$, where F_i is the marginal distribution on $\boldsymbol{\xi}_i$ and $F_{\mathbf{u}_i}$ is the marginal distribution on $\mathbf{u}(i)$. Because it may be difficult to find the distribution of $\mathbf{u}$, the generalized moment problem can be solved to obtain bounds on each integral in $\Re$. Generalized linear programming may solve this problem but can be inefficient. To simplify this process, in Birge and Dulá [1991], it is shown that a large class of functions requires only two points of support in the bounding distribution. A single line search can determine these points and give a bound on f over all distributions with bounded first and second moments for the marginals.

We develop bounds following Birge and Dulá [1991] on $\int \eta_i(x(i))$ $dF_i(x(i))$ by referring to g as a function on $\Re$ $(N = 1)$. We then consider the moment problem (5.27) with $s = 0$, and $M = 2$ and where the constraints correspond to known first and second moments. In other words, we wish to find:

$$
\begin{aligned}
U = \sup_{Q \in \mathcal{P}} \int_\Xi g(\xi) P(d\xi) \\
\int_\Xi \xi P(d\xi) &= \bar{\xi}, \\
\int_\Xi \xi^2 P(dx) &= \bar{\xi}^{(2)},
\end{aligned}
\tag{5.38}
$$

where $P \in \mathcal{P}$ is the set of probability measures on $(\Xi, \mathcal{B}^1)$, the first moment of the true distribution is $\bar{\xi}$, and the second moment is $\bar{\xi}^{(2)}$.

A generalization of Carathéodory's theorem (Valentine [1964]) for the convex hull of connected sets tells us that y^* can be expressed as a convex combination of at most three extreme points of C, giving us a special case of Theorem 9. Therefore, an optimal solution to (5.38) can be written, $\{\xi^*, p^*\}$, where the points of support, $\xi^* = \{\xi_1^*, \xi_2^*, \xi_3^*\}$ have probabilities, $p^* = \{p_1^*, p_2^*, p_3^*\}$. An optimal solution may, however, have two points of support. A function that has this property for a given instance of (5.27) is called a *two–point support* function. We will give sufficient conditions for a function to have this two-point support property. This property then allows a simplified solution of (5.38). It is given in the next theorem which is proven in Birge and Dulá [1991].

Theorem 10. *If g is convex with derivative g' defined as a convex function on $[a, c]$ and as a concave function on $(c, b]$ for $\Xi = [a, b]$ and $a \leq c \leq b$, then there exists an optimal solution to (5.38) with at most two support points, $\{\xi_1, \xi_2\}$, with positive probabilities, $\{p_1, p_2\}$.*

A corollary of Theorem 10 is that any function g that has a convex or concave derivative has the two-point support property. The class of functions that meets the criteria of Theorem 10 contains many useful examples, such as:

1. Polynomials defined over ranges with at most one third-derivative sign change.

2. Exponential functions of the form, $c_0 e^{c_1 \xi}, c_0 \geq 0$.

3. Logarithmic functions of the form, $\log_j(c\xi)$, for any $j \geq 0$.

4. Certain hyperbolic functions such as $\sinh(c\xi), c, \xi \geq 0, \cosh(cx)$.

5. Certain trigonometric and inverse trigonometric functions such as $\tan^{-1}(c\xi), c, \xi \geq 0$.

In fact, Theorem 10 can be applied to provide an upper bound on the expectation of any convex function with known third derivative when the distribution function has a known third moment, $\bar{\xi}^{(3)}$. Suppose $a > 0$ (if not, then this argument can be applied on $[a, 0]$ and $[0, b]$); then let $g(\xi) = \beta\xi^3 + g(\xi)$. The function g is still convex on $[0, b)$ for $\beta \geq 0$. By defining $\beta \geq (-1/6)\min(0, \inf_{\xi \in [a,b]} f'''(\xi))$, g' is convex on $[a, b]$, and an upper bound, $UB(g)$, on $E_g(\xi)$ has a two-point support. The expectation of g is then bounded by

$$E_g(\xi) \leq UB(g) - \beta\bar{\xi}^{(3)}. \tag{5.39}$$

The conditions in Theorem 10 are only sufficient for a two–point support function. They are not necessary (see Exercise 8). Note also that not all functions are two-point support functions (although bounds using (3.4) are available). A function requiring three support points, for example, is $g(\xi) = (1/2) - \sqrt{(1/4) - (\xi - (1/2))^2}$ (Exercise 9).

Given that a function is a two-point support function, the points $\{\xi_1, \xi_2\}$ can be found using a line search to find a maximum.

For the special case of piecewise linear functions, the points, ξ_1, ξ_2, can be found analytically. In this case, suppose that $g(\xi) = \psi^{SR}(h, \chi)$, the simple recourse function defined by:

$$\psi^{SR}(h, \chi) = \begin{cases} q^-(\chi - h) & \text{if } h \leq \chi, \\ q^+(h - \chi) & \text{if } h > \chi. \end{cases} \tag{5.40}$$

Consider the nonintersecting intervals, $A = (0, \bar{\xi}^{(2)}/(2\bar{\xi}))$, $B = [\bar{\xi}^{(2)}/(2\bar{\xi})$, $(1 - \bar{\xi}^{(2)})/(2(1 - \bar{\xi}))]$, and $C = ((1 - \bar{\xi}^{(2)})/(2(1 - \bar{\xi})), 1)$. The points of support for this semilinear, convex function defined on $[0, 1]$ are

$$\{\xi_1^*, \xi_2^*\} = \begin{cases} \{0, \bar{\xi}^{(2)}/\bar{\xi}\} & \text{if } \chi \in A, \\ \{\chi - d, \chi + d\} & \text{if } \chi \in B, \\ \{(\bar{\xi} - \bar{\xi}^{(2)})/(1 - \bar{\xi}), 1\} & \text{if } \chi \in C, \end{cases} \tag{5.41}$$

where $d = \sqrt{\chi^2 - 2\chi\bar{\xi} + \bar{\xi}^{(2)}}$. This result can be obviously extended to other finite intervals. Infinite intervals can also be solved analytically for these semilinear, convex functions. For $X = [0, \infty)$, the results are as in (5.41) with $B = [\bar{\xi}^{(2)}/(2\bar{\xi}), \infty)$ and $C = \emptyset$. For the interval $(-\infty, \infty)$, the points of support are those for interval B in (5.41). We note that special cases for these supports of semilinear, convex functions were considered in Jagganathan [1977] and Scarf [1958].

Other bounds are also possible using the generalized moment problem framework. One possible approach is to use piecewise linear constraints on the quadratic functions defining second-moment constraints as in (5.38). This approach is described in Birge and Wets [1987] which also considers

unbounded regions that lead to measures that are limiting solutions to
(5.27) but that may not actually be probability measures but are instead
nonnegative measures with weights on extreme directions of Ξ. An example
is given in Exercise 12.

To see how these bounds are constructed for unbounded regions, weights
can be placed on extreme recession directions, $r^j, j = 1, \ldots, J$, such that
$\xi + \beta r^j \in \Xi$ for all $\xi \in \Xi$ and r^j not expressable in non-negative multiples of
other recession directions. Then, if the recourse function Q has a recession
function, $rcQ(x, r^j) \geq \frac{Q(x, \xi + \beta r^j) - Q(x, \xi)}{\beta}$ for all $\beta > 0$, then $Q(x, \xi) \leq$
$\sum_{k=1,\ldots,K} \lambda^j Q(x, e^j) + \sum_{j=1,\ldots,J} \mu^j rcQ(x, r^j)$, when $\xi = \sum_{k=1,\ldots,K} \lambda^j e^j +$
$\sum_{j=1,\ldots,J} \mu^j r^j$, $\sum_{k=1,\ldots,K} \lambda^j = 1$, $\lambda^j, \mu^j \geq 0$. Now, an analogous result
to Theorem 1 can be constructed where $\lambda^j = \int_\Xi \lambda(\xi, e^j) P(d\xi)$ and $\mu^j =$
$\int_\Xi \mu(\xi, r^j) P(d\xi)$ are constructed from measures $\lambda(\xi, \cdot)$ and $\mu(\xi, \cdot)$ such that
$\xi = \sum_{k=1,\ldots,K} e^j \lambda(\xi, e^j) + \sum_{j=1,\ldots,J} r^j \mu^j(\xi, r^j)$ for all $\xi \in \Xi$.

With piecewise linear functions, $v_i(\xi) = \beta_{il}\xi + \beta_{il}$ on Ξ^l, $l = 1, \ldots, L$,
$P[\Xi^l] = p^l$,

$$\int_\Xi v_i(\xi) P(d\xi) = \sum_{l=1}^{L} \sum_{e \in ext\Xi_l} \beta_{il} e \lambda^l(e) + \sum_{r \in rc\Xi_l} \beta_{il} r \mu_l(r) - \beta_{il} p^l, \quad (5.42)$$

where $\lambda^l(e)$ is a weight on the extreme point e in Ξ^l and $\mu^l(r)$ is a weight
on extreme direction r of Ξ^l. From (5.42), we can use a piecewise linear v_i
to bound nonlinear v from below. If

$$\int_\Xi v(\xi) P(d\xi) \leq \bar{v}, \quad (5.43)$$

then

$$\int_\Xi v_i(\xi) P(d\xi) \leq \bar{v}. \quad (5.44)$$

Thus, we can use (5.44) in place of (5.43) to obtain an upper bound on
a moment problem. The advantage of (5.44) is that we need only use the
extreme values of the Ξ^l from (5.42) in (5.44).

Other types of bounds are also possible that depend on different types
of functions, such as lower piecewise linear functions (see Marti [1975] or
Birge and Wets [1986]). Stochastic dominance of probability distributions
can also be used to construct bounds, although this tends to be difficult in
higher dimensions (see Birge and Wets [1986, section 7]). Another alternative is to identify optimization procedures that improve among all possible
distributions (see, e.g., Marti [1988]). Still other procedures are possible
using conjugate function information directly; Birge and Teboulle [1989]
use nonlinear functions that are otherwise not easily evaluated.

We have not yet considered approximations based on sampling ideas.
Many possibilities exist in this area as well. We will describe these bounds
and algorithms based on them in Chapter 10.

Exercises

1. Verify the derivation of $\eta(\xi, \cdot)$ in (5.2).

2. Derive the result in (5.3).

3. Consider the sugar beet recourse function, $\mathcal{Q}_3$, in Section 1.1. Suppose that the selling price above 6000 is actually a random variable, $\mathbf{q}$, that has mean 10 and is distributed on $[5, 15]$. Suppose also that $E[\mathbf{qr}_3] = 250$. Use (5.9) to derive a lower bound on $\mathcal{Q}_3(300)$.

4. Verify the result of the integration in (5.5) given in (5.9).

5. Verify that $\sum_{\Lambda \in \mathcal{L}} c_{\Lambda}(e_l) m_{\Lambda} = 0$ implies $\sum_{\Lambda \in \mathcal{L}} c_{\Lambda}(e_l) m_{j,\Lambda} = 0$ and that, if both are nonzero, then $\dfrac{\sum_{\Lambda \in \mathcal{L}} c_{\Lambda}(e_l) m_{j,\Lambda}}{\sum_{\Lambda \in \mathcal{L}} c_{\Lambda}(e_l) m_{\Lambda}}$ is in the closure of the support of $\mathbf{h}$.

6. Find the functions Ψ_{D^i} as functions of χ for each i as in the example. Also find the optimal value function Ψ in terms of χ. Graph these functions as functions of χ_2 for values of $\chi_1 = 0.j, j = 0, \ldots, 9$. Compare the convex hulls of the approximations with the graph of Ψ.

7. Using the data for Example 1, solve (5.34) to determine an upper bound with the total second-moment constraint.

8. Construct a two-point support function that does not meet the conditions in Theorem 3.

9. Show that $g(\xi) = (1/2) - \sqrt{(1/4) - (\xi - (1/2))^2}$ requires three support points to obtain the best upper bound with mean of 0.5 and variance of $1/6$ on $\Xi = [0, 1]$.

10. Find the Edmundson-Madansky and two-moment bounds for the ξ uniform on $\Xi = [0, 1]$ and the following functions: $e^{-\xi}$, ξ^3, $sin(\pi(\xi + 1))$.

11. Use the results in Theorems 9 and 10 to bound the following nonlinear recourse function with the form in (5.38). We suppose in this case that

$$g(\xi_1, \xi_2) = \begin{cases} \min(\xi_1 - 1)^2 + (\xi_2 - 2)^2 \\ \text{s. t. } \xi_1^2 + \xi_2^2 - 1 \le \xi_1, \\ (\xi_1 - 1)^2 + \xi_2^2 - 1 \le \xi_2. \end{cases}$$

12. Suppose that it is known that the ξ_i are non-negative, that $\bar{\xi}_i = 1$, and that $\bar{\xi}_i^{(2)} = 1.25$. In this case, we would like an upper bound on

the expected performance $E(g(\boldsymbol{\xi})$. We construct a bound by first finding $\eta_i(\boldsymbol{\xi}_i)$ as in (5.37). This problem may correspond to determining a performance characteristic of a part machined by two circular motions centered at $(0,0)$ and $(1,0)$, respectively. Here, the performance characteristic is proportional to the distance from the finished part to another object at $(2,1)$. The square of the radii of the tool motions is $\boldsymbol{\xi}_i + 1$, where $\boldsymbol{\xi}_i$ is a non-negative random variable associated with the machines' precision.

13. As an example of using (5.41), consider Example 1, but assume that Ξ is the entire non-negative orthant and that each $\boldsymbol{\xi}_i$ is exponentially distributed with mean 0.5. Use a piecewise linear lower bound on the individual second moments that is zero for $0 \leq \xi_i \leq 0.5$, and $2\xi_i - 1$ for $\xi \geq 0.5$. Solve the moment problem using these regions to obtain an upper bound for all expected recourse functions with the same means and variances as the exponential. Also, solve the moment problem with only mean constraints and compare the results.

9.6 General Convergence Properties

For the following bounding discussions, we use a general function notation because these results hold quite broadly. The discussion in this section follows Birge and Qi [1995], which gives a variety of results on convergence of probability measures. Other references are Birge and Wets [1986] and King and Wets [1991]. This section is fundamental for theoretical properties of convergence of approximations.

We consider the *expectational functional* $E(g(\cdot)) = E\{g(\cdot,\boldsymbol{\xi}))\}$, where $\boldsymbol{\xi}$ is a random vector with support $\Xi \subseteq \Re^N$ and g is an extended real-valued function on $\Re^n \times \Xi$. Here,

$$E(g(x)) = \int g(x,\boldsymbol{\xi})P(d\boldsymbol{\xi}), \qquad (6.1)$$

where P is a probability measure defined on $\Re^n$.

We assume that $E(g(\cdot))$ (which represents the recourse function $\mathcal{Q}$) is difficult to evaluate because of the complications involved in g and the dimension of Ξ. The basic goal in most approximations is to approximate (6.1) by

$$E^\nu(g(x)) = \int g(x,\boldsymbol{\xi})P^\nu(d\boldsymbol{\xi}), \qquad (6.2)$$

where $\{P^\nu, \nu = 1, ...\}$ is a sequence of probability measures converging in distribution to the probability measure P. By *convergence in distribution*, we mean that $\int g(\boldsymbol{\xi})P^\nu(d\boldsymbol{\xi}) \to \int g(\boldsymbol{\xi})P(d\boldsymbol{\xi})$ for all bounded continuous g on Ξ. For more general information on convergence of distribution functions, we refer to Billingsley [1968].

In the following, we use E^0 and P^0 instead of E and P for convenience. If $C \subseteq \Re^n$ is a closed convex set, then Ψ_C^* is the support function of C, defined by $\Psi^*(g|C) = \sup\{< x, g >: x \in C\}$. A sequence of closed convex sets $\{C_\nu : \nu = 1, ...\}$ in $\Re^n$ is said to converge to a closed convex set C in $\Re^n$ if for any $g \in \Re^n$,

$$\lim_{\nu \to +\infty} \Psi^*(g|C_\nu) = \Psi^*(g|C).$$

One may easily prove the following proposition that is stated without proof.

Proposition 11. *Suppose that C and C^ν, for $\nu = 1, ...$, are closed convex sets in $\Re^n$. The following two statements are equivalent:*
(a) C_ν *converges to C as $\nu \to +\infty$;*
(b) *a point $x \in C$ if and only if there are $x^\nu \in C^\nu$ such that $x_\nu \to x$.* $\square$

This notion of set convergence is important in the study of convergence of functions. We say that a sequence of functions, $\{g^\nu; \nu = 1, ...\}$, *epi-converges* to function, g, if and only if the epigraphs, epi $g^\nu = \{(x, \beta)|\beta \geq g^\nu(x)\}$, of the functions converge as sets to the epigraph of g, epi $g = \{(x, \beta)|\beta \geq g(x)\}$. Epi-convergence has many important properties, which are explored in detail in Wets [1980a] and Attouch and Wets [1981]. A chief property (Exercise 1) is that any limit point of minima of g^ν is a minimum of g.

In the following, we restrict our attention to convex integrands g although extensions to nonconvex functions are also possible as in Birge and Qi [1995]. In this case, one can use the generalized subdifferential in the sense of Clarke [1983] or other definitions as in Michel and Penot [1984] or Mordukhovich [1988]. The next theorem appears in Birge and Wets [1986] with some extensions in Birge and Qi [1995]. Other results of this type appear in Kall [1987].

Theorem 12. *Suppose that*
(i) $\{P^\nu, \nu = 1, ...\}$ *converges in distribution to P;*
(ii) $g(x, \cdot)$ *is continuous on Ξ for each $x \in D$, where*

$$D = \{x : E(g(x)) < +\infty\} = \{x : g(x, \xi) < +\infty, a.s.\};$$

(iii) $g(\cdot, \xi)$ *is locally Lipschitz on D with Lipschitz constant independent of ξ;*
(iv) *for any $x \in D$ and $\epsilon > 0$, there exists a compact set S_ϵ and ν_ϵ such that for all $\nu \geq \nu_\epsilon$,*

$$\int_{\Xi \setminus S_\epsilon} |g(x, \xi)| P^\nu(d\xi) < \epsilon,$$

and with $V_x = \{\xi : g(x, \xi) = +\infty\}$, $P(V_x) > 0$ if and only if $P^\nu(V_x) > 0$ for $\nu = 0, 1, ...$
Then

(a) $E^\nu(g(\cdot))$ epi- and pointwise converges to $E(g(\cdot))$; if $x, x^\nu \in D$ for $\nu = 1, 2, ...$ and $x^\nu \to x$, then

$$\lim_{\nu \to \infty} E^\nu(g(x^\nu)) = E(g(x));$$

(b) $E^\nu(g(\cdot))$, where $\nu = 0, 1, ...$, is locally Lipschitz on D; furthermore, for each $x \in D$, $\{\partial E^\nu(g(x)) : \nu = 0, 1, ...\}$ is bounded;
(c) if $x^\nu \in D$ minimizes $E^\nu(g(x))$ for each ν and x is a limiting point of $\{x^\nu\}$, then x minimizes $E(g(x))$.

Proof. First, we establish pointwise convergence of the expectation functionals. Suppose $x \in D$ and consider S_ϵ as in the hypothesis. Let $M_\epsilon = \sup_{\xi \in S_\epsilon} |g(x, \xi)|$, which is finite for g continuous and S_ϵ compact. Construct a bounded and continuous function,

$$g^\epsilon(\xi) = \begin{cases} g(x, \xi) & \text{if } |g(x, \xi)| \leq M_\epsilon, \\ M_\epsilon & \text{if } |g(x, \xi)| > M_\epsilon, \\ -M_\epsilon & \text{if } |g(x, \xi)| < -M_\epsilon. \end{cases}$$

By convergence in distribution, $\beta_\epsilon^\nu \to \beta_\epsilon$, for $\beta_\epsilon^\nu = \int_\Xi g_\epsilon(\xi) P^\nu(d\xi)$ and $\beta_\epsilon = \int_\Xi g_\epsilon(\xi) P(d\xi)$. Let $\beta^\nu = \int_\Xi g(x, \xi) P^\nu(d\xi)$. Noting that for $\nu > \nu_\epsilon$, $\int_{\Xi \backslash S_\epsilon} g_\epsilon(\xi) P^\nu(d\xi) < \epsilon$,

$$|\beta^\nu - \beta_\epsilon^\nu| < 2\epsilon. \tag{6.3}$$

We also have that

$$|\beta - \beta_\epsilon| < 2\epsilon. \tag{6.4}$$

From the convergence of the β^ν, there exists some $\bar{\nu}_\epsilon$ such that for all $\nu \geq \bar{\nu}_\epsilon$,

$$|\beta_\epsilon^\nu - \beta_\epsilon| < 2\epsilon. \tag{6.5}$$

Combining (6.3), (6.4), and (6.5) for any $\nu > \max\{\bar{\nu}_\epsilon, \nu_\epsilon\}$,

$$|\beta - \beta^\nu| < 6\epsilon,$$

which establishes that $E^\nu(g(x)) \to E(g(x))$ for any $x \in D$.

To establish epi-convergence, from (b) of Proposition 11, we need to show that if $x \in D$ and $h \geq E(g(x))$, then there exists $x^\nu \in D$ and $h^\nu \geq E^\nu(g(x^\nu))$ such that $(x^\nu, h^\nu) \to (x, h)$, and, if $x^\nu \in D$ and $h^\nu \geq E^\nu(g(x^\nu))$ such that $(x^\nu, h^\nu) \to (x, h)$, then $x \in D$ and $h \geq E(g(x))$. The former follows by letting $x^\nu = x$ and $h^\nu = E^\nu(g(x)) + (h - E(g(x)))$ and using pointwise convergence. The latter follows from pointwise convergence and continuity because $\nu = \lim_\nu h^\nu \geq \lim_\nu E^\nu(g(x^\nu)) = \lim_\nu [(E^\nu(g(x^\nu)) - E^\nu(g(x)) + (E^\nu(g(x)) - E(g(x))) + E(g(x))] = E(g(x))$.

For (b), again let $x, x^\nu \in D, x^\nu \to x$. For any $x \in D, y$, and z close to x, $\nu = 0, 1, ...,$

$$|E^\nu(g(y)) - E^\nu(g(z))|$$

$$\leq \int |g(y, \boldsymbol{\xi}) - g(z, \boldsymbol{\xi})| P^\nu(d\boldsymbol{\xi})$$

$$\leq \int L_x \|y - z\| P^\nu(d\boldsymbol{\xi})$$

$$= L_x \|y - z\|,$$

where L_x is the Lipschitz constant of $g(\cdot, \boldsymbol{\xi})$ near x, which is independent of $\boldsymbol{\xi}$ by (iii). By (ii) and (iii), x is in the interior of the domain of $E^\nu(g(x))$. Hence, (see Theorem 23.4 in Rockafellar [1969]), the subdifferential $\partial E^\nu(g(x))$ is a nonempty, compact convex set, for each ν. The two-norms of subgradients in these subdifferentials are bounded by L_x.

By (b), $E^\nu(g(x))$ are lower semicontinuous functions. By (a), $E^\nu(g(x))$ epi-converges to $E(g(x))$. We get the conclusion of (c) from the statement in Exercise 1. This completes the proof. $\square$

This result also extends directly to nonconvex functions, as we mentioned earlier. In terms of stochastic programming computations, the most useful result may be (c), which implies convergence of optima for approximating distributions. Actually achieving optimality for each approximation may be time-consuming. One might, therefore, be interested in achieving convergence of subdifferentials. This may allow suboptimization for each approximating distribution.

In the case of closed convexity, Wets showed in Theorem 3 of Wets [1980a] that if $g, g^\nu : \Re^n \to \Re \cup \{+\infty\}, \nu = 1, 2, ...,$ are closed convex functions and $\{g^\nu\}$ epi-converges to g, then the graphs of the subdifferentials of g^ν converge to the graph of the subdifferential of g, i.e., for any convergent sequence $\{(x^\nu, u^\nu) : u^\nu \in \partial g^\nu(x^\nu)\}$ with (x, u) as its limit, one has $u \in \partial g(x)$; for any (x, u) with $u \in \partial g(x)$, there exists at least one such sequence $\{(x^\nu, u^\nu) : u^\nu \in \partial g^\nu(x^\nu)\}$ converging to it.

However, in general, it is not true that

$$\partial g(x) = \lim_{\nu \to \infty} \partial g^\nu(x) \tag{6.6}$$

even if $x \in \text{int}(\text{dom}(g))$ (See Exercise 2). However, if g is G-differentiable at x, (6.6) is true. This is the following result from Birge and Qi [1995].

Theorem 13. *Suppose that $g, g^\nu : \Re^n \to \Re \cup \{+\infty\}, \nu = 1, 2, ...,$ are closed convex functions and $\{g^\nu\}$ epi-converges to g. Suppose further that g is G-differentiable at x. Then*

$$\nabla g(x) = \lim_{\nu \to \infty} \partial g^\nu(x). \tag{6.7}$$

In fact, for any $x \in \text{int}(\text{dom}(g))$, there exists ν_x such that for any $\nu \geq \nu_x$, $\partial g^\nu(x)$ is nonempty, and $\{\partial g^\nu(x) : \nu \geq \nu_x\}$ is bounded. Thus, for any $x \in$

int(dom(g)), the right hand side of (6.7) is nonempty and always contained in the left-hand side of (6.7). But equality does not necessarily hold by our example. We also state the following result in Corollary 2.5 of Birge and Qi [1995].

Corollary 14. *Suppose the conditions of Theorem 2 and that $g(\cdot,\xi)$ is convex for each $\xi \in \Xi$. Then for $D = dom(E(g(\cdot)))$,*
(d) there is a Lebesgue zero-measure set $D_1 \subseteq D$ such that $E(g(x))$ is G-differentiable on $D \setminus D_1$, $E(g(x))$ is not G-differentiable on D_1, and for each $x \in D \setminus D_1$

$$\lim_{\nu \to \infty} \partial E^\nu(g(x)) = \nabla E(g(x));$$

(e) for each $x \in D$,

$$\partial E(g(x)) = \{ \lim_{\nu \to \infty} u^\nu : u^\nu \in \partial E^\nu(g(x^\nu)), x^\nu \to x \}.$$

Proof. By closed convexity of $g(\cdot,\xi)$, $E^\nu(g(x))$ are also closed convex for all ν. Now (d) follows Theorem 13 and the differentiability property of convex functions, and (e) follows Theorem 3 of Wets [1980a]. $\square$

Many other results are possible using Theorem 13 and results on epi-convergence. As an example, we consider convergence of sampled problem minima following King and Wets [1991]. Let P^ν be an empirical measure derived from an independent series of random observations $\{\xi^1, ..., \xi^\nu\}$ each with common distribution P. Then, for all x,

$$E^\nu(g(x)) = \frac{1}{\nu} \sum_{i=1}^{\nu} g(x, \xi_i).$$

Let $(\Xi, \mathcal{A}, P)$ be a probability space completed with respect to P. A closed-valued multifunction G mapping Ξ to $\Re^n$ is called *measurable* if for all closed subsets $C \subseteq \Re^n$, one has

$$G^{-1}(C) := \{\xi \in \Xi : G(\xi) \cap C \neq \emptyset\} \in \mathcal{A}.$$

In the following, "with probability one" refers to the sampling probability measure on $\{\xi^1, ..., \xi^\nu, ...\}$ that is consistent with P (see King and Wets [1991] for details). Applying Theorem 2.3 of King and Wets [1991] and Corollary 14, we have the following.

Corollary 15. *Suppose for each $\xi \in \Xi$, $g(\cdot,\xi)$ is closed convex and the epigraphical multifunction $\xi \mapsto epi\, g(\cdot,\xi)$ is measurable. Let $E^\nu(g(x))$ be calculated by (6.2). If there exists $\bar{x} \in dom(E^\nu(g(x)))$ and a measurable selection $\bar{u}(\xi) \in \partial g(\bar{x},\xi)$ with $\int \|\bar{u}(\xi)\| P(d\xi)$ finite, then the conclusions of Corollary 14 hold with probability one.* $\square$

King and Wets [1991] applied their results to the two-stage stochastic program with fixed recourse repeated here as

$$\min c^T x + \int Q(x,\xi)P(d\xi)$$

$$\text{s. t. } Ax = b, \tag{6.8}$$

$$x \geq 0,$$

where $x \in \Re^n$ and

$$Q(x,\xi) = \inf\{q(\xi)^T y | Wy = h(\xi) - T(\xi)x, y \in \Re_+^{n_2}\} \tag{6.9}$$

It is a fixed recourse problem because W is deterministic. Combining their Theorem 3.1 with our Corollary 14, we have the following.

Corollary 16. *Suppose that the stochastic program (6.8) has fixed recourse (6.9) and that for all i,j,k, the random variables $q_i\zeta_j$ and $q_i T_{jk}$ have finite first moments. If there exists a feasible point $\bar{x}$ of (6.9) with the objective function of (6.9) finite, then the conclusions of Corollary 14 hold with probability one for*

$$g(x,\xi) = c^T x + Q(x,\xi) + \delta(x),$$

where $\delta(x) = 0$ if $Ax = b, x \geq 0$, $\delta(x) = +\infty$ otherwise. $\square$

By Theorem 3.1 of King and Wets [1991], one may solve the approximation problem

$$\min c^T x + \frac{1}{\nu}\sum_{i=1}^{\nu} Q(x,\xi_i)$$

$$\text{s. t. } Ax = b, \tag{6.10}$$

$$x \geq 0,$$

instead of solving (6.8). If the solution of (6.10) converges as ν tends to infinity, then the limiting point is a solution of (6.8). Alternatively, by Corollary 16, one may directly solve (6.8) with a nonlinear programming method and use

$$c^T x + \frac{1}{\nu}\sum_{i=1}^{\nu} Q(x,\xi_i)$$

and

$$c + \frac{1}{\nu}\sum_{i=1}^{\nu} \partial_x Q(x,\xi_i)$$

as approximate objective function values and subdifferentials of (6.8) with $\nu = \nu(k)$ at the kth step. Notice that $-u^T T(\xi_i) \in \partial_x Q(x,\xi_i)$ if and only if u is an optimal dual solution of (6.9) with $\xi = \xi_i$. In this way, one may directly solve the original problem using the subgradients $-u^T T(\xi_i)$ and the probability that each is optimal (equivalently that the corresponding basis is primal feasible). The calculation is therefore reduced to obtaining the probability of satisfying a system of linear inequalities, which can be approximated well (see Prékopa [1988] and Section 4). This procedure may allow computation without calculating the actual objective value, which may involve a more difficult multiple integral.

These results give some general idea about the uses of approximations in stochastic programming. We can also introduce approximating functions, g^ν, such that g^ν converges to g pointwise in D. Similar convergence results are also obtained there. The general rule is that approximating distribution functions that converge in distribution (even with probability one) to the true distribution function lead to convergence of optima and, for differentiable points, convergence of subgradients.

Exercises

1. Prove that if g^ν epi-converges to g and x^* is a limit point of $\{x^\nu\}$, where $x^\nu \in arg\min g^\nu = \{x | g^\nu(x) \leq \inf g^\nu\}$, then $x^* \in arg\min g$.

2. Construct an example where g^ν epi-converges to g but $\partial g(x) \neq \lim_\nu \partial g^\nu(x)$.

3. Consider the basic bounding method in Section 2. Suppose that Ξ is compact and that for any $\epsilon > 0$, there exists some ν_ϵ such that for all $\nu \geq \nu_\epsilon$, $\text{diam} S_l \leq \epsilon$ for all $S_l \in \mathcal{S}^\nu$. Show that this implies that P^ν converges to P in distribution.

10
Monte Carlo Methods

Each function value in a stochastic program can involve a multidimensional integral with extremely high dimensions. Because Monte Carlo simulation appears to offer the best possibilities for higher dimensions (Deák [1988]), it seems to be the natural choice for use in stochastic programs. In this chapter, we describe some of the basic approaches built on sampling methods. The key feature is the use of statistical estimates to obtain confidence intervals on results. Some of the material uses probability measure theory which is necessary to develop the analytical results.

The first section describes the basic results for statistical analyses of stochastic programs. We consider a stochastic program formed with a set of random observations and the asymptotic properties of optimal solutions to those problems. In general, these problems can be solved using any technique that might apply to the sampled problems.

Section 2 considers methods based on the L-shaped method without resolving each problem in each period. We first consider possibilities for estimating the cuts using a large number of samples for each cut. We then consider the *stochastic decomposition method* (Higle and Sen [1991b]) that forms many cuts with few additional samples on each iteration.

Section 3 considers methods based on the stochastic quasi-gradient, which can be viewed as a generalization of the steepest descent method. These approaches have a wide variety of applications that extend beyond stochastic programming.

In Section 4, we consider extensions of Monte Carlo methods to include analytical evaluations exploiting problem structure in probabilistic constraint estimation and empirical sample information for methods that may use updated information in dynamic problems.

10.1   General Results for Sampled Problems

We begin by considering a stochastic program in the following basic form:

$$\inf_{x \in X} \int_{\Xi} g(x, \boldsymbol{\xi}) P(d\boldsymbol{\xi}), \qquad (1.1)$$

where $X \subset \Re^n$ and $\boldsymbol{\xi}$ is now defined on the probability space $(\Xi, \mathcal{B}, P)$ so that we can work directly with $\boldsymbol{\xi}$ instead of through ω. Suppose that (1.1) has an optimal solution, x^*.

A natural approach to solving (1.1) is to consider an approximate problem derived by taking ν samples from $\boldsymbol{\xi}$. The discrete distribution with these samples could be P^ν, which would allow us to apply the results in Chapter 9 to obtain convergence of the ν problem optimal solutions to the optimal solution in (1.1). We would like even more, however, to describe distributional properties of these solutions so that we can construct confidence intervals in place of the (probability one) bounds found in Chapter 9.

We therefore wish to consider a sample $\{\xi^i\}$ of independent observations of $\boldsymbol{\xi}$ that are used in the approximate problem:

$$\inf_{x \in X} (\frac{1}{\nu}) \sum_{i=1}^{\nu} g(x, \xi^i). \qquad (1.2)$$

Suppose that $\mathbf{x}^\nu$ is the random vector of solutions to (1.2) with independent random samples, $\boldsymbol{\xi}^i, i = 1, \ldots, \nu$. The general question considered in King and Rockafellar [1993] is to find a distribution $\mathbf{u}$ such that $\sqrt{\nu}(\mathbf{x}^\nu - x^*)$ converges to $\mathbf{u}$ in distribution. Properties of $\mathbf{u}$ can then be used to derive confidence intervals for x^* from an observation of $\mathbf{x}^\nu$.

We give the main result without proof. The interested reader can refer to King and Rockafellar [1993] and, for the statistical origin, Huber [1967].

Theorem 1. *Suppose that $g(\cdot, \xi)$ is convex and twice continuously differentiable, X is a convex polyhedron, $\nabla g : \Xi \times \Re^n \mapsto \Re^n$:*

- *i. is measurable for all $x \in X$;*
- *ii. satisfies the Lipschitz condition that there exists some $a :$ $\Xi \mapsto \Re$, $\int_{\Xi} |a(\boldsymbol{\xi})|^2 P(d\boldsymbol{\xi}) < \infty$, $|\nabla g(x_1, \xi) - \nabla g(x_2, \xi)| \leq a(\xi)|x_1 - x_2|$, for all $x_1, x_2 \in X$;*
- *iii. satisfies that there exists $x \in X$ such that $\int_{\Xi} |g(x, \boldsymbol{\xi})|^2 P(d\boldsymbol{\xi}) < \infty$; and, for $G^* = \int \nabla^2 g(x^*, \boldsymbol{\xi}) P(d\boldsymbol{\xi})$,*
- *iv. $(x_1 - x_2)^T G^* (x_1 - x_2) > 0, \forall x_1 \neq x_2, x_1, x_2 \in X.$*

Then the solution $\mathbf{x}^\nu$ *to (1.2) satisfies:*

$$\sqrt{\nu}(\mathbf{x}^\nu - x^*) \mapsto \mathbf{u}, \tag{1.3}$$

where $\mathbf{u}$ *is the solution to:*

$$\begin{array}{ll} \min & \frac{1}{2}u^T G^* u + \mathbf{c}^T u \\ \text{s. t.} & A_i.u_i \leq 0, i \in I(x^*), u^T \nabla \bar{g}^* = 0, \end{array} \tag{1.4}$$

$X = \{x | Ax \leq b\}$, (x^*, π^*) *solve* $\nabla \int_\Xi g(x^*, \boldsymbol{\xi}) P(d\boldsymbol{\xi}) + (\pi^*)^T A = 0$, $\pi^* \geq 0$, $Ax^* \leq b$, $I(x^*) = \{i | A_i.x^* = b_i\}$, $\nabla \bar{g}^* = \int \nabla g(x^*, \boldsymbol{\xi}) P(d\boldsymbol{\xi})$, *and* $\mathbf{c}$ *is distributed normally* $N(0, \Sigma^*)$ *with* $\Sigma^* = \int (\nabla g(x^*, \boldsymbol{\xi}) - \nabla \bar{g}^*)(\nabla g(x^*, \boldsymbol{\xi}) - \nabla \bar{g}^*)^T P(d\boldsymbol{\xi})$.

Proof: See King and Rockafellar [1993, Theorem 3.2]. $\square$

Example 1

Suppose that $X = [a, \infty)$, $\boldsymbol{\xi}$ is normally distributed $N(0, 1)$, and $g(x, \xi) = (x - \xi)^2$. Problem (1.1) then becomes:

$$\inf_{x \geq a} \int_\Xi \frac{(x - \xi)^2}{\sqrt{2\pi}} e^{-\frac{\xi^2}{2}} d\xi, \tag{1.5}$$

where we substituted for P the standard normal density with mean zero and unit standard deviation.

Because the expectation in (1.5) is just $x^2 + 1$, for $a \geq 0$, the clear solution is $x^* = a$. For $a < 0$, $x^* = 0$. In this case, $\nabla g(x^*, \xi) = -2(x - \xi)$, $G^* = 2$, $A = [-1]$, and $\nabla \bar{g}^* = -2x$. The variance of $\mathbf{c}$ is $\Sigma^* = E_{\boldsymbol{\xi}}[(2\boldsymbol{\xi})^2] = 4$. The asymptotic distribution $\mathbf{u}$ then solves:

$$\begin{array}{ll} \min & u^2 + \mathbf{c}^T u \\ \text{s. t.} & u \geq 0 \text{ if } x^* = a, u(-2x^*) = 0. \end{array} \tag{1.6}$$

For $a > 0$, the solution of (1.6) is $u^* = 0$ so that asymptotically $\sqrt{\nu}(\mathbf{x}^\nu - x^*) \mapsto 0$ in distribution. If $a = 0$, then note that because $\mathbf{c}/2$ is $N(0, 1)$, the overall result is that asymptotically the estimate, $\sqrt{\nu}\mathbf{x}^\nu$, for (1.5) approaches a distribution with a probability mass of 0.50 at 0 and a density for the normal distribution, $N(0, 1)$, over $(0, \infty)$. Exercise 1 asks the reader to find the asymptotic distribution for $a < 0$. In each case, the actual distribution of $\mathbf{x}^\nu$ can be found and compared to the asymptotic result (see Exercise 2).

Many other results along these lines are possible (see, e.g., Dupačová and Wets [1988]). They often concern the stability of the solutions with respect to the underlying probability distribution. For example, one might only have observations of some random parameter but may not know the parameter's distribution. This type of analysis appears in Dupačová [1984], Römisch and Schultz [1991a], and the survey in Dupačová [1990].

Another useful result is to have asymptotic properties of the optimal approximation value. For this, suppose that z^* is the optimal value of (1.1) and z^ν is the random optimal value of (1.2). We use properties of g and ξ^i so that each $g(x, \xi^i)$ is an independent and identically distributed observation of $g(x, \boldsymbol{\xi})$, and $g(x, \boldsymbol{\xi})$ has finite variance, $Var_g(x) = \int_\Xi |g(x, \boldsymbol{\xi})|^2 P(d\boldsymbol{\xi}) - (E_g(x))^2$. We can thus apply the central limit theorem to state that $\sqrt{\nu}[(\frac{1}{\nu})\sum_{i=1}^\nu g(x, \xi^i) - \int_\Xi g(x, \boldsymbol{\xi})P(d\boldsymbol{\xi})]$ converges to a random variable with distribution, $N(0, Var_g(x))$. Moreover, with the condition in Theorem 1, the random function on x defined by $\sqrt{\nu}[(\frac{1}{\nu})\sum_{i=1}^\nu g(x, \xi^i) - \int_\Xi g(x, \boldsymbol{\xi})P(d\boldsymbol{\xi})]$ is continuous. We can then derive the following result of Shapiro [1991, Theorem 3.3].

Theorem 2. *Suppose that X is compact and g satisfies the following conditions:*

   i. $g(x, \cdot)$ is measurable for all $x \in X$;

   ii. there exists some $a : \Xi \mapsto \Re$, $\int_\Xi |a(\boldsymbol{\xi})|^2 P(d\boldsymbol{\xi}) < \infty$, $|g(x_1, \boldsymbol{\xi}) - g(x_2, \boldsymbol{\xi})| \le a(\boldsymbol{\xi})|x_1 - x_2|$, for all $x_1, x_2 \in X$;

   iii. for some $x^0 \in X$, $\int g(x^0, \boldsymbol{\xi})P(d\boldsymbol{\xi}) < \infty$;

and $E_g(x)$ has a unique minimizer $x^0 \in X$. Then $\sqrt{\nu}[z^\nu - z^*]$ converges in distribution to a normal $N(0, Var_g(x^0))$.

Further results along these lines are possible using the specific structure of g for the recourse problem as in (3.1.1). For example, if K_1 is bounded and Q has a strong convexity property, Römisch and Schultz [1991b] show that the distance between the optimizing sets in (1.1) and (1.2) can be bounded.

Given the results in Theorems 1 and 2 and some bounds on the variances and covariances, one can construct asymptotic confidence intervals for solutions using (1.2). All the previous discrete methods can be applied to (1.2) to obtain solutions as ν increases. Various procedures can be used in incrementing ν and solving the resulting approximation (1.2).

In the next sections, this need to solve (1.2) completely is avoided so that the optimization and sampling are performed somewhat simultaneously. First, we describe methods for doing this with the L-shaped method, then we consider stochastic quasi-gradient approaches.

As a final note, we should mention that analogous procedures can be built around *quasi-random* sequences that seek to fill a region of integration with approximately uniformly spaced points. The result is that errors are asymptotically about of the order $log(\nu)/\nu$ instead of $1/\sqrt{\nu}$ (see Niederreiter [1978]). The difficulty is in the estimation of the constant term but quasi-Monte Carlo appears to work quite well in practice (see Fox [1986] and Birge [1994]). In terms of expected performance over broad function classes, quasi–Monte Carlo performs with the same order of complexity (Woźniakowski [1991]). For the following methods, we may substitute quasi-random sequences for pseudo-random sequences for practical implementations.

Exercises

1. For Example 1, find the asymptotic result from Theorem 1 for $\sqrt{\nu}(\mathbf{x}^\nu - x^*)$ for $a < 0$.

2. For Example 1, derive the actual distribution of $\sqrt{\nu}(\mathbf{x}^\nu - x^*)$ for a feasible region $x \geq a$ in each case of a, $a < 0$, $a = 0$ and $a > 0$. Find the limits of these distributions and verify the result from Theorem 1.

3. Consider a news vendor problem as in Section 1.1. Suppose this problem is solved using a sampling approach. The sampled problem with continuous cumulative distribution function F^ν has a solution at $(F^\nu)^{-1}(\frac{s-a}{s-r}) = x^\nu$. Find the distribution of this quantile and show how to construct a confidence interval around x^*.

10.2 Using Sampling in the L-Shaped Method

The disadvantage of sampling approaches that solve the νth approximation completely is that some effort might be wasted on optimizing when the approximation is not accurate. An approach to avoid these problems is to use sampling within another algorithm without complete optimization. A natural candidate is to embed sampling into the L-shaped method, which often works well for discrete distributions. We consider two such approaches in this section. The first uses importance sampling to reduce variance in deriving each cut based on a large sample (see Dantzig and Glynn [1990]). The second approach uses a single sample stream to derive many cuts that eventually drop away as iteration numbers increase (Higle and Sen [1991b]).

a. Importance sampling

The first approach, by Dantzig and Glynn, is to sample Q in the L-shaped method instead of actually computing it. Techniques to reduce the variance, called *importance sampling* (see, e.g., Rubinstein [1981] and Deák [1990]), can then be used to achieve converging results. Given an iterate x^s, the result is an estimate, $\bar{Q}^\nu(x^s) = (\frac{1}{\nu}(\sum_{i=1}^{\nu} Q(x^s, \xi^i)))$, and an estimate of $\nabla Q(x^s)$ as $\bar{\pi}_s^\nu = (\frac{1}{\nu} \sum_{i=1}^{\nu} \pi_s^i)$ where $\pi_s^i \in \partial Q(x^s, \xi^i)$. Now, for Q convex in x, one obtains

$$Q(x, \xi^i) \geq Q(x^s, \xi^i) + (\pi_s^i)^T (x - x^s) \tag{2.1}$$

for all x. We also have that

$$Q^\nu(x) = (\frac{1}{\nu})(\sum_{i=1}^{\nu} Q(x, \xi^i)) \geq Q^\nu(x^s) + (\bar{\pi}_s^\nu)^T(x - x^s) = LB_s^\nu(x), \quad (2.2)$$

where, by the central limit theorem, $\sqrt{\nu}$ times the right-hand side is asymptotically normally distributed with a mean value,

$$\sqrt{\nu}(Q(x^s) + \nabla Q(x^s)^T(x - x^s)), \quad (2.3)$$

which is a lower bound on $\sqrt{\nu}Q(x)$, and a variance, $\rho^s(x)$.

Note that the cut placed on $Q(x)$ as the right-hand side of (2.2) is a support of Q with some error,

$$Q(x) \geq Q^\nu(x^s) + (\bar{\pi}_s^\nu)^T(x - x^s) - \epsilon_s(x), \quad (2.4)$$

where $\epsilon_s(x)$ is an error term with zero mean and variance equal to $\frac{1}{\nu}\rho^s(x)$. Of course, the error term is not known. At iteration s, the L-shaped method involves the solution of:

$$
\begin{aligned}
\min \quad & c^T x \quad +\theta \\
\text{s. t.} \quad & Ax \quad = b, \\
& D_l x \quad \geq d_l, l = 1, \ldots, r, \\
& E_l x \quad +\theta \quad \geq e_l, l = 1, \ldots, s, \\
& x \geq 0,
\end{aligned} \quad (2.5)
$$

where D_l, d_l is a feasibility cut as in (5.1.7)–(5.1.8), $E_l = -\bar{\pi}_l$, and $e_l = Q^\nu(x^l) + (\bar{\pi}_l)^T(-x^l)$, where we count iterations only when a finite $Q^\nu(x^s)$ is found. Note that the generation of feasibility cuts occurs whenever ξ^i is sampled and $Q(x^l, \xi^i)$ is ∞.

We suppose that (2.5) is solved to yield x^{s+1} and θ^{s+1}, where

$$\theta^{s+1} = \max_l \{e_l - E_l x^{s+1}\}, \quad (2.6)$$

where each $e_l - E_l x^{s+1}$ can be viewed as a sample from a normally distributed random variable with mean at most $Q(x^{s+1})$ and variance at most $\frac{1}{\nu}(\sigma^{max}(x^{s+1}))^2 = \frac{1}{\nu}(\max_l \rho^l(x^{s+1}))$. Note that θ^{s+1} is a maximum of these random variables so the solution of (2.5) has a bias that may skew results for large s. Confidence intervals can, however, be developed based on certain assumptions about the functions and the supports.

If the variances become small, one can stop with a high confidence solution. Other approaches may also be used. Infanger [1991] makes several assumptions that can lead to tighter confidence intervals on the optimal value. Solutions of large problems appear in Dantzig and Infanger [1991] with these assumptions.

Variances and any form of confidence interval may be quite large when crude Monte Carlo samples are used as indicated earlier. Dantzig and Glynn

[1990] proposed the use of importance sampling to reduce the variance. Their results indicate that the variance can be reduced quite substantially with this technique.

In importance sampling, the goal is to replace a sample using the distribution of $\boldsymbol{\xi}$ with one that uses an alternative distribution that places more weight in the areas of importance. To see this, suppose that $\boldsymbol{\xi}$ has a density $f(\xi)$ over Ξ so that we are trying to find:

$$\mathcal{Q}(x) = \int_{\Xi} Q(x,\xi)f(\xi)d\xi. \tag{2.7}$$

The crude Monte Carlo technique generates each sample ξ^i according to the distribution given by density f.

In importance sampling, a new probability density $g(\xi)$ is introduced that is somewhat similar to $Q(x,\xi)$ and such that $g(\xi) > 0$ whenever $Q(x,\xi)f(\xi) \neq 0$. We then generate samples ξ^i according to this distribution while writing the integral as:

$$\mathcal{Q}(x) = \int_{\Xi} \frac{Q(x,\xi)f(\xi)}{g(\xi)} g(\xi)d\xi. \tag{2.8}$$

In this case, we generate random samples from the distribution with density $g(\xi)$. Note that if $g(\xi) = Q(x,\xi)f(\xi)/\mathcal{Q}(x)$, then every sample $\xi^i(imp)$ under importance sampling yields an importance sampling expectation, $\mathcal{Q}^1_{imp}(x) = \mathcal{Q}(x)$.

Of course, if we could generate samples from the density $Q(x,\xi)/\mathcal{Q}(x)$, we would already know $\mathcal{Q}(x)$. We can, however, use approximations such as the sublinear approximations in Section 9.5 that may be close to $\mathcal{Q}(x)$ and should result in lower variances for $\mathcal{Q}^\nu_{imp}$ over $\mathcal{Q}^\nu$. This approximation is the approach suggested in Infanger [1991].

In the sublinear approximation approach, the approximating density $g(\xi)$ is chosen as

$$g(\xi) = \sum_{i=1}^{m_2} \psi_{I(i)}((Tx)_{i\cdot}, h_i)/\Psi_I(Tx). \tag{2.9}$$

In this way, much lower variances can result in comparison to the crude Monte Carlo approach. One complication is, however, in generating a random sample from the density in (2.9). The general techniques for generating such random vectors is to generate sequentially from the marginal distributions conditionally, first choosing ξ_1 with the first marginal, $g_1(\xi_1) = \int_{\xi_2,\ldots,\xi_N} g(\xi)d\xi$. Then, sequentially, ξ_i is chosen with density, $g_i(\xi_i|\xi_1,\ldots, \xi_{i-1})$. Remember that in each case, a random sample with density $g_i(\xi_i)$ on an interval Ξ_i of $\Re$ can be found by choosing from a uniform random sample u from $[0,1]$ and then taking ξ such that $G(\xi) = u$ where $G(x) = \int_{-\infty}^{x} g_i(\xi_i)d\xi_i$.

Example 2

Consider Example 1 of Section 9.2 with $x_1 = x_2 = x$. We consider both the crude Monte Carlo approach and the importance sampling using the sublinear approximation for $g(\xi)$. In this case, $g(\xi)$ is actually chosen to depend on x as $g_x(\xi)$ defined by:

$$g_x(\xi) = \frac{|x - \xi_1| + |x - \xi_2|}{E_{\boldsymbol{\xi}}[|x - \boldsymbol{\xi}_1| + |x - \boldsymbol{\xi}_2|]}. \tag{2.10}$$

For comparison, we first consider the L-shaped method with ξ^i chosen by crude Monte Carlo from the original uniform density on $[0, 1] \times [0, 1]$ and by the importance sampling method with distribution $g_x(\xi)$ in (2.10). The results appear in Figure 1 for the solution x^s at each iteration s of the crude Monte Carlo and importance sampling L-shaped method with $\nu = 500$ on each L-shaped iteration. The figures show up to 101 L-shaped iterations, which involve more than 50,000 recourse problem solutions.

In Figure 1, the crude Monte Carlo iteration values x appear as $x(crude)$ while the importance sampling iterations appear as $x(imp)$. We also include the optimal solution $x^* = \sqrt{2} - 1$ on the graph. Note that $x(imp)$ is very close to x^* from just over 40 iterations while $x(crude)$ does not appear to approach this accuracy within 100 iterations. Note that $x(imp)$ begins to deteriorate after 80 iterations as the accumulation of cuts increases the probability that some cuts are actually above $\mathcal{Q}(x)$.

The advantage of importance sampling can also be seen in Figure 2, which compares the optimal value $\mathcal{Q}(x^*)$ with sample values, $\mathcal{Q}^\nu(x^\nu)$, with crude Monte Carlo denoted as $\mathcal{Q}(\text{crude})$ and $\mathcal{Q}^\nu_{imp}(x^\nu)$ with importance sampling denoted as $\mathcal{Q}(\text{imp})$. Note that the crude Monte Carlo values have a much wider variance, in fact, double the variance of the importance sampling results. Also note that in both sampling methods, the estimates have a mean close to the optimal value after 40 iterations.

The results in Figures 1 and 2 indicate that sampled cuts in the L-shaped method can produce fairly accurate results but that convergence to optimal values may require large numbers of samples for each cut even for small problems. One difficulty is that initial cuts with small numbers of samples may limit convergence unless they are removed in favor of more accurate cuts. One procedure to avoid this problem is gradually to remove initial cuts as the algorithm progresses. This is the intent of the next approach.

b. Stochastic decomposition

The alternative considered in Higle and Sen [1991b] is to generate many cuts with small numbers of additional samples on each cut and to adjust these cuts to drop as the algorithm continues processing. Their method is called *stochastic decomposition*. They assume complete recourse and a

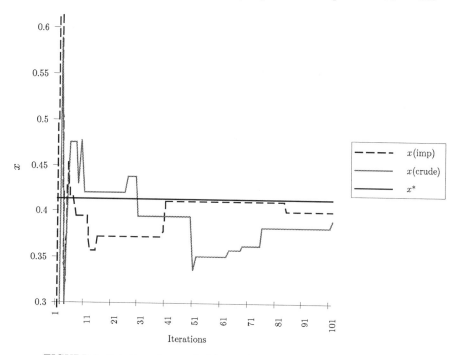

FIGURE 1. Solutions for crude Monte Carlo and importance sampling.

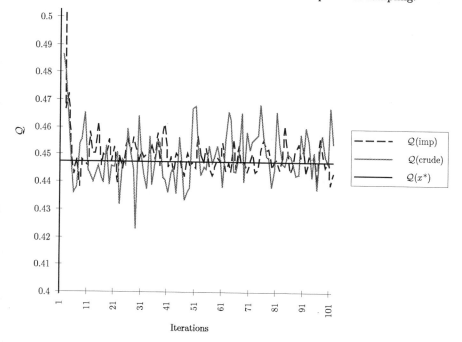

FIGURE 2. Objective values for crude Monte Carlo and importance sampling.

known (probability one) lower bound on $Q(x, \xi)$ (e.g., 0). They also assume that the set of dual solutions to the recourse problem (3.1.1) is bounded and that K_1 and Ξ are also compact.

With these assumptions, the basic stochastic decomposition method generates iterates, x^k, and observations, ξ^k. We can state the basic stochastic decomposition method in the following way.

Basic Stochastic Decomposition Method

Step 1. Set $\nu = 0$, $\xi^0 = \bar{\xi}$, and let x^1 solve

$$\min_{Ax=b,x\geq 0} \{c^T x + Q(x, \xi^0)\}. \tag{2.12}$$

Step 2. Let $\nu = \nu + 1$ and let ξ^ν be an independent sample generated from $\boldsymbol{\xi}$. Find $Q^\nu(x^\nu) = \frac{1}{\nu}\sum_{s=1}^{\nu} Q(x^\nu, \xi^s) = \frac{1}{\nu}\sum_{s=1}^{\nu}(\pi_s^\nu)^T(\xi^s - Tx^\nu)$. Let $E_\nu = \frac{1}{\nu}\sum_{s=1}^{\nu}(\pi_s^\nu)^T T$ and $e_\nu = \frac{1}{\nu}\sum_{s=1}^{\nu}(\pi_s^\nu)^T\xi^s$.

Step 3. Update all previous cuts by $E_s \leftarrow \frac{\nu-1}{\nu}E_s$ and $e_s \leftarrow \frac{\nu-1}{\nu}e_s$ for $s = 1, \ldots, \nu - 1$.

Step 4. Solve the L-shaped master problem as in (2.5) to obtain $x^{\nu+1}$. Go to Step 2.

This method differs slightly from the basic method in Higle and Sen [1991b] in our assuming π_s^ν to be optimal dual solutions in each iteration. Higle and Sen allow a restricted set of dual optima that may decrease the solution effort (with perhaps fewer effective cuts).

The main convergence result is contained in the following theorem.

Theorem 3. *Assuming complete recourse, $Q(x, \xi) \geq 0$, bounded dual solutions to (3.1.1), K_1 and Ξ compact, there exists a subsequence, $\{x^{\nu_j}\}$, of the iterates of the basic stochastic decomposition method such that every limit point of $\{x^{\nu_j}\}$ solves the recourse problem (3.1.1) with probability one.*

Proof. We follow the proof of Theorem 4 in Higle and Sen [1991b]. We use their Theorem 3 (Exercise 5), which gives the existence of a subsequence of $\{x^\nu\}$ such that

$$\lim_{\nu\to\infty} \theta^\nu - \max_{l=1,\ldots,\nu} (e_{\nu-1}^l - E_{\nu-1}^l x^\nu) = 0. \tag{2.12}$$

Suppose $\{x^{\nu_j}\}$ is a subsequence of the subsequence achieving (2.12) such that $\lim_j x^{\nu_j} = \hat{x}$ where $A\hat{x} = b, x \geq 0$. This occurs for some subsequence by compactness. From x^* optimal,

$$c^T x^* + \mathcal{Q}(x^*) \leq c^T \hat{x} + \mathcal{Q}(\hat{x}). \tag{2.13}$$

Note that because $Q(x,\xi) \geq 0$ for all $\xi \in \Xi$ and $Q(x,\xi^i) \geq \pi^T(h^i - Tx)$ for any $\pi^T W \leq q$ and any sample ξ^i, for any $1 \leq s \leq \nu$,

$$\sum_{i=1}^{\nu} Q(x,\xi^i) \geq \sum_{i=1}^{s} \pi^T(h^i - Tx), \qquad (2.14)$$

where π is any feasible multiplier in the recourse problem for ξ^i. From (2.14), it follows that $\frac{1}{\nu}\sum_{i=1}^{\nu} Q(x,\xi^i) \geq e_l^\nu - E_l^\nu x$ for all l and ν, where E_l^ν and e_l^ν are the components of Cut l on Iteration ν. Therefore,

$$c^T x + \max_{l=1,\dots,\nu} (e_l^\nu - E_l^\nu x) \leq c^T \hat{x} + \frac{1}{\nu}\sum_{i=1}^{\nu} Q(x,\xi^i). \qquad (2.15)$$

As ν increases, $\frac{1}{\nu}\sum_{i=1}^{\nu} Q(x,\xi^i) \to \mathcal{Q}(x)$, so

$$\limsup_{\nu}[c^T x^* + \max_{l=1,\dots,\nu} (e_l^\nu - E_l^\nu x^*)] \leq c^T x^* + \mathcal{Q}(x^*), \qquad (2.16)$$

with probability one. We can also show that (Exercise 6)

$$\lim_{j} c^T x^{\nu_j} + \max_{l=1,\dots,\nu} (e_l^\nu - E_l^\nu x^{\nu_j}) = c^T \hat{x} + \mathcal{Q}(\hat{x}), \qquad (2.17)$$

with probability one. Thus, (2.16), (2.17), and the fact that x^{ν_j} minimizes $c^T x + \max_{l=1,\dots,\nu-1}(e_l^{\nu-1} - E_l^{\nu-1}x)$ over feasible x yield

$$c^T x^* + \mathcal{Q}(x^*) \leq c^T \hat{x} + \mathcal{Q}(\hat{x})$$
$$\leq \limsup_{\nu}[c^T x^* + \max_{l=1,\dots,\nu} (e_l^\nu - E_l^\nu x^*)] \leq c^T x^* + \mathcal{Q}(x^*), (2.18)$$

which proves the result. $\square$

One difficulty in this basic method is that convergence to an optimum may only occur on a subsequence. To remedy this, Higle and Sen suggest retaining an incumbent solution that changes whenever the objective function falls below the best known value so far. The incumbent is updated each time a sufficient decrease in the νth iteration objective value is obtained. They also show that the sequence of incumbents contains a subsequence with optimal limit points, and then show how this subsequence can be identified. Various approaches may be used for practical stopping conditions, such as the statistical verification tests for optimality conditions in Higle and Sen [1991a].

Example 2 (continued)

We again consider Example 1 from Section 9.2. The basic stochastic decomposition method results appear in Figures 3 and 4. In Figure 3, both the basic result x^ν and the incumbent solution, x^ν(incumbent), which is adjusted whenever a solution after the first 100 iterations improves the previous best estimate by 1%. Figure 3 also gives the optimal solution, x^*. The total number of iterations yields about 50,000 subproblem solutions, which is approximately equal to the total number of iterations in Figures 1 and 2. Note that the raw solutions x^ν oscillate rapidly, while the incumbent solutions settle close to x^* quite quickly after their initiation at $\nu = 100$.

The objective value estimates, θ^ν, $\mathcal{Q}^\nu$, and $\mathcal{Q}^\nu(x^\nu(\text{inc}))$ for the incumbent, and the optimal objective value, $\mathcal{Q}(x^*)$, appear in Figure 4. Note that the θ^ν values from the master problem have wide oscillations. The $\mathcal{Q}^\nu(x^\nu)$ values have lower but significant variation. The incumbent objective values, however, show low variation that begins to approach the optimum.

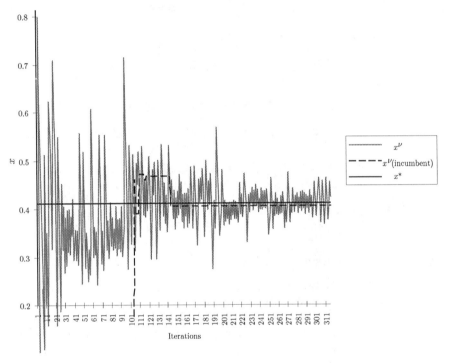

FIGURE 3. Solutions for the stochastic decomposition method.

Exercises

1. Prove Theorem 3. Show first that eventually the objective value of (2.5) for x^{ν_n} at iteration ν_n is the same as the objective value of (2.5) for x^{ν_n} at iteration ν_{n-1}.

2. Show how to sample from the density $g_x(\xi)$ as the sum of the absolute values $|x - \xi_i|, i = 1, 2$ for Example 2.

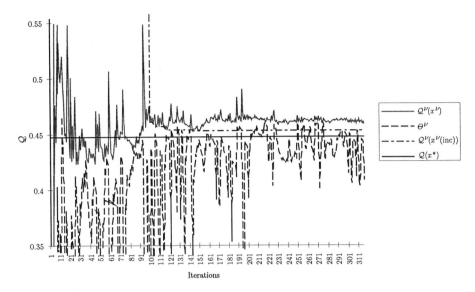

FIGURE 4. Objective values for the stochastic decomposition method.

3. Consider Example 1 in Section 5.1 with ξ uniformly distributed on $[1, 5]$. Apply the crude Monte Carlo L-shaped method to this problem for 100 iterations with 100 samples per cut. What would the result be with importance sampling in this case?

4. Apply both the crude Monte Carlo and importance sampling approaches to Example 3 of Section 9 with both x_1 and x_2 decision variables. First, use 100 samples for each cut for 100 iterations and then compare to results with an increase to 500 samples per cut.

5. Prove that there exists a subsequence of iterates $\{x^\nu\}$ in the basic stochastic decomposition method with the assumptions so that (2.12) holds.

6. Suppose a subsequence of iterates $x^{\nu_j} \to \hat{x}$ in the basic stochastic decomposition method. Prove that (2.17) holds.

7. Apply the basic stochastic decomposition method to Example 1 in Section 5.1 with ξ uniformly distributed on $[1, 5]$. Record the sequence of iterations until 10 consecutive iterations are within 1% of the optimal objective value.

10.3   Stochastic Quasi-Gradient Methods

Stochastic quasi-gradient methods represent one of the first computational developments in stochastic programming. They apply to a broad class of

problems and represent extensions of stochastic approximation methods
(see, e.g., Dupač [1965] and Kushner [1971]). Our treatment will be brief because the emphasis in this book is on methods that exploit the structure of
deterministic equivalent or approximation problems. Ermoliev [1988] provides a more complete survey of these methods.

Stochastic quasi-gradient methods (SQG) apply to a general mathematical program of the form:

$$\min_{x \in X \subset \Re^n} g_o(x)$$

$$\text{s.t. } g_i(x) \le 0, \ i = 1, \ldots, m, \tag{3.1}$$

where we assume that each g_i is convex. We suppose that an initial point,
$x^o \in X$, is given. The method generates a sequence of points, $\{x^\nu\}$, that
converges to an optimal solution of (3.1).

Given a history at time ν, $(x^o, \ldots, x^\nu)$, the method selects function estimates, $\boldsymbol{\eta}_i(\nu)$, and subgradient estimates, $\boldsymbol{\beta}_i(\nu)$, such that

$$E[\boldsymbol{\eta}_i(\nu) \mid (x^o, \ldots, x^\nu)] = g_i(x^\nu) + a_i(\nu) \tag{3.2}$$

and

$$E[\boldsymbol{\beta}_i(\nu) \mid (x^o, \ldots, x^\nu)] + b_i(\nu) \subset \partial g_i(x^\nu), \tag{3.3}$$

where $a_i(\nu), b_i(\nu)$ may depend on $(x^o, \ldots, x^\nu)$ but must satisfy

$$a_i(\nu) \to 0 \text{ and } \| b_i(\nu) \| \to 0. \tag{3.4}$$

When $b_i(\nu) \ne 0$, $\boldsymbol{\beta}_i(\nu)$ is called a *stochastic quasi-gradient*. Otherwise,
$\boldsymbol{\beta}_i(\nu)$ is a *stochastic subgradient*.

We first consider the method when all constraints are deterministic and
represented in X. Thus, Problem (3.1) becomes

$$\min_{x \in X \subset \Re^n} g_o(x). \tag{3.5}$$

The method requires a projection onto X represented by

$$\Pi_X(y) = \arg\min_x \{\| x - y \|^2 \mid x \in X\}.$$

In the basic method, a sequence of step sizes $\{\rho^\nu\}$ is given. The stochastic
quasi-gradient method defines a stochastic sequence of iterates, $\{\mathbf{x}^\nu\}$, by

$$\mathbf{x}^{\nu+1} = \Pi_X[\mathbf{x}^\nu - \rho^\nu \boldsymbol{\beta}_o(\nu)], \tag{3.6}$$

where we interpret the projection as operating separately on each element
$\omega \in \Omega$, so that $x^{\nu+1}(\omega) = \Pi_X[(x(\omega)^\nu - \rho^\nu \beta_o(\omega)(\nu))]$.

To place all these results into the two-stage recourse problem as in (1.1.2),
let $X = \{x | Ax = b, \ x \ge 0\}$, $g_o(x) = \int g^o(x, \boldsymbol{\xi}) P(d\boldsymbol{\xi})$ where

$$g^o(x, \boldsymbol{\xi}) = \inf_{\mathbf{y}} \{\mathbf{q}^T \mathbf{y} \mid W\mathbf{y} = \mathbf{h} - \mathbf{T}x, \ \mathbf{y} \ge 0\}.$$

Thus, we can use $\boldsymbol{\beta}_0^i(x)$ such that $\boldsymbol{\beta}_0^i(x)^T(\mathbf{h}^i - \mathbf{T}^i x) = g^o(x, \boldsymbol{\xi}^i), \boldsymbol{\beta}_0^i(x)^T \omega \leq \mathbf{q}^i$ for a sample $\boldsymbol{\xi}^i$ composed of the components, $\mathbf{h}^i$, $\mathbf{T}^i$, and $\mathbf{q}^i$. The stochastic quasi-gradient method takes a step in this direction and then projects back onto K_1. In the following example and the exercises, we explore the use of this approach.

For these examples, we use an estimate of the objective value by taking a moving average of the last 500 samples, $Q^{\nu-ave}(x^\nu) = \sum_{i=0}^{499} Q(x^\nu - i, \boldsymbol{\xi}^{\nu-i})/500$. Changes in this estimate (or the lack thereof) can be used to evaluate the convergence of stochastic quasi-gradient methods. Gaivoronski [1988] discusses various practical approaches in this regard.

Example 2 (continued)

We consider the same example and apply the stochastic quasi-gradient method. On each step ν, a random sample $\boldsymbol{\xi}^\nu$ is taken with $\beta_o(\nu) \in \nabla Q(x^\nu, \boldsymbol{\xi}^\nu)$. For $X = \{x | 0 \leq x \leq 1\}$, the projection operation yields $x^{\nu+1} = \min(1, \max(x^\nu + \rho^\nu \beta_o(\nu), 0))$. Figures 5 and 6 show these iterations for solutions x^ν and objective estimates, $Q^{\nu-ave}$ for every multiple of 500 iterations up to 50,000 so that total numbers of recourse problem solutions are the same as in Figures 1 to 4.

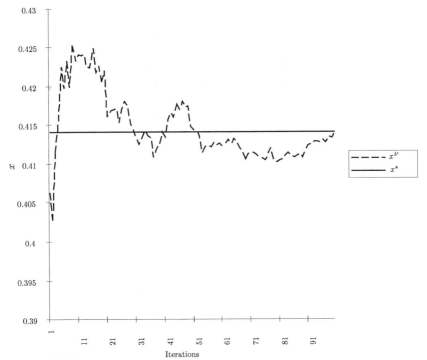

FIGURE 5. Solutions for the stochastic quasi-gradient method.

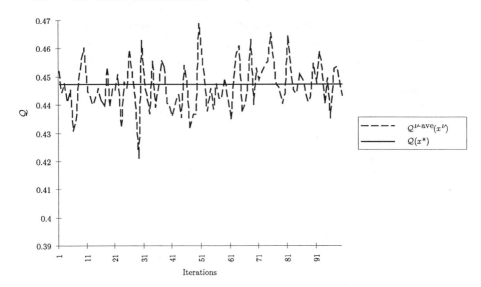

FIGURE 6. Objective values for the stochastic quasi-gradient method.

Note that the iterations in Figure 5 appear to approach x^* much more quickly than the results in Figures 1 to 4. They also seem to show lower variances in the objective estimates in Figure 6, although these results are not converging to zero variance because the sample length 500 is not changing. To achieve convergence or greater confidence in a solution, the number of samples in the estimate must increase.

While the results in Figures 5 and 6 indicate that stochastic quasi-gradient methods may be more effective than the decomposition methods, we should note that this example is quite low in dimension. For higher dimensions, the results are often quite different. In general, stochastic quasi-gradient methods exhibit similar behavior to subgradient optimization methods that often have slow convergence properties in higher dimension problems. They are, nonetheless, easy to implement and can give good results, especially in small problems.

In the rest of this section, we discuss the theory behind the stochastic quasi-gradient method convergence. The exercises consider examples for using SQG.

The basic method in (3.6) traces back to the unconstrained methods of Robbins-Monro [1951]. The main device in demonstrating convergence of $\{x^\nu\}$ to a point in X^* is the use of a *stochastic quasi-Feyer sequence* (see Ermoliev [1969]), a sequence of random vectors, $\{w^\nu\}$, defined in (Ω, Σ, P) such that for a set $W \subset \Re^n$, $E[\| w^o \|^2] < +\infty$, and any $w \in W$,

$$E\{\| w - w^{\nu+1} \|^2 | w^o, \dots, w^\nu\}$$

$$\leq \| w - w^\nu \|^2 + \gamma_\nu, \quad \nu = 0, 1, \dots,$$

$$\gamma_\nu \geq 0, \ \sum_{\nu=0}^{\infty} E[\gamma_\nu] < +\infty. \tag{3.7}$$

The following result shown in Ermoliev [1976] is the basis for the convergence results.

Theorem 4. *If $\{\mathbf{w}^\nu\}$ is a stochastic quasi-Feyer sequence for a set W, then*

(a) $\{\| w - \mathbf{w}^{\nu+1} \|^2\}$ *converges with probability one for any set $w \in W$, and $E[\| w - w^\nu \|^2] < c < +\infty$,*

(b) *the set of limit points of $\{w^\nu(\omega)\}$ is nonempty for almost all $\omega \in \Omega$.*

(c) *if $\bar{w}^1(\omega)$ and $\bar{w}^2(\omega)$ are two distinct limit points of $\{w^\nu(\omega)\}$ such that $\bar{w}^1(\omega) \notin W, \bar{w}^2(\omega) \notin W$, then $W \subset H$, a hyperplane such that $\eta = \{w \mid \alpha^T w = \alpha_0\}$, $\| \bar{w}^1(\omega) - \Pi_\eta(\bar{w}^1(\omega)) \| = \| \bar{w}^2(\omega) - \Pi_\eta(\bar{w}^2(\omega)) \|$, where Π_η denotes projections onto η.*

We can now state the most basic convergence result.

Theorem 5. *Given the following:*

(a) $g_o(x)$ *is convex and continuous,*

(b) X *is a convex compact set,*

(c) *the parameters, ρ_ν, and $\gamma^0(\nu) = \inf_{x^* \in X^*} \beta_0(\nu)^T(\mathbf{x}^\nu - x^*)$, satisfy with probability one,*

$$\rho^\nu > 0, \ \sum_{\nu=0}^{\infty} \rho^\nu = +\infty, \ \sum_{\nu=0}^{\infty} E[\rho^\nu \mid \gamma^0(\nu) \mid +$$

$$(\rho^\nu)^2 \| \beta_0(\nu) \|^2] < \infty, \tag{3.8}$$

then, with probability one, for any $\bar{x}(\omega) = \lim_{\nu_i} x^{\nu_i}(\omega)$, $\bar{x}(\omega) \in X^*$.

Proof. First note that for $-u(\omega) \equiv x^\nu(\omega) + \rho^\nu \beta_0^\nu(\omega) - x^{\nu+1}(\omega)$, $u^T(\omega)(x^{\nu+1}(\omega) - x) \leq 0$ for any $x \in X$ by the nearest point property (see, e.g., Bazaraa and Shetty [1979, Theorem 2.3.1]).

We can write

$$\| x^* - \mathbf{x}^{\nu+1} \|^2 = \| x^* - \mathbf{x}^\nu \|^2 - 2\rho^\nu \beta_0^{\nu T}(x^* - x^\nu)$$

$$+\rho^{\nu 2} \| \beta_0^\nu \|^2 - 2u^T(x^* - \mathbf{x}^\nu - \rho^\nu \beta_0^\nu) + u^T u, \tag{3.9}$$

where $\mathbf{u}^T(x^* - \mathbf{x}^\nu - \rho^\nu \beta_0^\nu) = u^T(x^* - \mathbf{x}^{\nu+1}) + u^T u$. Thus,

$$\| x^* - \mathbf{x}^{\nu+1} \| = \| x^* - \mathbf{x}^\nu \|^2 - 2\rho^\nu \beta_0^{\nu T}(x^* - x^\nu)$$

$$+\rho^{\nu 2} \parallel \boldsymbol{\beta}_0^\nu \parallel^2 -2\mathbf{u}^T(x^* - \mathbf{x}^{\nu+1})$$

$$-\mathbf{u}^T\mathbf{u}$$

$$\leq \parallel x^* - \mathbf{x}^\nu \parallel^2 -2\rho^\nu \boldsymbol{\beta}_0^{\nu T}(x^* - \mathbf{x}^\nu)$$

$$+\rho^{\nu 2} \parallel \boldsymbol{\beta}_0^\nu \parallel^2 . \tag{3.10}$$

Taking expectations in (3.10) yields

$$E\left[\parallel x^* - \mathbf{x}^{\nu+1} \parallel\right] \leq E\left[\parallel x^* - \mathbf{x}^\nu \parallel^2\right]$$

$$-2\rho^\nu E[\boldsymbol{\beta}_0^{\nu T}(x^* - \mathbf{x}^\nu)]$$

$$+\rho^{\nu 2} E\left[\parallel \boldsymbol{\beta}_0^\nu \parallel^2\right], \tag{3.11}$$

$$\leq E\left[\parallel x^* - \mathbf{x}^\nu \parallel^2\right] + \nu E\left[(\rho^\nu \mid \boldsymbol{\gamma}(\nu) \mid +(\rho^\nu)^2) \parallel \boldsymbol{\beta}_0^\nu \parallel)\right],$$

where ν is a constant. Now (3.11) implies that $\{\mathbf{x}^\nu\}$ is a stochastic quasi-Feyer sequence for X^*. Applying Theorem 4, we have $\parallel x^* - \mathbf{x}^{\nu+1} \parallel^2$ converging with probability one for any $x^* \in X^*$ and $\{\mathbf{x}^\nu\}$ almost always has a limit point. We just need to show that one limit point is in X^* by Theorem 4c.

To show this, note that

$$E\left[\parallel x^* - \mathbf{x}^{\nu+1} \parallel^2\right] \leq E\left[\parallel x^* - \mathbf{x}^o \parallel\right] + 2E\left[\sum_{\nu=0}^{\nu} \rho_i(\boldsymbol{\beta}_0(i)^T(x^* - \mathbf{x}^i))\right]$$

$$+\nu \sum_{i=0}^{\nu} E\rho_i^2 \parallel \boldsymbol{\beta}_0(i) \parallel^2$$

$$\leq E\left[\parallel x^* - x^o \parallel\right] + 2E\left[\sum_{i=0}^{\nu} \rho_\nu(g^o(x^*) - g_o(\mathbf{x}^i))\right]$$

$$+\nu \sum_{i=0}^{\nu} E\left[\rho_i \mid \boldsymbol{\gamma}(i) \mid +\rho_\nu^2 \parallel \boldsymbol{\beta}_0(i) \parallel^2\right], \tag{3.12}$$

where we note that $g_o(x^*) \leq g_o(\mathbf{x}^i)$. Hence $\sum_{i=0}^{\infty} g_o(\mathbf{x}^i) - g_o(x^*) < \infty$, or there must exist a subsequence $\{\mathbf{x}^{\nu_i}\}$ such that $x^{\nu_i}(\omega) \to x^i \in X^*$ for almost all $\omega \in \Omega.$ $\square$

The general method can be amplified in a variety of ways. Condition (c) can be relaxed to remove the finiteness of $\sum_{\nu=0}^{\infty} \rho_\nu^2$ when $\gamma(\nu) = 0$ for all ν, but the convergence is for $\dfrac{\sum_\nu \mathbf{x}^\nu \rho^\nu}{\sum_\nu \rho^\nu}$ (see Uriasiev [1988]).

Two important aspects of stochastic quasi-gradient implementations are the determinations of step sizes and stopping rules. Various adaptive step sizes are considered by Mirzoachmedov and Uriasiev [1983]. For stopping rules, we refer to Pflug [1988], where details appear.

The results describe the use of stopping times $\{\tau_\epsilon\}$, to yield *uniform asymptotic level α confidence regions*, defined by

$$\lim_{\epsilon \to 0} \inf_{x^0} P\{\| \mathbf{x}_{\tau_\epsilon} - x^* \| \le \epsilon\} \ge 1 - \alpha. \qquad (3.13)$$

Deterministic step size rules do not, unfortunately, produce such uniform confidence intervals. Instead, Pflug shows that an oscillation test stopping rule does obtain such confidence regions. In this rule, a test is performed to check whether the iterates are oscillating without objective improvement. The key is building consistent estimates of the objective Hessian at x^* and the covariance matrix of objective errors. For other issues concerning implementation, we refer to Gaivoronski [1988].

Exercises

1. Consider Example 1 in Section 5.1. Find the projection of a point onto $X = \{x | 0 \le x \le 10\}$. Solve this problem using the stochastic quasi-gradient method until 20 consecutive iterations are within 1% of the optimal solution.

2. Consider Example 2, where both x_1 and x_2 can be chosen instead of $x = x_1 = x_2$. Follow the stochastic quasi-gradient method again until 20 consecutive iterations are within 1% of the optimal solution.

3. Prove Theorem 4.

4. Consider Example 1 in Section 5.1. Find the projection of a point onto $X = \{x | 0 \le x \le 10\}$. Solve this problem using the stochastic quasi-gradient method until three consecutive iterations are within 1% of the optimal solution.

10.4   Sampling Extensions: Uses with Analytical and Empirical Observations

Monte Carlo sampling procedures can be enhanced by including some forms of analytical computations or allowing for empirical sampling to improve distribution information. These approaches have a wide variety of uses in probabilistically constrained problems. The goal is to produce efficient estimates of probabilities that involve multidimensional integrals with useful decompositions. In general, the goal is to use problem structure to obtain significant decreases in sample variance. Various approaches have been suggested. As examples, we refer to Deák [1980], Gassmann [1988], and Szántai [1986].

We briefly describe Szántai's method here. The basic idea is to use Bonferroni-type inequalities to write the probability of a set with many constraints in terms of sums and differences of integrals of subsets of the constraints, as we described in Section 9.5. In sampling procedures, these alternative estimates allow for significant variance reduction.

For Szántai's approach, suppose we wish to find

$$p = P[A = A_1 \cap \ldots \cap A_m] = \int_A dF(\boldsymbol{\xi}). \tag{4.1}$$

Szántai takes three estimates of p:

1. $\hat{p}^1$—a direct Monte Carlo sample;

2. $\hat{p}^2$—finding the first-order Bonferroni terms, $1 - \sum_{i=1}^m P(\hat{A}_i)$, directly and sampling from higher-order terms;

3. $\hat{p}^3$—Calculating the first- and second-order terms explicitly, $1 - \sum_{i=1}^m P(\hat{A}_i) + \sum_{i<j} P(\hat{A}_i \cap \hat{A}_j)$, and sampling from higher order terms.

Sampling from all higher order terms may be difficult, but Szántai shows that the effort may be reduced at each sample ξ^j to finding $\hat{n}(j)$ defined as the number of constraints violated by ξ^j, i.e., $\hat{n}(j) = \sum_{i=1}^N 1_{\{\xi^j \notin A_i\}}$. With this quantity defined, we can define *unbiased estimates*, i.e., estimates whose expectations have no error, using the following:

$$\gamma^1 = \frac{1}{\nu} \sum_{j=1}^\nu \max\{0, 1 - \hat{n}(j)\}, \tag{4.2}$$

$$\gamma^2 = \frac{1}{\nu} \sum_{j=1}^\nu \max\{0, \hat{n}(j) - 1\}, \tag{4.3}$$

and

$$\gamma^3 = \frac{1}{\nu} \sum_{j=1}^\nu \frac{\max\{0, \hat{n}(j) - 1\}\hat{n}(j) - 2)}{2}. \tag{4.4}$$

These quantities are then used to form unbiased estimates:

$$\hat{p}^1 = \gamma^1, \tag{4.5}$$

$$\hat{p}^2 = 1 - \sum_{i=1}^{m} P[\hat{A}_i] + \gamma^2, \tag{4.6}$$

$$\hat{p}^3 = 1 - \sum_{i=1}^{m} P[\hat{A}_i] + \sum \sum_{i<j} P[\hat{A}_i \cap \hat{A}_j] - \gamma^3. \tag{4.7}$$

These three estimators are combined to form

$$\hat{p}^4 = \lambda_1 \hat{p}^1 + \lambda_2 \hat{p}^2 + (1 - \lambda_1 - \lambda_2)\hat{p}^3, \tag{4.8}$$

where the weights λ_1 and λ_2 are chosen to minimize the variance of $\hat{p}^4$. They are calculated using the sample covariance matrix of $(\gamma^1, \gamma^2, \gamma^3)$, which we denote as $C = [c_{ij}]$. In this case, $\lambda_1 = \frac{\mu_1}{\mu_1 + \mu_2 + \mu_3}$, $\lambda_1 = \frac{\mu_2}{\mu_1 + \mu_2 + \mu_3}$, where

$$\mu_1 = c_{12}(c_{33} - c_{23}) + c_{22}(c_{13} - c_{33}) + c_{23}(c_{23} - c_{13}), \tag{4.9}$$

$$\mu_2 = c_{11}(c_{23} - c_{33}) + c_{12}(c_{33} - c_{13}) + c_{13}(c_{123} - c_{23}), \tag{4.10}$$

and

$$\mu_3 = c_{11}(c_{23} - c_{22}) + c_{12}(c_{12} - c_{23}) + c_{13}(c_{22} - c_{12}). \tag{4.11}$$

The result is that $\hat{p}^4$ can have significantly lower variance than standard Monte Carlo. In fact, Szántai obtains efficiencies (variance ratios) of 100 and higher, implying that the same error can be obtained with $\hat{p}^4$ in 1% of the number of samples for using $\hat{p}^1$ alone.

This approach combines analytical techniques with simulation to produce lower variance. Another approach is to use empirical sample information. This is the area studied in Jagganathan [1985], where some sample information can be used in a Bayesian framework to determine probabilities of underlying distributions. These may be used for probabilistic constraints, for recourse functions, or for both.

As an example, consider the basic two-stage model in (3.1.1), where the distribution function of $\boldsymbol{\xi}$ is $F(\boldsymbol{\xi}, \boldsymbol{\eta})$, where $\boldsymbol{\eta}$ is a k-vector of unknown parameters with prior distribution function, $G(\cdot)$. Given an observation, $\hat{\xi}^l = (\xi^1, \dots, \xi^l)$, we can define a posterior distribution, $G_l(\cdot|\hat{\xi}^l)$. Using this, we may obtain an improved solution.

Without sample information, we would have the solution to (3.1.1) as

$$R(G) = \min_{x \in K_1} \{c^T x + \int_{\eta} \int_{\xi} Q(x, \xi) F(\xi, \eta) G(d\eta)\}. \tag{4.12}$$

However, using $\hat{\boldsymbol{\xi}}^l$, which we assume has a conditional distribution given by $W(\hat{\boldsymbol{\xi}}^l, \eta)$ for some value η of $\boldsymbol{\eta}$, we obtain a value with sample information as

$$R^l(G) = \int_{\eta} [\min_{x \in K_1} \{c^T x + \int_{\eta} \int_{\xi} Q(x, \xi) F(d\xi, \eta) G_l(d\eta|\hat{\xi}^l)\}] W(d\hat{\boldsymbol{\xi}}^l, \eta) G(d\eta). \tag{4.13}$$

The difference $R^l(G) - R(G)$ is the *expected value of sample information.* This represents the additional expected value from observing the sample information. This type of analysis can also be extended to problems with probabilistic constraints.

A different use of sample information is for dynamic problems that may change over time. In these cases, future characteristics, such as product demand, may not be known with certainty but they can be predicted roughly using past experience. These problems were examined by Cipra [1991], who also considered the possibility that more recent information might be more valuable than older information.

For example, consider the news vendor problem in Section 1.1. Suppose the demand occurs as ξ_t for periods $t = 1, \ldots, H$. At time H, suppose that $\xi^H = (\xi_1, \ldots, \xi_H)$ have been observed. The news vendor wishes to place an order based on these observations. One solution might be to use a discount factor, $\beta \in (0, 1)$, to choose $x(H)$ to

$$\min_{x \geq 0} \left(\sum_{i=0}^{H-1} \beta^i ((a - s)x + (s - r)(x - \xi_{H-i})^+) \right). \tag{4.14}$$

The solution of this problem is straightforward (Exercise 5). Alternative perspectives on the value of empirical observations can also be introduced, as could Bayesian approaches as in (4.13). For another view of decisions made over time, refer to Jagganathan [1991].

Exercises

1. Show that $\hat{p}^i$ are unbiased estimators of the probability p in (4.1).

2. Suppose that γ_i, $i = 1, 2, 3$, are independent standard gamma random variables with parameters, η_i, $i = 1, 2, 3$. Let $x_i = \gamma_1 + \gamma_{i+1}$ for $i = 1, 2$. Give a one dimensional integral that represents $P[x_i \leq w_i, i = 1, 2]$ using cumulative gamma distribution functions in the integrand.

3. The result in Exercise 2 allows calculations of $\hat{p}^2$. For example, suppose that y_i, $i = 1, 2, 3, 4$ in Exercise 2 and $x_i = y_1 + y_{i+1}$ for $i = 1, 2, 3$. Find $\hat{p}_4$ for $p = P[x_i \leq z_i, i = 1, 2, 3]$ when $z_i = 6$, $i = 1, 2, 3$, and $\eta_i = 3$, $i = 1, 2, 3, 4$. Also, find sample variances for increasing sample sizes and compare to the sample variances for $\hat{p}_1$.

4. Suppose that ξ is known to take on a finite number K of possible values but the probabilities η^i of these values are not known but have a Dirichlet prior distribution. Show how to find $R(G)$ and $R^l(G)$ in this case.

5. Find the solution to (4.14). (Hint: Order the observed demands.)

11
Multistage Approximations

Most decision problems involve effects that carry over from one time to another. Sometimes, as in the power expansion problem of Section 1.3, random effects can be confined to a single period so that recourse is block separable. In other cases, however, this separation is not possible. For example, power may be stored by pumping water into the reservoir of a hydroelectric station when demand is low. In this way, decisions in one period are influenced by decisions in previous periods.

Problems with this type of linkage among periods are the subject of this chapter. We again wish to derive approximations that can be used to bound the error involved in any decision based on the approximate problem solution. In Chapter 10, we saw that the number of random variables can lead to rapidly growing problems. In this chapter, we have the additional effect that the number of periods leads to exponential increases in problem size even if the number of realizations in each period remains constant (see Figure 3.3).

We can again construct bounds based on the properties of the multistage recourse functions. These analogues of the basic Jensen and Edmundson-Madansky bounds are given in Section 1. They correspond to fixing values at means or extreme values of the support of the random vectors in each period.

Keeping the number of periods fixed may not lead to sufficient reductions in problem size, especially if no time is clearly the end of the problem. This case would mean facing either an uncertain or an infinite horizon decision problem. These problems can also be approximated by aggregating several periods together. Section 2 describes this procedure to obtain both upper and lower bounds.

The bounds of Sections 1 and 2 can all be viewed as discretization procedures. We can also construct separable bounds that do not require discretization as in Chapter 9. These bounds correspond to separable responses to any changes in the problem. They are described in Section 3.

In multistage problems, specific problem forms and structures can again lead to substantial savings. We describe uses of the form of production problems and vehicle allocation problems in Section 4.

11.1   Bounds Based on the Jensen and Edmundson-Madansky Inequalities

The basic Jensen and Edmundson-Madansky inequalities can be extended to multiple periods directly. The principle is to use Jensen's inequality (or a feasible dual solution) to derive the lower bound and construct a feasible primal solution using extreme points to construct the upper bound. To present these results, we consider the linear case first, although extensions to nonlinear, convex problems are directly possible. We use concepts from measure theory in the following discussion. Readers without this background may skip to the declarations to find the major results for actual implementations.

The multistage stochastic linear program is to find $\mathbf{x} = (x_0^T, \mathbf{x}_1^T, \ldots, \mathbf{x}_H^T)^T$ to

$$
\begin{aligned}
\min \ & c_0^T x_0 + E_\Omega[\mathbf{c}_1^T \mathbf{x}_1 + \cdots + \mathbf{c}_H^T \mathbf{x}_H] \\
\text{s. t. } & A_0 x_0 = b_0, \\
& \mathbf{B}_{t-1}\mathbf{x}_{t-1} + A_t \mathbf{x}_t = \mathbf{h}_t, \\
& \qquad t = 1, \ldots, H, a.s., \\
& \mathbf{x}_t - E_{\Omega^t}[\mathbf{x}_t] = 0, \\
& \qquad t = 1, \ldots, H, a.s., \\
& \mathbf{x}_t \geq 0, \\
& \qquad t = 1, \ldots, H, a.s.,
\end{aligned}
\tag{1.1}
$$

where we have used the explicit nonanticipativity constraints as in (3.5.9). We have also assumed that the recourse within each period A_t is known and not random.

The basic Jensen bound again follows by assuming a partition of Ω, the support vector of all random components. Here, we write Ω as $\Omega = \Omega_1 \times \cdots \times \Omega_H$. We suppose that $\Omega^t = \{\omega^t = (\omega_1, \ldots, \omega_t) | \omega_i \in \Omega_i, i = 1, \ldots, t\}$. In this way, we can characterize all events up to time t by measurability with respect to the Borel field defined by Ω^t, Σ^t. We assume that Ω^t is partitioned as $\Omega^t = S_{t,1} \cup \cdots \cup S_{t,\nu_t}$ and that $S_{t,i} = \cup_{j \in \mathcal{D}_{t+1}(i)} \{\omega^t | (\omega^t, \omega_{t+1}) \in S_{t+1,j}\}$ so that the partitions are consistent from one period to another. We construct measurable decisions at time t if they are constant over each $S_{t,j}$.

Next, assume that $p_{t,i} = P[S_{t,i}]$, $\mathbf{c}_t = c_t$, and that $E_{S_{t,i}}[(\mathbf{h}_t, \mathbf{B}_t)] = (\bar{h}_{t,i}, \bar{B}_{t,i})$ for all t and i. The problem then is to find:

$$
\min c_0^T x_0 + \sum_{i=1}^{\nu_1} p_{1,i} c_1^T x_{1,i} + \cdots + \sum_{i=1}^{\nu_H} p_{H,i} c_H^T x_{H,i}
$$

$$
\text{s. t. } A_0 x_0 = b_0,
$$
$$
\bar{B}_{t-1,i} x_{t-1,i} + A_t x_{t,j} = \bar{h}_{t,j},
$$
$$
t = 1, \ldots, H, i = 1, \ldots, l_{t-1}, \tag{1.2}
$$
$$
j \in \mathcal{D}_{t+1}(i),
$$
$$
x_{t,i} \geq 0,
$$
$$
i = 1, \ldots, \nu_t, t = 1, \ldots, H.
$$

The first result is that (1.2) provides a lower bound on the optimal solution in (1.1) provided the expectations of $(\bar{h}_{t,i}, \bar{B}_{t,i})$ are independent of the past. If not, then the conditional expectation form in (1.2) may not actually achieve a bound.

Theorem 1. *Given that $E_{S_{t,i}}[(\mathbf{h}_t, \mathbf{B}_t)] = (\bar{h}_{t,i}, \bar{B}_{t,i}) = E_{S_{t,j}}[(\mathbf{h}_t, \mathbf{B}_t)]$ for all $S_{t,i}$ and $S_{t,j}$ that have a common outcome at time t, i.e., such that $(\omega^{t-1}, \omega_t) \in S_{t,j}$ if and only if there exist some $(\hat{\omega}^{t-1}, \omega_t) \in S_{t,j}$. The optimal value of (1.2) with the definitions given earlier provides a lower bound on the optimal value of (1.1).*

Proof: Suppose an optimal solution x^* to (1.2) with optimal dual variables $(\pi^*_{t,i})$ corresponding to constraints in (1.2) with right-hand sides, $\bar{h}_{t,i}$ (b_0 if $t = 0$). By dual feasibility in (1.2),

$$
p_{t,i} c_t \geq \pi^{*T}_{t,i} A_t + \sum_{j \in \mathcal{D}_{t+1}(i)} \pi^{*T}_{t+1,j} \bar{B}_{t+1,j}, \tag{1.3}
$$

for every (t, i). Let $\pi_t(\omega) = \sum_{i=1}^{\nu_t} 1_{\omega^t \in S_{t,i}} [\pi^*_{t,i}/p_{t,i}]$, where $1_S(w) = 1$ if $w \in S$ and 0 otherwise. We also have $\rho_t(\omega) = -\sum_{i=1}^{\nu_t} 1_{\omega^t \in S_{t,i}} [\pi^{*T}_{t,i} B_{t,i}(\omega)/p_{t,i}] + \sum_{i'|i' \in \mathcal{D}_{t-1}(A_{t-1}(i))} [\pi^{*T}_{t,i'} \bar{B}_{t,i}/p_{t-1,A_{t-1}(i)}]$. Note how the ρ variables represent nonanticipativity.

The condition for dual feasibility from the multistage version of Theorem 3.13 (see Exercise 1) is that

$$
c_t(\omega) - \pi_t^T(\omega) A_t - \pi_{t+1}^T(\omega) B_{t+1}(\omega) - \rho_{t+1}(\omega) \geq 0, a.s., \tag{1.4}
$$

and

$$
E_{\Sigma_t}[\rho_{t+1}(\omega)] = 0. \tag{1.5}
$$

Substituting in the right-hand side of (1.4) yields:

$$c_t - [\pi_{t,i}^{*T}/p_{t,i}]A_t - [\pi_{t+1,j}^{*T}/p_{t+1,j}]B_{t+1,j}(\omega) + [\pi_{t+1,j}^{*T}/p_{t+1,j}]B_{t+1,j}(\omega)$$
$$- \sum_{j|j \in \mathcal{D}_t(i)} [\pi_{t+1,j}^{*T}\bar{B}_{t+1,j}/p_{t,i}]$$

$$(1.6)$$

for each $S_{t,i}$ and $j \in \mathcal{D}_t(i)$, which is non-negative from (1.3). Also, by their definition and the assumption that integration of $B_{t+1,j}(\omega)$ over varying $S_{t,i}$ does not change its conditional outcome,

$$E_{\Sigma_t}[\rho_{t+1}(\omega)] = \sum_{j \in \mathcal{D}_t(i)} [\pi_{t+1,j}^{*T}/p_{t+1,j}]\bar{B}_{t+1,j}p_{t+1,j}$$
$$- \sum_{j|j \in \mathcal{D}_t(i)} p_{t,1}[\pi_{t+1,j}^{*T}\bar{B}_{t+1,j}/p_{t,i}] = 0 \qquad (1.7)$$

yielding (1.5). Hence, we have constructed a dual feasible solution whose objective value is a lower bound on the objective value of (1.1) by the multistage version of Theorem 3.13. Because this value is the same as the optimal value in (1.2), we obtain the result. $\square$

Thus, lower bounds can be constructed in the same way for multistage problems as for two-stage problems, provided the data have serial independence. Such independence is not necessary if only right-hand sides vary because the dual feasibility is not affected in that case. The key procedure is in developing a dual feasible solution (lower bounding support function). Upper bounds can follow as before by constructing primal feasible solutions. These bounds can also be used in conjunction with the lower bounds to obtain bounds when objective coefficients (c_t) are also random.

To develop the upper bounds, the basic result is an extension of Theorem 9.2. We assume the following general form in which the decision variables x are explicit functions of the random outcome parameters, $\boldsymbol{\xi}$:

$$\inf_{\mathbf{x} \in \mathcal{N}} E_\Xi[\sum_{t=0}^{T} f_t(x_t(\boldsymbol{\xi}), x_{t+1}(\boldsymbol{\xi}), \boldsymbol{\xi}_{t+1})], \qquad (1.8)$$

where the random vector $\boldsymbol{\xi} = (\boldsymbol{\xi}_1, \ldots, \boldsymbol{\xi}_{H+1})$ has an associated probability space, (Ξ, Σ, P), $\mathcal{N}$ is the space of nonanticipative decisions, f_t is convex, and $\boldsymbol{\xi}_{t+1}$ is measurable with respect to Σ_{t+1} and $\boldsymbol{\xi}_{t+1} \in \Xi_{t+1}$, which is compact, convex, and has extreme points, $ext\Xi_t$, with Borel field, $\mathcal{E}_{t+1}$. In this representation, $\mathbf{x}$ *nonanticipative* means that $x_t(\xi(\omega))$ is Σ_t-measurable for all t. It could also be described in terms of measurability with respect to Σ^t, the Borel field defined by the history process $\boldsymbol{\xi}^t = (\boldsymbol{\xi}_1, \ldots, \boldsymbol{\xi}_t)$.

Suppose that $e = (e_1, \ldots, e_H)^T$ where each $e_t \in ext\Xi_t$. The set of all such extreme points is written $ext\Xi$. Suppose $x' = (x'_1, \ldots, x'_H)$, where

$x'_t : ext\Xi_t \to \Re^{n_t}$. We say that x' is *extreme point nonanticipative*, or $x' \in \mathcal{N}'$, if x'_t is measurable with respect to the Borel field, $\mathcal{E}^t$, on $ext\Xi$, defined by $(e_1, \ldots, e_t)$, where $e_j \in ext\Xi_j$ (for $t = 0$, this will be with respect to $\{\emptyset, ext\Xi\}$). With these definitions, we obtain the following result.

Theorem 2. *Suppose that $\xi \mapsto f_t(x_t, x_{t+1}, \xi_{t+1})$ is convex for $t = 0, \ldots, H$, Ξ_t is compact, convex, and has extreme points, $ext\Xi_t$. For all $\xi_t \in \Xi_t$, let $\phi(\xi, \cdot)$ be a probability measure on $(ext\Xi, \mathcal{E})$ where $\mathcal{E}$ is the Borel field of $ext\Xi$, such that*

$$\int_{e \in ext\Xi} e\phi(\xi, de) = \xi, \qquad (1.9)$$

and $\xi \mapsto \phi(\xi, A)$ is measurable with respect to Σ^t for all $A \in \mathcal{E}^t$. Then there exists, $\mathbf{x} \in \mathcal{N}$, such that $x_t(\xi) = \int_{e \in ext\Xi} x'_t(e)\phi(\xi, de)$,

$$E[\sum_{t=0}^{T} f_t(\mathbf{x}_t, \mathbf{x}_{t+1}, \boldsymbol{\xi}_{t+1})] \le \int_{e \in ext\Xi} \sum_{t=0}^{T} f_t(x'_t, x'_{t+1}, e_{t+1})]\lambda(de), \qquad (1.10)$$

where x' is extreme point nonanticipative and λ is the probability measure on $\mathcal{E}$ defined by

$$\lambda(A) = \int_{\Xi} \nu(\xi, A)P(d\xi). \qquad (1.11)$$

Proof. We must first show that $\mathbf{x}$ as defined in the theorem is nonanticipative, or that $x_t(\xi)$ is Σ^t-measurable. This follows because $x'_t(e)$ is $\mathcal{E}^t$-measurable, and, for any $A \in \mathcal{E}^t$, $\phi(\xi, A)$ is Σ^t-measurable. Because each f_t is convex, for any ξ,

$$f_t(x_t(\xi), x_{t+1}(\xi), \xi_{t+1}) = f_t(\int_{e \in ext\Xi} x'_t(e)\phi(\xi, de), \int_{e \in ext\Xi} x'_{t+1}(e)\phi(\xi, de),$$

$$\int_{e \in ext\Xi} e_{t+1}\phi(\xi, de))$$

$$\le \int_{e \in ext\Xi} f_t(x'_t(e), x'_{t+1}(e), e_{t+1})\phi(\xi, de).$$

$$(1.12)$$

Integrating with respect to P, the result in (1.10) is obtained. $\square$

As in Chapter 9, we implement the result in Theorem 2 by finding an appropriate ϕ and then solving the following approximation problem.

$$\inf_{x' \in \mathcal{N}'} \int_{ext\Xi} [\sum_{t=0}^{H} f_t(x_t(e), x_{t+1}(e), e_{t+1})]\lambda(de) \qquad (1.13)$$

to find an upper bound on the value in (1.8). One can also refine these bounds by taking partitions of Ξ.

The simplest type of bound from Theorem 2 is the extension of the Edmundson-Madansky bound on rectangular regions with independent components. For this bound, we assume that all components, $\xi_t(i)$, are stochastically independent and distributed on $[a_t(i), b_t(i)]$. In this case, we can define

$$\nu^{EM-I}(\xi, e) = \Pi_{t=1}^{H} \Pi_{i=1}^{m_t} \frac{|\xi_t(i) - e_t(i)|}{(b_t(i) - a_t(i))}, \tag{1.14}$$

so that

$$\lambda^{EM-I}(e) = \Pi_{t=1}^{H} \Pi_{i=1}^{m_t} \frac{|\bar{\xi}_t(i) - e_t(i)|}{(b_t(i) - a_t(i))}. \tag{1.15}$$

It is easy to check that this ν meets the nonanticipative measurability requirements. Problem 1.13 now can be written as:

$$\inf_{x} \left[\sum_{t=0}^{H} \left(\sum_{i_1=1}^{I_1} \cdots \sum_{i_{t+1}=1}^{I_{t+1}} \left[\sum_{i_{t+2}=1}^{I_{t+2}} + \cdots + \sum_{i_{H+1}=1}^{I_{H+1}} \lambda(e_{i_1}, \ldots, e_{i_{H+1}}) \right] \right. \right.$$
$$f_t(x_t(i_1, \ldots, i_t), x_{t+1}(i_1, \ldots, i_{t+1}), e_{i_{t+1}})], \tag{1.16}$$

where $x_t(i_1, \ldots, i_t)$ corresponds to the tth-period decision depending on the outcomes in extreme point combination e_{i_s} from each period $s = 1, \ldots, H$. This places the nonanticipativity back into the problem implicitly.

Example 1

To see how this bound might be implemented, consider Example 1 in Section 7.1. Suppose that demand is uniformly and independently distributed on $[1, 3]$ in each period. In this case, we obtain a decision vector (x_s^t, w_s^t, y_s^t) in period t for scenario $s = 2^{i_1} + i_2$ for i_1 and i_2 in $\{1, 2\}$. Problem (7.1.7) is, therefore, the upper bounding problem (1.16) for this uniform distribution case, yielding an upper bound of 6.25. In this case, the lower bound using the expected demand value of two in each period is three. In Exercise 2, you are asked to refine these bounds until they are within 25% of each other.

Other extreme point combinations are clearly also possible in multiperiod problems as they are in single-period problems. Extensions to dependent random variables and f_t concave in some arguments can also be made.

The bounds given in this section so far apply only to fixed numbers of periods. When periods are combined, we call the resulting problem an *aggregated problem*. These problems are described in the next section.

Exercises

1. Consider the multistage stochastic linear program in the form of (1.1). Prove the multistage version of Theorem 3.13.

2. Refine the extreme point (Edmundson-Madansky) and conditional expectation (Jensen) bounds on partitions for Example 1 from Section 7.1 until the upper bound is within 25% of the lower bound.

11.2 Bounds Based on Aggregation

The main motivation for aggregation bounds is to deal with problems with many (perhaps an infinite number of) periods by combining periods to obtain a simpler approximate problem with fewer periods. The basic procedures in this chapter appear in Birge [1985a] and Birge [1984]. They follow the general aggregation results in Zipkin [1980a, 1980b]. Similar methods, especially for dealing with infinite horizon problems, appear in Grinold ([1976, 1983, 1986]). Generalizations appear in Wright [1994].

To derive both upper and lower bounds in this framework, we consider a special form for the multistage problem in (3.5.1). We allow feasibility by adding a penalty variable y_t that can achieve feasibility in each period. This notion of model robustness is quite common, although the penalty parameter q may be quite high. The form of the multistage stochastic linear program in this case is:

$$\min z = c^T x^1 + E_{\boldsymbol{\xi}}[\sum_{t=2}^{H} \rho^{t-1}(c^T x^t(\boldsymbol{\xi}_2,\ldots,\boldsymbol{\xi}_t)+q^T y^t(\boldsymbol{\xi}_2,\ldots,\boldsymbol{\xi}_t))]$$

$$\text{s. t. } Wx^1 \geq h^1,$$
$$T x^{t-1}(\boldsymbol{\xi}_2,\ldots,\boldsymbol{\xi}_{t-1}) + W x^t(\boldsymbol{\xi}_2,\ldots,\boldsymbol{\xi}_t) + y^t(\boldsymbol{\xi}_2,\ldots,\boldsymbol{\xi}_t) \geq \boldsymbol{\xi}^t,$$
$$t = 2,\ldots,H,$$
$$x^1 \geq 0; x^t(\omega) \geq 0, a.s., t = 2,\ldots,H,$$
$$y^t(\omega) \geq 0, a.s., t = 2,\ldots,H,$$

$$(2.1)$$

where c is a known vector in $\Re^{n_1}$, h^1 is a known vector in $\Re^{m_1}$, $\xi^t(\omega) = h^t(\omega)$ is a random m-vector defined on (Ω, Σ^t, P) (where $\Sigma^t \subset \Sigma^{t+1}$) for all $t = 2,\ldots,H$, and T and W are known $m \times n$ matrices. We also suppose that Ξ^t is the support of ξ^t. The parameter ρ is a discount factor.

Note that in (2.1), we assume that the parameters T, W, c, and q are all constant across time (with objective coefficients varying only with the discount factor). This assumption is basically made to simplify the following presentation. Varying parameters are possible with little additional work.

The key observation for these bounds is that an optimal solution in (2.1) is no lower than

$$\pi^1 h^1 + E_{\boldsymbol{\xi}}[\sum_{t=2}^{H} (\pi^t(\xi^2,\ldots,\xi^t))^T \xi^t]$$

$$(2.2)$$

for any $(\pi^1, \ldots, \pi^t(\boldsymbol{\xi}^2, \ldots, \boldsymbol{\xi}^t), \ldots, \pi^T(\boldsymbol{\xi}^2, \ldots, \boldsymbol{\xi}^T)) \geq 0$ a.s. that satisfies

$$(\pi^1)^T W + E_{\boldsymbol{\xi}}[\pi^2(\boldsymbol{\xi}_2)]^T T \leq c^T,$$

$$\pi^t(\boldsymbol{\xi}^2, \ldots, \boldsymbol{\xi}^t)^T W + E_{\boldsymbol{\xi}_{t}(\boldsymbol{\xi}_2, \ldots, \boldsymbol{\xi}_t)}[\pi^{t+1}(\boldsymbol{\xi}^2, \ldots, \boldsymbol{\xi}^{t+1})]^T T \leq \rho^{t-1} c^T,$$

$$t = 2, \ldots, H - 1, \qquad (2.3)$$

$$\pi(\boldsymbol{\xi}^2, \ldots, \boldsymbol{\xi}^H)^T W \leq \rho^{H-1} c^T,$$

$$\pi(\boldsymbol{\xi}^2, \ldots, \boldsymbol{\xi}^H)^T W \leq \rho^{H-1} q^T.$$

You are asked to show that (2.2) subject to (2.3) provides a bound in Exercise 1.

The basic idea behind the aggregation bounds is that we can either construct solutions $(\mathbf{x}, \mathbf{y})$ that are feasible in (2.1) or solutions $\boldsymbol{\pi}$ that are feasible in (2.3). As before, the former provide upper bounds, while the latter provide lower bounds.

The other assumption we make is that some set of finite upper bounds exists in $\mathbf{x}_t$ so that for any $\mathbf{x}^*$ optimal in (2.1):

$$\mathbf{x}_t^*(\boldsymbol{\xi}_2, \ldots, \boldsymbol{\xi}_t) \leq u_t(\boldsymbol{\xi}_2, \ldots, \boldsymbol{\xi}_t). \qquad (2.4)$$

In most problems, some form of bound satisfying (2.4) can be found. The tightness of this bound may, however, significantly affect the bounding results.

The basic bound is first to assume that the Jensen type of conditional expectation bound has been applied in each period. We illustrate this with a single partition, although finer partitions are possible. We also collapse everything into a two-period problem. Less aggregated models are constructed in the same way. Note in the following that H is quite arbitrary and, assuming finite sums, could even be infinite.

The problem is formed by defining aggregate variables, $\hat{X}^1$, $\hat{X}^2$, and $\hat{Y}^2$, and parameters,

$$\hat{W} = (\sum_{t=2}^{H} \rho^{t-2}) W + (\sum_{t=2}^{H} \rho^{t-2}) T, \hat{I} = (\sum_{t=2}^{H} \rho^{t-2} I),$$

$$\hat{c} = (\sum_{t=2}^{H} \rho^{t-1}) c, \hat{q} = (\sum_{t=2}^{H} \rho^{t-1}) q, \hat{\xi} = (\sum_{t=2}^{H} \rho^{t-2} \bar{\xi}^t).$$

The resulting aggregate approximation problem is:

$$\begin{aligned}
\min \, & c^T \hat{X}^1 + \hat{c}^T \hat{X}^2 \hat{q}^T \hat{Y}^2 \\
\text{s. t. } & W \hat{X}^1 \geq h^1, \\
& T \hat{X}^1 + \hat{W} \hat{X}^2 + \hat{T} \hat{Y}^2 \geq \hat{\xi}, \\
& \hat{X}^1, \hat{X}^2, \hat{Y}^2 \geq 0.
\end{aligned} \qquad (2.5)$$

Suppose (2.5) has an optimal solution (X_1^*, X_2^*, Y_w^*) with multipliers, Π^*. These solutions are not directly feasible in (2.1) or (2.3), but feasible solutions can be easily constructed from them. To do so, we need only let $\hat{x}_1 = X_1^*$, $\hat{x}_t(\xi_2, \ldots, \xi_t) = X_2^*$, and $\hat{y}_t(\xi_2, \ldots, \xi_t) = Y_2^*$ for all t and ξ. We also let $\hat{\pi}_1 = \Pi_1^*$ and $\hat{\pi}_t(\xi_2, \ldots, \xi_t) = \rho^{t-2}\Pi_2^*$ for all t and ξ. In this way, the value of (2.5) is the same as

$$\hat{z} = c^T \hat{x}^1 + E_{\Xi}[\sum_{t=2}^{H} \rho^{t-1}(c^T \hat{x}^t(\xi_2, \ldots, \xi_t) + q^T \hat{x}^t(\xi_2, \ldots, \xi_t))], \qquad (2.6)$$

which forms the basis for our bounds. The result is contained in the following theorem.

Theorem 3. *Let z^* be a finite optimal value for (2.1). Then*

$$\hat{z} + \epsilon^+ \geq z^* \geq \hat{z} - \epsilon^-, \qquad (2.7)$$

where

$$\epsilon^- = -\sum_{t=?}^{H}\sum_{j=1}^{n}[\int_{\Xi} [\min\{\rho^{t-1}c_j - \rho^{t-2}\Pi_2^* W_{.j} - \rho^{t-1}\Pi_2^* T_{.j}, 0\}u_t(j)(\xi)]P(d\xi)]$$

and

$$\epsilon^+ = \sum_{t=2}^{H}\sum_{j=1}^{n}[\int_{\Xi} [\max\{-W_{.j}X_2^* - T_{.j}X_2^* - Y_2^*(j) + \xi^t, 0\}\rho^{t-1}q(j)]P(d\xi)].$$

The proof of this theorem is Exercise 2. The basic idea is to write out z^* in terms of $(\mathbf{x}^*, \mathbf{y}^*)$ and to add on $\hat{\pi}^t(\xi)^T(\xi_t - Wx_t^*(\xi) - y_t^* - Tx_{t-1}^*(\xi))$ terms, which are all nonpositive. This yields ϵ^-. The upper bound comes from showing that $\{\hat{x}^t(\boldsymbol{\xi}), \hat{y}^t(\boldsymbol{\xi}) + \max\{-W_{.j}X_2^* - T_{.j}X_2^* - Y_2^*(j) + \boldsymbol{\xi}^t, 0\}\}$ is always feasible in (2.1).

These bounds can be quite useful, but the penalty and variable bound assumptions may not be apparent in many problems. Sometimes bounds on groups of variables are possible and can be useful. In other cases, properties of the constraint matrices can be exploited to obtain other bounds similar to those in Theorem 3. Several of these ideas are presented in Birge [1985a].

Example 2

In production/inventory problems, these values are especially easy to find, as in Birge [1984]. Consider a basic problem of the form

$$\min z \quad = E_{\boldsymbol{\xi}}[\sum_{t=1}^{H} \rho^{t-1}(-c^t x^t(\boldsymbol{\xi}) + q^t y_+^t(\boldsymbol{\xi}) + r^t s^t(\boldsymbol{\xi}))]$$

$$\text{s. t.} \quad \mathbf{x}^t - \mathbf{s}^t \leq \mathbf{k}^t, a.s.,$$
$$\mathbf{w}^{t-1} + \mathbf{x}^t - \mathbf{w}^t = 0, a.s.,$$
$$\mathbf{z}^t \geq \mathbf{b}^t, a.s.,$$
$$\mathbf{y}_+^{t-1} + \mathbf{x}^t - \mathbf{y}_+^t = \boldsymbol{\xi}_t, a.s.,$$
$$t = 1, \ldots, H,$$
$$\mathbf{y}_+^{t-1}, \mathbf{x}^t, \mathbf{s}^t, \mathbf{w}^t \geq 0, a.s.;$$
$$t = 1, \ldots, H;$$
$$\mathbf{y}_+^t, \mathbf{y}_-^t, \mathbf{x}^t, \mathbf{s}^t, \mathbf{w}^t, \text{all } \Sigma^t \text{ measurable}$$
$$t = 1, \ldots, H, \tag{2.8}$$

where x^t represents total production, s^t represents overtime production, w^t is cumulative production, b^t is a lower bound to achieve a service reliability criterion (see Bitran and Yanasse [1984]), c^t, q^t, and r^t are cost parameters, and $\boldsymbol{\xi}^t$ is the random demand.

For problems with the form in (2.7), it is possible to find bounds on all primal and dual variables for an optimal solution. These bounds can then be used with Theorem 3. Exercises 3, 4, and 5 explore the aggregation bounds in this context more fully.

Exercises

1. Verify that a non-negative π satisfying the conditions in (2.3) provides a bound on (2.1)'s optimal value through (2.2).

2. Prove Theorem 3.

3. Find bounds on all optimal variable values in (2.7) as functions of the parameters and previous realizations.

4. Using the bounds in (2.3), construct bounds based on Theorem 3 for a problem with four periods, uniform demand on $[8000, 10,000]$, $b^t = t(9500)$, $p = 19$, $h = 0.4$, $q = 9.5$, $\rho = 0.9$, and $k = 9000$.

5. It is not necessary to take expectations before aggregating periods. Using the example in (2.7), construct bounds with a two-period problem that uses a weighted sum of future demands in the first period. What type of stochastic program is this?

11.3 Bounds Based on Separable Responses

In this section, we extend the basic separable bounds presented in Section 9.5b to multistage problems. The main idea is to use the two-stage method repeatedly to approximate the objective function by separable functions. For linear problems, this leads to sublinear or piecewise linear functions as in Section 9.5b. Functions without recession directions (e.g., quadratic functions) would require some type of nonlinear (e.g., quadratic) function that should again be easily integrable, requiring, for example, limited moment information (second moments for quadratic functions). We consider the linear case (following Birge [1989]).

The goal is to construct a problem that is separable in the components of the random vector. In each period t, a decision, x_t, is made subject to the constraints, $A_t x_t = \xi_t - B_{t-1} x_{t-1}, x_t \geq 0$, where ξ_t is the realization of random constraints and x_{t-1} was the decision in period $t-1$. The objective contribution from this decision is $c_t^T x_t$. We can view this decision as a response to the input, $\eta_t = \xi_t - B_{t-1} x_{t-1}$. The period t decision, x_t, then becomes a function of this input, so $x_t(\omega)$ becomes $x_t(\eta_t)$. Problem (2.2) becomes

$$\min c_1^T x_1 + \mathrm{E}[c_2^T x_2(\eta_2) + \ldots + c_H^T x_H(\eta_H)]$$
$$\text{s. t. } A_1 x_1 = b_1,$$
$$A_t x_t(\eta_t) = \eta_t,$$
$$t = 2, \ldots, H, \text{ a.s.},$$
$$\eta_t = \xi_t - B_{t-1} x_{t-1}(\eta_{t-1}),$$
$$t = 2, \ldots, H, \text{ a.s.},$$
$$x_t(\eta) \geq 0, t = 1, \ldots, H.$$

The optimization problem is to determine the correct response to η_t. The two-stage method given in Section 9.5b gives a response that is separable in the components of $\xi = \eta_2$. In multiple stages, ξ is replaced by η_t for period t. The response must consider future actions and costs, so it is no longer simply optimization of the second-period problem.

The dimension of $\eta = (\eta_2, \ldots, \eta_H)$ makes direct solution difficult in general. An upper bound is, however, obtained for any feasible response, i.e., decision vectors, $x_t(\eta_t)$, that satisfy $A_t x_t(\eta_t) = \eta_t, x_t(\eta_t) \geq 0$, a.s., where $\eta_t = \xi_t - B_{t-1} x_{t-1}(\eta_{t-1})$ for all t. The two-stage method can be used to obtain feasible responses that are separable in the components of η_t, i.e., where $x_t(\eta_t) = \sum_i x_t^i(\eta_t(i))$.

One choice is to let $x_t^i(\eta_t(i))$ solve

$$\min c_t^T x_t \text{ s. t. } A_t x_t = \eta_t(i) e(i), x_t \geq \beta, \tag{3.1}$$

where $e(i)$ is the ith unit vector and β depends on choices for the other x_t^i. Program (3.1) is a parametric linear program in $\eta_t(i)$. It is particularly easy to solve if $\beta = 0$. In this case, $x_t^i(\eta_t(i))$ is linear for positive and negative $\eta_t(i)$. We suppose this case and let the optimal solutions be $x_t^{\pm i}$ when $\eta_t(i) = \pm 1$.

A solution can be obtained if we can find the distribution of the $\eta_t(i)$ given responses determined by solutions of (3.1). The resulting problem to solve is

$$\min \; c_1 x_1 + \sum_{t=2}^{T} \sum_{i=1}^{m_t} \int \psi_t^i(\boldsymbol{\eta}_t(i)) P(d\boldsymbol{\eta}_t(i))$$

$$\text{s. t. } A_1 x_1 = b_1, x_1 \geq 0, \tag{SL}$$

where $\psi_t^i(\boldsymbol{\eta}_t(i)) = c_t x_t^{+i} \eta_t(i)$ if $\eta_t(i) \geq 0$, and $\psi_t^i(\boldsymbol{\eta}_t(i)) = c_t x_t^{-i}(-\eta_t(i))$ if $\eta_t(i) \leq 0$. Assuming that the distribution of $\boldsymbol{\eta}_t$ is known in this approximation, we can find $\boldsymbol{\eta}_{t+1}$. Initially, $\boldsymbol{\eta}_2 = \boldsymbol{\xi}_2 - B_1 x_1$, which has the same distributional form as $\boldsymbol{\xi}_2$. In general, $\boldsymbol{\eta}_{t+1}(j)$ is given by:

$$\boldsymbol{\eta}_{t+1}(j) = \boldsymbol{\xi}_{t+1}(j) - B_t(j, \cdot)[\sum_{i=1}^{m_t}(x_t^{+i}\mathbf{1}_{\boldsymbol{\eta}_t(i) \geq 0} + x_t^{-i}\mathbf{1}_{\boldsymbol{\eta}_t(i) < 0})(|\boldsymbol{\eta}_t(i)|)]. \tag{3.2}$$

Note that the values in (3.2) are linear functions of $\boldsymbol{\eta}_t$ on the regions where $\boldsymbol{\eta}_t$ has constant sign. We can, therefore, construct $\boldsymbol{\eta}_{t+1}$ as a function of $\boldsymbol{\eta}_t$ by overlaying these linear transformations of random variables. For normally distributed data, this may be possible because the transformation does not affect the distribution class. For other distributions, it is more difficult. Even in the normal case, however, we have different distribution parameters for all possible sign combinations of all random variables in previous period inputs. Exponential growth of the calculations in the number of periods is not avoided.

Because the approximation given earlier may be difficult to compute even with normal distributions, it may be necessary to approximate the distribution of $\boldsymbol{\eta}_{t+1}$. We can use bounds on the $P\{\boldsymbol{\eta}_t(i) \geq 0\}$ and on the moments conditional on $\boldsymbol{\eta}_t(i) \geq$ or < 0. Given these values, moment problems can be solved to calculate corresponding values for $\boldsymbol{\eta}_{t+1}$ and to bound ψ_t^i (see Birge and Wets [1989]). Any other bounds on the input $(B_t x_t)$ from period t to period $t + 1$ can also be used to obtain crude bounds on the ψ values.

Note that certain problems, such as networks, may have very few nonzeros in the $B_t(j, \cdot)$ term in (3.1) (because they have a close-to-simple recourse structure). The random input vector $\boldsymbol{\eta}_{t+1}$ may be easily calculable for these problems.

Another looser but more implementable bound can be obtained by forcing a feasible and separable response in all future periods depending on a single random variable in the current period. This eliminates the problem of characterizing the distribution of inputs to all periods. It does, however, force a dependency in future periods that may increase the bound.

To develop this response function, let $X_t(\pm i)$ be an optimal solution, $(x_t, \ldots, x_H)$, $(t > 1)$, to:

$$
\begin{aligned}
\min \quad & c_t^T x_t \quad + \ldots \quad + c_H^T x_H \\
\text{s. t.} \quad & A_t x_t & = \pm e_i, \\
& B_t x_t + A_{t+1} x_{t+1} & = 0, \\
& \qquad \cdots & \vdots \\
& A_H x_H & = 0, \\
& x_\tau \geq 0, \quad \tau = t, \ldots, H.
\end{aligned} \tag{3.3}
$$

Now define

$$
\begin{aligned}
z_t^i(\hat{\xi}_t(i)) = & \int_{\boldsymbol{\xi}_t - \hat{\xi}_t(i) > 0} \{ C_t^T X_t(+i)(\boldsymbol{\xi}_t(i) - \hat{\xi}_t(i)) \\
& + \int_{\boldsymbol{\xi}_t - \hat{\xi}_t(i) \leq 0} \{ C_t^T X_t(-i)(-\boldsymbol{\xi}_t(i) + \hat{\xi}_t(i)),
\end{aligned}
$$

where $C_t = (c_t, \ldots, c_H)$. An *upper bound* on the objective value of (3.1) is obtained by solving the separable nonlinear program:

$$
\begin{aligned}
\min \quad & c_1^T x_1 \quad \ldots + c_H^T x_H \quad + \sum_t \sum_i (z_t^i(\hat{\xi}_t(i))) \\
\text{s. t.} \quad & A_1 x_1 & = b, \\
& B_t x_t + A_{t+1} x_{t+1} \quad - \hat{\xi}_{t+1} & = 0, \\
& & t = 1, \ldots, H-1, \\
& x_t \geq 0, \quad t = 1, \ldots, H. \\
& \hat{\xi} \in \Xi,
\end{aligned} \tag{3.4}
$$

where Ξ is the support set of the random variables. Note that if we drop the nonlinear term in the objective and replace $\hat{\xi}$ in the constraints with a fixed valued of $E[\boldsymbol{\xi}]$, then we can obtain a *lower bound* on the optimal objective value in (3.1) (see Birge and Wets [1986]). We should note that in some cases, we may not have a solution to (3.3) for $\pm e_i$ but may only have a solution for $+e_i$, e.g. In this case, $\hat{\xi}_{t+1}(i)$ could be constrained to be less than the minimum possible value of $\boldsymbol{\xi}_{t(i)}$.

In (3.4), we are solving to determine a *centering point*, $\hat{\xi}$, that obtains minimum cost if we assume the response to any variation from $\hat{\xi}$ is a solution of (3.3). By allowing some variation of the choice of centering point, a "best" approximation of this type is found. The value of (3.4) is an upper bound because the composition of the x_t solutions from (3.4) and the X_t values used in the z terms yield a feasible solution for all ξ.

This procedure may also be implemented as responses to several scenarios. In this case, the random vectors are partitioned as in Section 1. The partitions may also be part of the higher-level optimization problem so that in some way a "best" partition can be found.

The points used within the partitions may be chosen as expected values, in which case the solution without penalty terms is again a lower bound on the optimal objective value. For an upper bound, this vector may be allowed to take on any value in the partition.

The use of multiple scenarios enters directly into the progressive hedging approach of Rockafellar and Wets (see Section 6.2). This can be used to solve the top-level problem and to approach a solution that is optimal for a given set of partitions and the piecewise linear penalty structure presented here.

Computations are then restricted to optimizing separable nonlinear functions subject to linear constraints. Implementations can be based on previous procedures (such as decomposition).

The basic framework for the upper bounding procedures given earlier is to construct a feasible solution that is easily integrated. Other procedures for constructing such feasible responses are possible. For example, Wallace and Yan [1993] suppose two types of restrictions of the set of solutions to obtain bounds. The first is to suppose only a subset of variables is used within a period, as, for example, with the penalty terms used for aggregation bounds in Section 2. The other approach is to suppose that all realizations from period to period must meet some common constraint on values passed between periods. This procedure effectively divides the multistage problem into a sequence of two-stage problems. It appears to work well on problems with many stages.

Exercise

1. Use the separable function approach and (3.4) to construct an upper bound on Example 1 with uniform demand distributions.

11.4   Bounds for Specific Problem Structures

To achieve further improvements in bounding multistage stochastic programs requires taking advantage of the specific structure of the problem considered. For Example 2 in Section 2, we considered a basic production problem that allows the construction of bounds on optimal primal and dual variables that can then be used in constructing optimal objective value bounds as in (2.7). Other bounds and approximations using similar production problem structures are also possible. We explore some of those bounds developed by Ashford [1984], following Beale, Forrest, and Taylor [1980], and Bitran and Yanasse [1984], and Bitran and Sarkar [1988]. These bounds can be viewed as extensions of the aggregation-type bounds in Section 2.

Extensions of the separable approximation bounds in Section 3 for specially structured problems are also possible. We consider these possibilities in the context of vehicle allocation problems as developed by Powell [1988] and Frantzeskakis and Powell [1990, 1993].

a. Production problems

The first type of extension of the production problem we consider is the model used in Ashford [1984] which is a slight generalization of (2.8). It is also an extension of similar work by Beale, Forrest, and Taylor [1980] on a production problem similar to (2.8). The model is to

$$\min z = E_\xi [\sum_{t=1}^{T} (-c^t x^t(\xi) - q^t y^t(\xi))]$$
$$\text{s. t. } A^t y^{t-1} + B^t x^t - y^t \leq \xi^t, a.s.,$$
$$t = 1, \ldots, H, \qquad (4.1)$$
$$y_t \geq l_t, u_t \geq x^t \geq 0, a.s., t = 1, \ldots, H,$$

where x^t represents production and related variables and y^t represents the state (e.g., inventory) after realizing demands, ξ_t. Both variables are bounded, although y^t may only have trivial bounds. One upper bound directly analogous to that in Theorem 3 can be constructed using this structure (see Exercise 1).

A lower bound on the optimal value of (4.1) can be obtained simply by substituting expected values for the random elements in (4.1). Ashford also presents an improved lower bound, however, that forms the basis for an approximation procedure. This bound consists of solving a *reduced problem*:

$$\min z^{RED}(G_1, \ldots, G_H) = \sum_{t=1}^{T} (-c^t x^t - q^t y^t)$$
$$\text{s. t. } A^t y^{t-1} + B^t x^t - w^t = \bar{\xi}^t, t = 1, \ldots, H, \qquad (4.2)$$
$$-y_t - w_t \leq -f_t(w_t - l_t),$$
$$t = 1, \ldots, H,$$
$$u_t \geq x^t \geq 0, a.s., t = 1, \ldots, H,$$

where the G_t are m_t-vectors of given distribution functions, G_{it}, $i = 1, \ldots, m_t$, and $f_t = (f_{1t}, \ldots, f_{m_t,t})$, with

$$f_{it}(\eta_i) = \int_{\infty}^{-\eta_i} (\eta_i + \zeta) dG_{it}(\zeta), \qquad (4.3)$$

for $i = 1, \ldots, m_t$.

The bound in (4.2) is chosen by first determining the distribution function, G_{it}. If G_t^* is the vector of distribution functions of $A^t y^{t-1,*} + B^t x^{t,*} - \xi^t$ for an optimal solution (y^*, x^*) of (4.1), then the following theorem holds.

Theorem 4. *The solution* $z^{RED}(G_1^*, \ldots, G_H^*)$ *provides a lower bound on the optimal solution* z^* *in (4.1) and* $z^{RED}(G_1^*, \ldots, G_H^*) \geq z(\bar{\xi})$, *the solution* *of the expected value problem, i.e., (4.1) with all random variables replaced by their expectations.*

Proof: Exercise 2. $\square$

It is possible to make the approximation in (4.2) into a deterministic equivalent of (4.1) if appropriate penalties are placed on the violation of bound constraints on x^t, but the calculation of this and of the bound given by Theorem 1 requires information about the optimal solutions which is not known. Another bound is, however, obtainable by substituting $G^\xi(t)$, the distribution function vector, corresponding to $(\xi^t - \bar{\xi}^t)$ (see Exercise 3). This represents the beginning of an approximation when the ξ^t vectors are normally distributed. The approximation successively estimates parameters of a normal approximation of the distribution of $A^t y^{t-1,*} + B^t x^{t,*} - \xi^t$ from t to $t+1$. This procedure continues until little improvement occurs in this updating procedure. Computational results with this procedure show significant savings over dynamic programming calculations.

This process can be viewed as a form of dynamic programming approximation using the input to each period's decisions as the quantity, $A^t y^{t-1,*} + B^t x^{t,*} - \xi^t$. In this way, it is also similar to the response method given in Section 3. An alternative approach is to build approximations of the value function from period to period. One application to problems with uncertainties in the B_{t-1} matrix in (4.1) appears in Beale, Dantzig, and Watson [1986]. The bounds developed by Bitran et al. follow these production examples closely. The model is again of the form in (2.8).

b. Vehicle allocation problems

Vehicle allocation problems provide a different problem structure that allows specific bound construction. These problems can be represented as multistage network problems with only arc capacities random. A formulation would then be the same as (1.1). The matrices A_t correspond to flows leaving nodes in period t while B_t corresponds to flow entering nodes in period $t+1$. The only exception is in the last period for which A_H just gathers flow into ending nodes. For simplicity, this model assumes that all flow requires one period to move between nodes.

The $x_t(ij)$ decisions are then flows from i in period t to j in period $t+1$. The randomness involves the demand from i to j in period t. We assume that $x_t(ij) = x_t^f(ij) + x_t^e(ij)$, where $x_t^f(ij)$ represents *full* loads (or vehicles) and $x_t(ij)^e$ represents *empty* vehicles (assuming that fractional vehicle loads are feasible). For demand of $\xi^t(ij)$, we would have $x_t^f(ij) \leq \xi^t(ij)$. The costs $c_t^f(ij)$ and $c_t^e(ij)$ then correspond to the unit values of moving full and empty vehicles from i to j at t. The result is that vehicles are conserved in (4.4). The decisions generally depend on the locations of vehicles at any point in time.

Frantzeskakis and Powell [1993] consider several alternative approximations of (4.4). First, one could solve the expected value problem to obtain
$\hat{x}_t$ values. These corresponding decisions can be used regardless of realized
demand (as, e.g., in Bitran and Yanasse [1984]). Then the x_t values could
be split into full and empty parts, $\mathbf{x}_t = \bar{x}_t$, $\mathbf{x}_t^f(ij) = \max\{\bar{x}_t(ij), \boldsymbol{\xi}_t(ij)\}$,
according to realized demand to produce both upper and lower bounds.
This could be viewed as a generalization of a simple recourse strategy;
hence Powell and Frantzeskakis refer to it as the *simple recourse* strategy.

Another approach is simply to solve the mean value problem, but only
actually to send a vehicle from i to j at t if there is sufficient demand. In
this way, $\mathbf{x}_t^f(ij) = \max\{\bar{x}_t(ij), \boldsymbol{\xi}_t(ij)\}$, but $\mathbf{x}_t(ij) = \mathbf{x}_t^f(ij)$ whenever $i \neq j$.
This strategy is called *null recourse*.

A further strategy is called *nodal recourse*, in which a set of decisions or
a policy, $\delta^t(i)$, is defined for each node i at all times t. This policy would
be a list of options for flow from i at t. The list would be a ranking of full
loads (i.e., preferred nodes, $j_1(i), \ldots j_k(i)$) if capacity is available followed
by an alternative for any remaining empty vehicles.

This preference structure can be constructed using a separable approximation from period $t+1$ to H. In period H, we can begin by assigning
some salvage/final value $-c_H(i)$ to vehicles on the arcs corresponding to
travel from one node to itself.

At period $H-1$, the value of sending a full load from i to j is simply
$-c_{H-1}^f(ij) - c_H(j)$. Including empty loads in the obvious way and ordering
in decreasing orders for each p determines the strategy at $H-1$. Now, given
the distributions of $\boldsymbol{\xi}_{H-1}$, these values yield an expected value function
for vehicles at i at t. The argument of this function is a state variable,
$y_{H-1}(i)$. With the function defined, similar decisions on expected values of
loads from i to j can be made in period $H-2$. A dynamic programming
recursion would be to find $\mathcal{Q}_t(\mathbf{y}_t) = E_{\boldsymbol{\xi}}[Q_t(\mathbf{y}_t, \boldsymbol{\xi}_t)]$ where:

$$Q_t(\mathbf{y}_t, \boldsymbol{\xi}_t) = \min_{\mathbf{x}_t, \mathbf{y}_t} \quad -c_t^T \mathbf{x}_t + \mathcal{Q}_{t+1}(\mathbf{y}_{t+1})$$

$$\text{s. t. } A_t \mathbf{x}_t = \mathbf{y}_t, \tag{4.4}$$
$$B_t \mathbf{x}_t - \mathbf{y}_{t+1} = 0,$$
$$\boldsymbol{\xi}_t \geq \mathbf{x}_t \geq 0.$$

If $\mathcal{Q}_{t+1}(\mathbf{y}_{t+1})$ is linear with coefficients, $\bar{Q}_{t+1}(i)$ in each component i of
$\mathbf{y}_{t+1}$ as it is for $t = H-1$, then the optimal solution to (4.4) is given by
the increasing ordering of $c_t^f(ij) + \hat{Q}_{t+1}(j)$ with each successive $x_t^f(ij)$ used
up to the minimum of $y_t(i)$ and $\xi_t(ij)$ according to this realization of $\boldsymbol{\xi}_t$.
The key is then to construct a linear approximation to $\mathcal{Q}_{t+1}(\mathbf{y}_{t+1})$.

With a linearization, the entire strategy can be simply carried back to
the first period. Overall, this represents a feasible but not optimal strategy
because it avoids calculating the full nonlinear value of each $y_{H-1}(i)$. The
objective function values are not, however, calculated fully because of the
linearization.

One way to compute a linearization is to assume an input value $\hat{y}_t(i)$ and to find the probability of each option multiplied by the expected linearized value of that option. Using this to determine the recourse value at each stage can lead to a lower bound at each stage and overall when the first-period problem is solved (see Exercise 4). An upper bounding linearization is also possible. This is analogous to the Edmundson-Madansky approach (Exercise 5).

Frantzeskakis and Powell [1993] mention that extensions of nodal recourse can apply to general network problems. These procedures are similar to the separable bounding procedures presented in Section 3. They again rely on building responses to random variation that depend separately on the random components and that are also feasible.

Other types of network structure can also yield bounds in specific cases. For PERT networks (see, e.g., Taha [1992]), for example, a typical problem is to balance the benefits of early completion against the possible penalty costs of exceeding a due date or promise date. In these problems, a natural separation occurs that allows calculation despite the interconnected structure of paths and possibly correlated times. Klein Haneveld [1986] considers bounds on expected tardiness penalties with mean constraints. Maddox and Birge extend this analysis to bounds with second moment information (Birge and Maddox [1995, 1996]) and to bounding probabilities of tardiness (Maddox and Birge [1991]).

The basic principle throughout this and Chapter 10 is to use convexity of objective and constraints. Relax the problem and substitute expectations properly to obtain a lower bound. Restrict the problem and maintain a feasible solution (as perhaps a combination of extremal solutions) to obtain an upper bound. Many more bounding approximations are possible based on these fundamental observations.

Exercises

1. Let A^{t+} be the matrix composed of the positive elements of A^t in (4.1) (with zeros elsewhere). Use this to construct a bound on any feasible dual variable value with $\beta_t = \sum_{\tau=t}^{H}(\prod_{s=t}^{\tau-1}(A^{s+})^T)q^\tau$, where $\prod_{s=t}^{t-1}(A^{s+})^T = I$. Combine this with Theorem 3 to obtain an upper bound on the optimal objective value using the solution to the mean value problem.

2. Prove Theorem 4.

3. Show that $z^{RED}(G_1^\xi, \ldots, G_H^\xi) \leq z^*$.

4. To construct a lower bound for nodal recourse, assume a projected value, $\hat{y}_t(i)$ of $\mathbf{y}_t(i)$ (as, e.g., an average of incoming and outgoing loads). Find an expression (in terms of the demand distributions on

the ranked full load alternatives) for the expected value (assuming linearized future costs) of an additional vehicle beyond $\hat{y}_t(i)$. Show that this procedure gives a lower bound on (4.4) when $t = 1$.

5. Show how an upper bounding linearization can be constructed for (4.4) using a linearization of $Q_{t+1}(y_{t+1})$. (Note: You can assume a constant number of total vehicles.)

6. Consider a three-period example with five total vehicles, three nodes (cities), and salvage values, $c_3(1) = 2$, $c_3(2) = 1$, and $c_3(3) = 4$. Currently, two vehicles are at A, two vehicles are at B, and one vehicle is at C. Suppose demand in each period is uniform on the integers from zero to $\xi^{\max}(ij)$, where $\xi^{\max}(ij)$ has the following values:

$$
\begin{array}{ccc}
\text{To } j = & 1 \quad 2 \quad 3 \\
\text{From } i = & \\
1 & 0 \quad 2 \quad 3 \\
2 & 2 \quad 0 \quad 2 \\
3 & 3 \quad 3 \quad 0.
\end{array}
$$

Suppose the costs (negative of profits) on each route for a full truck are

$$
\begin{array}{ccc}
\text{To } j = & 1 \quad\; 2 \quad\;\; 3 \\
\text{From } i = & \\
1 & 0 \quad -1 \quad -2 \\
2 & -1 \quad\; 0 \quad -3 \\
3 & -2 \quad -3 \quad\; 0.
\end{array}
$$

Empty load costs are

$$
\begin{array}{ccc}
\text{To } j = & 1 \quad 2 \quad 3 \\
\text{From } i = & \\
1 & 0 \quad 1 \quad 2 \\
2 & 1 \quad 0 \quad 3 \\
3 & 2 \quad 3 \quad 0.
\end{array}
$$

Use the lower and upper bounding procedures in Exercises 4 and 5 to construct upper and lower bounds on (4.4) for these data.

Part V

A Case Study

12

Capacity Expansion

This chapter presents a case study of a stochastic programming study for a manufacturing firm. The study was to determine a method for allocating capacity for different products in different plants. This process is known as *adding flexible capacity*. In general, flexible capacity represents primarily option value and cannot be evaluated by standard methods of discounting cash flows (see, e.g., Myers [1984]). Because of correlations among the random variables, it is necessary to evaluate recourse actions, which are then used in the analysis.

This chapter's study is similar to the earlier study by Eppen, Martin, and Schrage [1989] in terms of our development. We follow the same steps although our risk characterization is somewhat different from that of Eppen and coworkers.

The chapter is divided into four sections. In the first section, we describe the model development and the use of an option value model perspective for the evaluation of flexible capacity. The second section describes the method for approximating the distribution of the random variables. A key feature here is that the distribution is not completely known so that methods based on the moment problem solutions in Section 9.5 are necessary. The third section compares solution methods for varying numbers of scenarios, while the fourth section presents analysis of the results and suggestions for future studies.

12.1   Model Development

The manufacturer faces the situation of having certain plants with installed capacity to produce specific products. The problem is to determine whether additional capacity should be installed at a plant where no capacity for a product currently exists. This additional capacity will allow the plant to continue production if demand for the new product is higher than existing capacity at other plants and if the demand for products at the new plant is lower than the existing plant capacity.

As an example, consider Figure 1. Here, there are two products, A and B, and three plants, 1, 2, and 3. The solid lines in the diagram indicate that each plant currently produces only a single product. We could assume that each of the plants is built to meet the mean demand exactly. In that case, if demand for a product exceeds the mean, potential sales are lost.

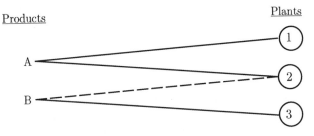

FIGURE 1. Adding flexible capacity at Plant 2.

By building additional capacity at Plant 2 for Product B (the dotted line in Figure 1), if demand for product A is lower than the mean and demand for product B is higher than its mean, then the excess Product B demand can be produced at Plant 2 and fewer sales would be lost. That is the basic goal of flexible capacity. Other relevant measures (apart from net expected value gains) are the use of the plants (the fraction of installed capacity actually used) and the number of sales lost for each product type.

The decision problem is to trade off the costs of adding capacity against the potential revenue from additional sales due to the extra capacity. As mentioned in the introduction, this basic problem has been considered by Eppen et al., who applied a mixed integer, stochastic linear programming model. A more general study in the context of flexible capacity appears in Fine and Freund [1990].

The model considered here is the form used in the analysis by Jordan and Graves [1991]. This model includes eight plants and sixteen products. The original installed capacities for the products at each of the plants are indicated in Figure 2. The total capacity in these plants and the expected demands are given in Table 1.

Given this network characterization, the model is to determine where to install additional capacity to maximize the value added to the firm by

Products   Plants

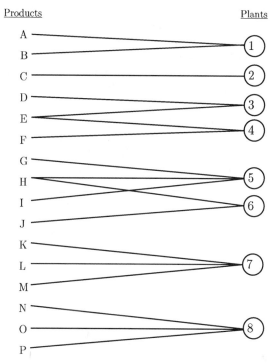

FIGURE 2. Original product-plant installed capacity.

TABLE 1. Total plant capacities and expected demands.

Plant	Capacity (1000s)	Product	Mean Demand (1000s)	Product	Mean Demand (1000s)
1	380	A	320	I	140
2	230	B	150	J	160
3	250	C	270	K	60
4	230	D	110	L	35
5	240	E	220	M	40
6	230	F	110	N	35
7	230	G	120	O	30
8	240	H	80	P	180

these capacity decisions. A standard net present value approach for this is, however, inadequate. In such an approach, the expected revenue for a capacity decision would be calculated and then discounted to the present to determine a net present value. Difficulties with doing this, however, are that the correlations among the products and other capacity decisions make a sequential analysis of independent capacity decisions inadequate and asymmetric revenue character of a flexible capacity return does not match the requirements for a net present value.

For this latter reasoning, consider Figure 3. In this figure, we show potential demand and expected revenues as linearly related so that each product has some fixed operating profit (excluding fixed capacity costs). The distribution of the potential demand is also indicated in Figure 3. The result is that revenues are not symmetric but are essentially truncated at the capacity limit. With revenues of this form, a direct use of the capital asset pricing model (see, e.g., Sharpe [1964]) to determine a discount factor cannot be fully justified. Instead we need to use an option value approach. (For basic background, see Jarrow and Rudd [1983]. A detailed explanation of the use of options in capacity planning appears in Birge [1995].)

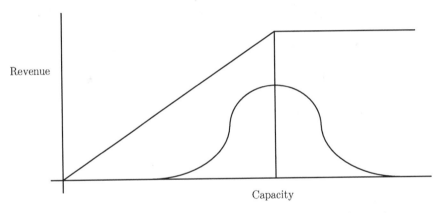

FIGURE 3. Demand and revenue relationship with limited capacity.

The diagram of returns in Figure 3 has the basic form of the position diagram of holding an asset and selling a call option at the value of the capacity. The call option would allow an investor to purchase the revenue from potential sales at the fixed price of the revenue from sales at full capacity. The call, therefore, has value whenever the demand exceeds the capacity.

With multiple products, the revenue from combinations of the sales becomes a more complicated type of option (or, in general, contingency claim), but it can still be evaluated. In some cases, in fact (see Andreou [1990]), analytical formulas may exist to evaluate the benefit of flexible capacity.

Our general approach is to assume, as in the Black-Scholes model (see Black and Scholes [1973] and Cox and Ross [1976]), that a riskless *hedging* strategy is possible. This assumption requires the ability to trade both the option and the underlying asset in perfectly divisible amounts. While such a market is most likely not possible in all asset categories, it does appear to represent a fairly consistent approximation of value calculations and is used for the purposes here.

Given the *risk-neutral* assumption and other basic assumptions (see Jarrow and Rudd [1983], Hull [1997]), an option value can be calculated by assuming that the value of an option in a risk-neutral environment is the same as an option value in a setting with risk. In this way, we can evaluate the asset as if it grew at the riskfree rate and find an option value under this situation that will be equivalent to the option value in the situation with risk aversion.

To carry out this analysis for the situation here, we must calculate the current value of all potential sales (revenue from any potential demand without capacity restrictions) and then assume that this value grows at the riskfree rate. To do so, we would discount sales without capacity restriction at a discount rate consistent with the total market risk (variance) of sales. Suppose this rate is r; then revenue in time t is discounted by a factor $(\frac{1}{r})^t$. Now, assuming risk neutrality, the value of these sales would increase at an annual rate of r_f, the riskfree rate.

The net effect of these two adjustment is that the revenue from Period t sales corresponds to a demand distribution that is shifted by a multiplier, $(\frac{1+r_f}{1+r})^t$ (see Figure 4). Note that if the capacity is high, the revenue effect is also just a shift by $(\frac{1+r_f}{1+r})^t$, but this is now discounted at the riskfree rate by $(\frac{1}{1+r_f})^t$. The original discount factor, $(\frac{1}{r})^t$, is then retained.

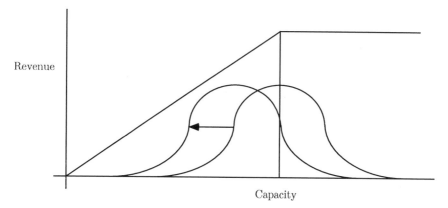

FIGURE 4. Demand distribution shift.

The more interesting effect of this shift is when capacity is low. In fact, if capacity is very low, the revenue is effectively deterministic. In this case, when the demand distribution is shifted, it has no effect on revenue. However, in this case, the revenues are again discounted with the riskfree factor, $(\frac{1}{1+r_f})^t$. The result is that the low-capacity (and hence low-risk) revenue stream is discounted at the riskfree rate.

This model clearly is consistent in these two extremes of abundant or very limited capacity. When capacity limits some but not all potential revenue, the effect is a moderate discounting. In this way, risk can be captured in terms of overall market effect. Other risk measures, such as downside risk, can also be used but generally require more complicated models.

In our model, instead of shifting the demand distribution and discounting at the riskfree rate, it is often simpler to increase the capacity by multiplying the capacity by $(\frac{1+r}{1+r_f})^t$ and then discounting with $(\frac{1}{r})^t$. This approach is used in this example because fewer data needed modification than would be required with demand shifts. Capacities in the following formulations are assumed to be adjusted in this way.

With this framework for assessing risk, we can now describe the stochastic programming formulation of this model. The basic decisions are to determine whether to equip plants with new fixtures. This decision corresponds to determining whether a product is produced at a plant.

We describe the model in the AMPL formulation that appears in Figure 5. The "SETS" in that model are the names of the products and plants. The parameters include the number of scenarios, the regular and overtime capacity of the plants, the demand for each product under each scenario, the cost of additional fixtures for each product to be produced at each plant, the capital budget, the operating profit on each product, the additional cost for overtime production, and the probabilities of the scenarios. We represent discounting with an amortization factor. This factor is varied according to the lifetime of each product.

The current model assumes that each product will be produced with the same demand pattern through the same time horizon. The model also assumes that capacity decisions only occur now and not in the future. In this way, we essentially build a two-stage model although it may actually encompass several periods of demand. In this way, sales in each period are discounted by some factor $\rho(H)$ that depends on the horizon length H and the discount rate r. We divide the objective by $\gamma(H)$ so that the first-stage decision is multiplied by $amort = \frac{1}{\gamma(H)}$ and the second-stage decision corresponds to a single year's revenues. This was again used to modify as few data items as necessary.

The most fundamental decision is to determine where each product may be produced. We designate this binary variable variable as $y(plant, product)$. The next variables determine regular and overtime production of each product at each plant and in each scenario.

The objective to maximize is simply the expected revenues with a regular operating profit factor minus the additional costs for overtime production

```
# This is a capacity planning model for assigning products to plants.
# SETS
set product;
set plant;
# PARAMETERS
param no_scenarios;
param capacity{plant}; # CAPACITY OF PLANT

param ot_capacity{plant}; # OVERTIME CAPACITY

param demand{product,1..no_scenarios}; # DEMAND FOR PRODUCT
# UNDER SCENARIO
param cap_cost{plant,product}; # CAPITAL COST
param cap_budget; # CAPITAL BUDGET
param profit{product}; # PROFIT FROM EACH PRODUCT

param otcost{product}; # OVERTIME OPERATING COSTS

param prob{1..no_scenarios}; # PROBABILITY OF SCENARIO
param amort; # AMORTIZATION FACTOR
#
var y{plant,product}>= 0; # 1 = j IS PRODUCED AT i, 0 OTHERWISE
#integer y
var reg_prod{plant,product,1..no_scenarios}>= 0; #REGULAR
# PRODUCTION
var ot_prod{plant,product,1..no_scenarios} >= 0; #OVERTIME
# PRODUCTION
# OBJECTIVE
maximize opt_val:
- amort*cap_cost[i,j]*y[i,j] + sum{i in plant} sum{j in product} sum{k in

1..no_scenarios} (prob[k]*profit[j]*(reg_prod[i,j,k] + ot_prod[i,j,k])
- otcost[j]*prob[k]*ot_prod[i,j,k])
;
# CONSTRAINTS
subject to
#
under_bud{k in 1..no_scenarios}:
sum{i in plant}sum{j in product} y[i,j]*cap_cost[i,j] <= cap_budget;
#
no_gdemand{j in product, k in 1..no_scenarios}:
sum{i in plant} (reg_prod[i,j,k] + ot_prod[i,j,k]) <= demand[j,k];
#
j_atplant{i in plant,j in product, k in 1..no_scenarios}:
(reg_prod[i,j,k] + ot_prod[i,j,k]) <= y[i,j]*(capacity[i]
+ ot_capacity[i]);
#
r_eq_rcap{i in plant, k in 1..no_scenarios}:
sum{j in product} reg_prod[i,j,k] <= capacity[i];
#
ot_eq_otc{i in plant, k in 1..no_scenarios}:
sum{j in product} ot_prod[i,j,k] <= ot_capacity[i];
```

FIGURE 5. AMPL model of capacity problem.

and the stage-one capital costs. The constraints state that the capital expenditure is under budget. They also limit production of specific products by demand and plant capacity multiplied by the binary variable of whether the product can be produced at a given plant. Total regular and overtime production within a plant are also limited.

Given the model in Figure 5, the next problem is to determine the data to drive the model. The basic assumption is that the data concerning costs and operating profits are relatively well known. Some variations might occur, but it is generally decided that these changes can be accommodated by varying the parameters after solving an initial model. The major driving random factor is considered to be demand. Thus, much of the stochastic modeling effort considers demand scenario generation as described in the next section.

12.2   Demand Distribution Modeling

Demand forecasting in this situation is a clear case of limited distribution information. Historical data may exist on some older products but newer product forecasts must be based on other techniques. In general, the level of information is at most an expected value for demand on each product, an estimate of a variance, and some general guidelines for correlation coefficients. No other distributional information would be available.

The general variance information followed Jordan and Graves [1991] in assuming 40% of the expected demand as a standard deviation for a single product demand. Correlation coefficients were assumed to be 0.3 for products in groups, A–F, G–M, and N–P. For products from different groups, a 0.0 correlation coefficient was assumed.

To accommodate this limited distribution information, the basic approximation framework of a generalized moment problem as in Section 9.5 is used. Instead of building a bound, however, the goal was to find a fixed a number of points that match the given moments as closely as possible. The basic generalized linear programming is used to first pick the set of points and then find weights to minimize deviations from the moment values.

New points are generated by a local optimization procedure (which is not necessarily optimal because the problem is not convex). Full optimization of the scenario set is not considered necessary due to the lack of precision in the moment estimates. The resulting solutions, with as few as six scenarios, gives fairly consistent results, although fewer scenarios lead to some loss of the hedging nature of the stochastic solution. Other types of sensitivity analyses are described in Section 4.

TABLE 2. Capacity planning solution times.

Stages	Scenarios	Rows	Columns	OSL (CPUs)	ND-UM (CPUs)
2	4	478	966	4.7	1.7
2	8	894	1.8E03	11.9	2.5
2	16	1.7E03	3.5E03	35.4	3.6
3	36	4.4E03	9.0E03	230.7	15.2
3	256	2.8E04	5.7E04	12361	140.5
4	4096	4.5E05	9.2E05	Failed	5024

12.3 Computational Comparisons

For the computational experiments, the basic procedure uses the integer programming capability of IBM's OSL (IBM [1991]). This software enabled solutions of relatively large capacity expansion models, although most analyses considered the six-scenario problem in order to have relatively quick turnaround times for varying parameter combinations on the RS6000 processors.

The linear programming relaxations of the basic model in Figure 5 were also solved using the nested decomposition method, ND-UM (Birge et al. [1994]), which uses the OSL solver for each subproblem. Because the OSL solver also follows the integer programming path, we only present the linear programming times for varying numbers of scenarios. For this experiment we also consider multiple-stage versions of the problem, although we did not solve the integer versions of these multiple-stage stochastic programs. In the multistage problems, the decisions in each period can consider previous capacity installations with possible learning about the form of demand and inventories carried from one period to the next. These interperiod ties mean that block separability does not apply.

Table 2 indicates a clear advantage for decomposition in these problems. The results on large problems represent orders of magnitude speed-ups over the straightforward simplex method approach in OSL. While these results do not occur in all practical problems, they are illustrative of the potential for special-purpose stochastic programming codes. The next section considers the analysis of the solutions.

12.4 Results and Extensions

The solution process began by assuming a given value of the amortization factor and then trying to establish a stable scenario set with which the solution did not change as more scenarios are included. With this set established, the amortization factor was varied to observe possible effects from early product retirement or lengthy product lifetimes.

At high amortization factors, products have short lifetimes and little incentive for flexible capacity exists. In this model, only a single additional capacity (J at 7) was added for amortization close to one. For amortizations of one-half, two additional capabilities are added (P at 5 and G at 3). As the amortization factor declines, more capacity is added. When *amort* is 0.2, a total of nine capacities are added. At *amort*= 0.1, 11 capabilities are added.

These results, in some sense, rank the priority for adding additional capacity. This information was considered for the actual capacity decision.

Other factors are also considered to evaluate the capacity decision completely. The expected number of lost sales and expected utilizations are also considered. These values give the decision makers more information about the value of flexible capacity. In general, flexible capacity increases both values. In this model, utilizations increased to 98% and lost sales declined by 80% as flexibility was added up to amortization factors of 0.1.

This model represents an important application of stochastic programming in capacity planning. The actual model was provided to a manufacturer for use in evaluating the effect of flexible capacity. Some other extensions, such as intermediate-stage capacity additions and variable product lifetimes, could represent the actual situation more fully, but this simple model gives much of the information used for actual decision making.

Appendix A
Sample Distribution Functions

This appendix gives the basic distributions used in the text. We provide their means and variances.

A.1 Discrete Random Variables

Uniform: U[1,n]

$$P(\boldsymbol{\xi} = i) = \frac{1}{n}, i = 1, \ldots, n, n \geq 1,$$

with $E[\boldsymbol{\xi}] = \frac{n+1}{2}$ and $Var[\boldsymbol{\xi}] = \frac{n^2-1}{12}$.

Binomial: Bi(n,p)

$$P(\boldsymbol{\xi} = i) = \binom{n}{i} p^i (1-p)^{n-i}, i = 0, 1, \ldots, n; 0 < p < 1,$$

with $E[\boldsymbol{\xi}] = np$ and $Var[\boldsymbol{\xi}] = np(1-p)$.

Poisson: P(λ)

$$P(\boldsymbol{\xi} = i) = e^{-\lambda} \frac{\lambda^i}{i!}, \lambda > 0, i = 0, 1, \ldots ,$$

with $E[\boldsymbol{\xi}] = \lambda$ and $Var[\boldsymbol{\xi}] = \lambda$.

A.2   Continuous Random Variables

Uniform: U[0,a]

$$f(\xi) = \frac{1}{a}, 0 \le \xi \le a, a > 0,$$

with $E[\xi] = \frac{a}{2}$ and $Var[\xi] = \frac{a^2}{12}$.

Exponential: exp(λ)

$$f(\xi) = \lambda e^{-\lambda \xi}, 0 \le \xi, \lambda > 0,$$

with $E[\xi] = \frac{1}{\lambda}$ and $Var[\xi] = (\frac{1}{\lambda})^2$.

Normal: N(μ,σ^2)

$$f(\xi) = \frac{1}{\sqrt{2\pi\sigma^2}} e^{-\frac{(\xi-\mu)^2}{2\sigma^2}}, \sigma > 0,$$

with $E[\xi] = \mu$ and $Var[\xi] = \sigma^2$.

Gamma: G(α, β)

$$f(\xi) = \frac{1}{\beta^2 \Gamma(\alpha)} \xi^{\alpha-1} e^{\frac{-\xi}{\beta}}, \alpha > 0, \beta > 0,$$

where $\Gamma(\alpha) = \int_0^\infty x^{\alpha-1} e^{-x} dx, \alpha > 0$, $E[\xi] = \alpha\beta$ and $Var[\xi] = \alpha\beta^2$.

References

[1] P.G. Abrahamson, "A Nested Decomposition Approach for Solving Staircase Linear Programs," Ph.D. Dissertation, Stanford University (1983).

[2] S.A. Andreou, "A capital budgeting model for product-mix flexibility," *Journal of Manufacturing and Operations Management* 3 (1990) pp. 5–23.

[3] K.M. Anstreicher, "A combined Phase I–Phase II projective algorithm for linear programming," *Mathematical Programming* 43 (1989) pp. 209–223.

[4] K.M. Anstreicher, "A standard form variant, and safeguarded linesearch, for the modified Karmarkar algorithm," *Mathematical Programming* 47 (1990) pp. 337–351.

[5] K.M. Anstreicher, "Strict monotonicity and improved complexity in the standard form projective algorithm for linear programming," *Mathematical Programming* 62 (1993) pp. 517–536.

[6] K.A. Ariyawansa and D.D. Hudson, "Performance of a benchmark parallel implementation of the Van Slyke and Wets algorithm for two-stage stochastic programs on the Sequent/Balance," *Concurrency Practice and Experience* 3 (1991) pp. 109–128.

[7] R. Ashford, "Bounds and an approximate solution method for multi-stage stochastic production problems," Warwick Papers in Industry, Business and Administration, No. 15, University of Warwick, Coventry, UK (1984).

[8] H. Attouch and R.J Wets, "Approximation and convergence in nonlinear optimization" in: O.L. Mangasarian, R.R. Meyer and S.M. Robinson, Eds., *Nonlinear programming, 4* (Academic Press, New York–London, 1981) pp. 367–394.

[9] M. Avriel and A.C. Williams, "The value of information and stochastic programming," *Operations Research* 18 (1970) pp. 947–954.

[10] E.R. Barnes, "A variation on Karmarkar's algorithm for solving linear programming problems," *Mathematical Programming* 36 (1986) pp. 174–182.

[11] M.S. Bazaraa and C.M. Shetty, *Nonlinear Programming: Theory and Algorithms* (John Wiley, Inc., New York, NY, 1979).

[12] M.S. Bazaraa, J.J. Jarvis, and H.D. Sherali, *Linear Programming and Network Flows* (John Wiley, Inc., New York, NY, 1990).

[13] E.M.L. Beale, "On minimizing a convex function subject to linear inequalities," *J. Royal Statistical Society, Series B* 17 (1955) pp. 173–184.

[14] E.M.L. Beale, "The use of quadratic programming in stochastic linear programming," Rand Report P-2404-1, The Rand Corporation (1961).

[15] E.M.L. Beale, J.J.H. Forrest, and C.J. Taylor, "Multi-time-period stochastic programming" in: M.A.H. Dempster, Ed., *Stochastic Programming* (Academic Press, New York, NY, 1980) pp. 387–402.

[16] E.M.L. Beale, G.B. Dantzig, and R.D. Watson, "A first order approach to a class of multi-time-period stochastic programming problems," *Mathematical Programming Study* 27 (1986) pp. 103–117.

[17] R. Bellman, *Dynamic Programming* (Princeton University Press, Princeton, NJ, 1957).

[18] A. Ben-Tal and M. Teboulle, "Expected utility, penalty functions, and duality in stochastic nonlinear programming," *Management Science* 32 (1986) pp. 1445–1466.

[19] J. F. Benders, "Partitioning procedures for solving mixed-variables programming problems," *Numerische Mathematik* 4 (1962) pp. 238–252.

[20] B. Bereanu, "Some numerical methods in stochastic linear programming under risk and uncertainty" in: M.A.H. Dempster, Ed., *Stochastic Programming* (Academic Press, New York, NY, 1980) pp. 169–205.

[21] J.O. Berger, *Statistical Decision Theory and Bayesian Analysis* (Springer-Verlag, New York, NY, 1985).

[22] O. Berman, R.C. Larson, and S.S. Chiu, "Optimal server location on a network operating as a M/G/1 queue," *Operations Research* 33 (1985) pp. 746–770.

[23] D. Bertsimas, P. Jaillet, and A. Odoni, "A priori optimization," *Operations Research* 38 (1990) pp. 1019–1033.

[24] D. Bienstock and J.F. Shapiro, "Optimizing resource acquisition decisions by stochastic programming," *Management Science* 34 (1988) pp. 215–229.

[25] P. Billingsley, *Convergence of Probability Measures* (John Wiley, Inc., New York, NY, 1968).

[26] J.R. Birge, "Solution Methods for Stochastic Dynamic Linear Programs," Ph.D. Dissertation and Technical Report SOL 80-29, Systems Optimization Laboratory, Stanford University (Stanford, CA 94305, 1980).

[27] J.R. Birge, "The value of the stochastic solution in stochastic linear programs with fixed recourse," *Mathematical Programming* 24 (1982) pp. 314–325.

[28] J.R. Birge, "Using sequential approximations in the L-shaped and generalized programming algorithms for stochastic linear programs," Technical Report 83-12, Department of Industrial and Operations Engineering, University of Michigan (Ann Arbor, MI, 1983).

[29] J.R. Birge, "Aggregation in stochastic production problems," *Proceedings of the 11th IFIP Conference on System Modelling and Optimization* (Springer-Verlag, New York, 1984).

[30] J.R. Birge, "Aggregation in stochastic linear programming," *Mathematical Programming* 31 (1985a) pp. 25–41.

[31] J.R. Birge, "Decomposition and partitioning methods for multi-stage stochastic linear programs," *Operations Research* 33 (1985b) pp. 989–1007.

[32] J.R. Birge, "Exhaustible recourse models with uncertain returns from exploration investment" in: Y. Ermoliev and R. Wets, Eds., *Numerical Techniques for Stochastic Optimization* (Springer-Verlag, Berlin, 1988a) pp. 481–488.

[33] J.R. Birge, "The relationship between the L-shaped method and dual basis factorization for stochastic linear programming" in: Y. Ermoliev and R. Wets, Eds., *Numerical Techniques for Stochastic Optimization* (Springer-Verlag, Berlin, 1988b) pp. 267–272.

[34] J.R. Birge, "Multistage stochastic planning models using piecewise linear response functions" in: G. Dantzig and P. Glynn, Eds., *Resource Planning under Uncertainty for Electric Power Systems* (NSF, 1989).

[35] J.R. Birge, "Quasi-Monte Carlo methods for option evaluation," Technical Report, Department of Industrial and Operations Engineering, University of Michigan (Ann Arbor, MI, 1994).

[36] J.R. Birge, "Option methods for incorporating risk into linear planning models," Technical Report 95-8, Department of Industrial and Operations Engineering, University of Michigan (Ann Arbor, MI, 1995).

[37] J.R. Birge and M.A.H. Dempster, "Optimality conditions for match-up strategies in stochastic scheduling and related dynamic stochastic optimization problems," Technical Report 92-58, Department of Industrial and Operations Engineering, University of Michigan (Ann Arbor, MI, 1992).

[38] J.R. Birge, C.J. Donohue, D.F. Holmes, and O.G. Svintsiski, "A parallel implementation of the nested decomposition algorithm for multistage stochastic linear programs," Technical Report 94-1, Department of Industrial and Operations Engineering, University of Michigan (Ann Arbor, MI, 1994), also *Mathematical Programming* 75 (1996), pp. 327–352.

[39] J.R. Birge and J. Dulá, "Bounding separable recourse functions with limited distribution information," *Annals of Operations Research* 30 (1991) pp. 277–298.

[40] J.R. Birge, R.M. Freund, and R.J. Vanderbei, "Prior reduced fill-in in the solution of equations in interior point algorithms," *Operations Research Letters* 11 (1992) pp. 195–198.

[41] J.R. Birge and D.F. Holmes, "Efficient solution of two-stage stochastic linear programs using interior point methods," *Computational Optimization and Applications* 1 (1992) pp. 245–276.

[42] J.R. Birge and F.V. Louveaux, "A multicut algorithm for two-stage stochastic linear programs," *European Journal of Operations Research* 34 (1988) pp. 384–392.

[43] J.R. Birge and M.J. Maddox, "Bounds on Expected Project Tardiness," *Operations Research* 43 (1995) pp. 838–850.

[44] J.R. Birge and M.J. Maddox, "Using second moment information in stochastic scheduling" in: G. Yin and Q. Zhang, Eds., *Recent Advances in Control and Manufacturing Systems* (Springer-Verlag, New York, NY, 1996) pp. 99–120.

[45] J.R. Birge and L. Qi, "Computing block-angular Karmarkar projections with applications to stochastic programming," *Management Science* 34 (1988) pp. 1472–1479.

[46] J.R. Birge and L. Qi, "Semiregularity and generalized subdifferentials with applications to optimization," *Mathematics of Operations Research* 18 (1993) pp. 982–1006.

[47] J.R. Birge and L. Qi, "Subdifferential convergence in stochastic programs," *SIAM J. Optimization* 5 (1995) pp. 436–453.

[48] J.R. Birge and M. Teboulle, "Upper bounds on the expected value of a convex function using subgradient and conjugate function information," *Mathematics of Operations Research* 14 (1989) pp. 745–759.

[49] J.R. Birge and S.W. Wallace, "Refining bounds for stochastic linear programs with linearly transformed independent random variables," *Operations Research Letters* 5 (1986) pp. 73–77.

[50] J.R. Birge and S.W. Wallace, "A separable piecewise linear upper bound for stochastic linear programs," *SIAM Journal on Control and Optimization* 26 (1988) pp. 725–739.

[51] J.R. Birge and R.J-B Wets, "Approximations and error bounds in stochastic programming" in: Y. Tong, Ed., *Inequalities in Statistics and Probability* (IMS Lecture Notes—Monograph Series, 1984) pp. 178–186.

[52] J.R. Birge and R.J-B Wets, "Designing approximation schemes for stochastic optimization problems, in particular, for stochastic programs with recourse," *Mathematical Programming Study* 27 (1986) pp. 54–102.

[53] J.R. Birge and R.J-B Wets, "Computing bounds for stochastic programming problems by means of a generalized moment problem," *Mathematics of Operations Research* 12 (1987) pp. 49–162.

[54] J.R. Birge and R.J-B Wets, "Sublinear upper bounds for stochastic programs with recourse," *Mathematical Programming* 43 (1989) pp. 131–149.

[55] G.R. Bitran and D. Sarkar, "On upper bounds of sequential stochastic production planning problems," *European Journal of Operational Research* 34 (1988) pp. 191–207.

[56] G.R. Bitran and H. Yanasse, "Deterministic approximations to stochastic production problems," *Operations Research* 32 (1984) pp. 999–1018.

[57] C.E. Blair and R.G. Jeroslow, "The value function of an integer program," *Mathematical Programming* 23 (1982) pp. 237–273.

[58] F. Black and M. Scholes, "The pricing of options and corporate liabilities," *Journal of Political Economy* 81 (1973) pp. 737–654.

[59] D. Blackwell, "Discounted dynamic programming," *Annals of Mathematical Statistics* 36 (1965) pp. 226–235.

[60] C. Borell, "Convex set functions in d-spaces," *Periodica Mathematica Jungarica* 6 (1975) pp. 111–136.

[61] S.L. Brumelle and J.I. McGill, "Airline seat allocation with multiple nested fare classes," *Operations Research* 41 (1993) pp. 127–137.

[62] C.C. Carøe and J. Tind, "L-shaped decomposition of two-stage stochastic programs with integer recourse," *Mathematical Programming* 83 (1998) pp. 407–424; Technical Report, Institute of Mathematics, University of Copenhagen (Copenhagen, Denmark, 1996).

[63] T. Carpenter, I. Lustig, and J. Mulvey, "Formulating stochastic programs for interior point methods," *Operations Research* 39 (1991) pp. 757–770.

[64] H.P. Chao, "Exhaustible resource models: the value of information," *Operations Research* 29 (1981) pp. 903–923.

[65] A. Charnes and W.W. Cooper, "Chance-constrained programming," *Management Science* 5 (1959) pp. 73–79.

[66] A. Charnes and W.W. Cooper, "Deterministic equivalents for optimizing and satisficing under chance constraints," *Operations Research* 11 (1963) pp. 18–39.

[67] A. Charnes and W.W. Cooper, "Response to 'Decision problems under risk and chance constrained programming: dilemmas in the transition'," *Management Science* 29 (1983) pp. 750–753.

[68] A. Charnes, W.W. Cooper, and G.H. Symonds, "Cost horizons and certainty equivalents: an approach to stochastic programming of heating oil," *Management Science* 6 (1958) pp. 235–263.

[69] I.C. Choi, C.L. Monma, and D.F. Shanno, "Further development of a primal-dual interior point method," *ORSA Journal on Computing* 2 (1990) pp. 304–311.

[70] E. Chu, A. George, J. Liu, and E. Ng, "SPARSPAK: Waterloo sparse matrix package user's guide for SPARSPAK-A," Research Report CS-84-36, Department of Computer Science, University of Waterloo (Waterloo, Ontario, 1984).

[71] K. L. Chung, *A Course in Probability Theory* (Academic Press, New York, NY, 1974).

[72] V. Chvátal, *Linear Programming* (Freeman, New York/San Francisco, CA, 1980).

[73] T. Cipra, "Moment problem with given covariance structure in stochastic programming," *Ekonom.-Mat. Obzor* 21 (1985) pp. 66–77.

[74] T. Cipra, "Stochastic programming with random processes," *Annals of Operations Research* 30 (1991) pp. 95–105.

[75] F. Clarke, *Optimization and Nonsmooth Analysis* (John Wiley, Inc., New York, NY, 1983).

[76] J. Cox and S. Ross, "The valuation of options for alternative stochastic processing," *Journal of Financial Economics* 3 (1976) pp. 145–166.

[77] J. Czyzyk, R. Fourer, and S. Mehrotra, "A study of the augmented system and column-splitting approaches for solving two-stage stochastic linear programs by interior-point methods," *ORSA Journal on Computing* 7 (1995) pp. 474–490.

[78] G.B. Dantzig, "Linear programming under uncertainty," *Management Science* 1 (1955) pp. 197–206.

[79] G.B. Dantzig, *Linear Programming and Extensions* (Princeton University Press, Princeton, NJ, 1963).

[80] G.B. Dantzig and P. Glynn, "Parallel processors for planning under uncertainty," *Annals of Operations Research* 22 (1990) pp. 1–21.

[81] G.B. Dantzig and G. Infanger, "Large-scale stochastic linear programs— Importance sampling and Benders decomposition" in: C. Brezinski and U. Kulisch, Eds., *Computational and applied mathematics, I (Dublin, 1991)* (North-Holland, Amsterdam, 1991) pp. 111–120.

[82] G.B. Dantzig and A. Madansky, "On the solution of two–stage linear programs under uncertainty," Proceedings of the Fourth Berkeley Symposium on Mathematical Statistics and Probability, (University of California Press, Berkeley, CA, 1961).

[83] G.B. Dantzig and A. Wald, "On the fundamental lemma of Neyman and Pearson," *The Annals of Mathematical Statistics* 22 (1951) pp. 87–93.

[84] G.B. Dantzig and P. Wolfe, "The decomposition principle for linear programs," *Operations Research* 8 (1960) pp. 101–111.

[85] D. Dawson and A. Sankoff, "An inequality for probabilities," *Proceedings of the American Mathematical Society* 18 (1967) pp. 504–507.

[86] I. Deák, "Three-digit accurate multiple normal probabilities," *Numerische Mathematik* 35 (1980) pp. 369–380.

[87] I. Deák, "Multidimensional integration and stochastic programming," in: Y. Ermoliev and R. Wets, Eds., *Numerical Techniques for Stochastic Optimization* (Springer-Verlag, Berlin, 1988) pp. 187–200.

[88] I. Deák, *Random Number Generators and Simulation* (Akadémiai Kiadó, Budapest, 1990).

[89] M.H. DeGroot, *Optimal Statistical Decisions* (McGraw-Hill, New York, NY, 1970).

[90] M.A.H. Dempster, "Introduction to Stochastic Programming" in: M.A.H. Dempster, Ed., *Stochastic Programming* (Academic Press, New York, NY, 1980) pp. 3–59.

[91] M.A.H. Dempster, "The expected value of perfect information in the optimal evolution of stochastic problems" in: M. Arato, D. Vermes, and A.V. Balakrishnan, Eds., *Stochastic Differential Systems* (Lecture Notes in Information and Control, Vol. 36, 1981) pp. 25–40.

[92] M.A.H. Dempster, "On stochastic programming II: dynamic problems under risk," *Stochastics* 25 (1988) pp. 15–42.

[93] M.A.H. Dempster and Papagaki-Papoulias, "Computational experience with an approximate method for the distribution problem" in: M.A.H. Dempster, Ed., *Stochastic Programming* (Academic Press, New York, NY, 1980) pp. 223–243.

[94] V.F. Demyanov and L.V. Vasiliev, *Nedifferentsiruemaya optimizatsiya (Nondifferentiable optimization)* (Nauka, Moscow, 1981).

[95] I.I. Dikin, "Iterative solution of problems of linear and quadratic programming," *Soviet Mathematics Doklady* 8 (1967) pp. 674–675.

[96] J.H. Dulá, "An upper bound on the expectation of simplicial functions of multivariate random variables," *Mathematical Programming* 55 (1991) pp. 69–80.

[97] V. Dupač, "A dynamic stochastic approximation method," *Annals of Mathematical Statistics* 6 (1965) pp. 1695–1702.

[98] J. Dupačová, "Minimax stochastic programs with nonconvex nonseparable penalty functions" in: A. Prékopa, Ed., *Progress in Operations Research* (Janos Bolyai Math. Soc., 1976) pp. 303–316.

[99] J. Dupačová, "The minimax approach to stochastic linear programming and the moment problem," *Ekonom.-Mat. Obzor* 13 (1977) pp. 297–307.

[100] J. Dupačová, "Stability in stochastic programming with recourse-contaminated distributions," *Mathematical Programming Study* 28 (1984) pp. 72–83.

[101] J. Dupačová, "Stability and sensitivity analysis for stochastic programming," *Annals of Operations Research* 27 (1990) pp. 115–142.

[102] J. Dupačová and R.J-B Wets, "Asymptotic behavior of statistical estimators and of optimal solutions of stochastic optimization problems," *Annals of Statistics* 16 (1988) pp. 1517–1549.

[103] B.C. Eaves and W.I. Zangwill, "Generalized cutting plane algorithms," *SIAM J. Control* 9 (1971) pp. 529–542.

[104] N.C.P. Edirisinghe, "Essays on Bounding Stochastic Programming Problems," Ph.D. Dissertation, The University of British Columbia (1991).

[105] N.C.P. Edirisinghe, "New second-order bounds on the expectation of saddle functions with applications to stochastic linear programming," *Operations Research* 44 (1996) pp. 909–922.

[106] H.P. Edmundson, "Bounds on the expectation of a convex function of a random variable," RAND Corporation Paper 982, Santa Monica, CA (1956).

[107] M. Eisner and P. Olsen, "Duality for stochastic programming interpreted as l.p. in L_p-space," *SIAM Journal of Applied Mathematics* 28 (1975) pp. 779–792.

[108] G.D. Eppen, R.K. Martin, and L. Schrage, "A scenario approach to capacity planning," *Operations Research* 37 (1989) pp. 517–527.

[109] Y. Ermoliev, "On the stochastic quasigradient method and quasi-Feyer sequences," *Kibernetika* 5 (2) (1969) pp. 73–83 (in Russian; also published in English as *Cybernetics* 5 (1969) pp. 208–220).

[110] Y. Ermoliev, *Methods of Stochastic Programming* (Nauka, Moscow (in Russian) 1976).

[111] Y. Ermoliev, "Stochastic quasigradient methods and their applications to systems optimization," *Stochastics* 9 (1983) pp. 1–36.

[112] Y. Ermoliev, "Stochastic quasigradient methods." (SC) in: Y. Ermoliev and R. Wets, Eds., *Numerical Techniques for Stochastic Optimization* (Springer-Verlag, Berlin, 1988) pp. 141–186.

[113] Y. Ermoliev, A. Gaivoronski, and C. Nedeva, "Stochastic optimization problems with partially known distribution functions," *SIAM Journal on Control and Optimization* 23 (1985) pp. 377–394.

[114] Y. Ermoliev and R. Wets, "Introduction" in: Y. Ermoliev and R. Wets, Eds., *Numerical Techniques for Stochastic Optimization* (Springer-Verlag, Berlin, 1988).

[115] L.F. Escudero, P.V. Kamesam, A.J. King, and R.J-B Wets, "Production planning via scenario modeling," *Annals of Operations Research* 43 (1993) pp. 311–335.

[116] W. Feller, *An Introduction to Probability Theory and Its Applications* (John Wiley, Inc., New York, NY, 1971).

[117] A. Ferguson and G.B. Dantzig, "The allocation of aircraft to routes: an example of linear programming under uncertain demands," *Management Science* 3 (1956) pp. 45–73.

[118] C.H. Fine and R.M. Freund, "Optimal investment in product-flexible manufacturing capacity," *Management Science* 36 (1990) pp. 449–466.

[119] S.D. Flåm, "Nonanticipativity in stochastic programming," *Journal of Optimization Theory and Applications* 46 (1985) pp. 23–30.

[120] S.D. Flåm, "Asymptotically stable solutions to stochastic problems of Bolza" in: F. Archetti, G. Di Pillo, and M Lucertini, Eds., *Stochastic Programming* (Lecture Notes in Information and Control 76, 1986) pp. 184–193.

[121] W. Fleming and R. Rischel, *Deterministic and Stochastic Control* (Springer-Verlag, New York, NY, 1975).

[122] R. Fourer, "A simplex algorithm for piecewise-linear programming. I: derivation and proof," *Mathematical Programming* 33 (1985) pp. 204–233.

[123] R. Fourer, "A simplex algorithm for piecewise-linear programming. II: finiteness, feasibility, and degeneracy," *Mathematical Programming* 41 (1988) pp. 281–315.

[124] R. Fourer, D.M. Gay, and B.W. Kernighan, *AMPL: A Modeling Language for Mathematical Programming* (Scientific Press, South San Francisco, CA, 1993).

[125] B. Fox, "Implementation and relative efficiency of quasirandom sequence generators," *ACM Transactions on Mathematical Software* 12 (1986) pp. 362–376.

[126] L. Frantzeskakis and W. Powell, "A successive linear approximation procedure for stochastic, dynamic vehicle allocation problems," *Transportation Science* 24 (1990) pp. 40–57.

[127] L.F. Frantzeskakis and W.B. Powell, "Bounding procedures for multistage stochastic dynamic networks," *Networks* 23 (1993) pp. 575–595.

[128] K. Frauendorfer, "Solving SLP recourse problems:The case of stochastic technology matrix, RHS, and objective," *Proceedings of 13th IFIP Conference on System Modelling and Optimization* (Springer-Verlag, Berlin, 1988a).

[129] K. Frauendorfer, "Solving S.L.P. recourse problems with arbitrary multivariate distributions – the dependent case," *Mathematics of Operations Research* 13 (1988b) pp. 377–394.

[130] K. Frauendorfer, "A simplicial approximation scheme for convex two-stage stochastic programming problems," Manuskripte, Institut für Operations Research, University of Zurich (Zurich, 1989).

[131] K. Frauendorfer, *Stochastic Two-Stage Programming* (Lecture Notes in Economics and Mathematical Systems 392, 1992).

[132] K. Frauendorfer and P. Kall, "A solution method for SLP recourse problems with arbitrary multivariate distributions—the independent case," *Problems in Control and Information Theory* 17 (1988) pp. 177–205.

[133] A.A. Gaivoronski, "Implementation of stochastic quasigradient methods" in: Y. Ermoliev and R. Wets, Eds., *Numerical Techniques for Stochastic Optimization* (Springer-Verlag, Berlin, 1988) pp. 313–352.

[134] J. Galambos, *The Asymptotic Theory of Extreme Order Statistics* (John Wiley, Inc., New York, 1978).

[135] S.J. Gartska, "An economic interpretation of stochastic programs," *Mathematical Programming* 18 (1980) pp. 62–67.

[136] S.J. Gartska and D. Rutenberg, "Computation in discrete stochastic programs with recourse," *Operations Research* 21 (1973) pp. 112–122.

[137] S.J. Gartska and R.J-B Wets, "On decision rules in stochastic programming," *Mathematical Programming* 7 (1974) pp. 117–143.

[138] H.I. Gassmann, "Conditional probability and conditional expectation of a random vector" in: Y. Ermoliev and R. Wets, Eds., *Numerical Techniques for Stochastic Optimization* (Springer-Verlag, Berlin, 1988) pp. 237–254.

[139] H.I. Gassmann, "Optimal harvest of a forest in the presence of uncertainty," *Canadian Journal of Forest Research* 19 (1989) pp. 1267–1274.

[140] H.I. Gassmann, "MSLiP: a computer code for the multistage stochastic linear programming problem," *Mathematical Programming* 47 (1990) pp. 407–423.

[141] H.I. Gassmann and W.T. Ziemba, "A tight upper bound for the expectation of a convex function of a multivariate random variable," *Mathematical Programming Study* 27 (1986) pp. 39–53.

[142] D.M. Gay, "A variant of Karmarkar's linear programming algorithm for problems in standard form," *Mathematical Programming* 37 (1987) pp. 81–90.

[143] M. Gendreau, G. Laporte, and R. Séguin, "Stochastic vehicle routing," *European Journal of Operational Research* 88 (1996) pp. 3–12.

[144] A.M. Geoffrion, "Elements of large-scale mathematical programming," *Management Science* 16 (1970) pp. 652–675.

[145] A.M. Geoffrion, "Duality in nonlinear programming: a simplified applications-oriented development," *SIAM Rev.* 13 (1971) pp. 1–37.

[146] C.R. Glassey, "Nested decomposition and multistage linear programs," *Management Science* 20 (1973) pp. 282–292.

[147] R.C. Grinold, "A new approach to multi-stage stochastic linear programs," *Mathematical Programming Study* 6 (1976) pp. 19–29.

[148] R.C. Grinold, "Model building techniques for the correction of end effects in multistage convex programs," *Operations Research* 31 (1983) pp. 407–431.

[149] R.C. Grinold, "Infinite horizon stochastic programs," *SIAM Journal on Control and Optimization* 24 (1986) pp. 1246–1260.

[150] J.M. Harrison, *Brownian Motion and Stochastic Flow Systems* (John Wiley, Inc., New York, NY, 1985).

[151] J.M. Harrison and L.M. Wein, "Scheduling networks of queues:Heavy traffic analysis of a two-station closed network," *Operations Research* 38 (1990) pp. 1052–1064.

[152] D. Haugland and S.W. Wallace, "Solving many linear programs that differ only in the righthand side," *European Journal of Operational Research* 37 (1988) pp. 318–324.

[153] D.P. Heyman and M.J. Sobel, *Stochastic Models in Operations Research, Volume II, Stochastic Optimization* (McGraw-Hill, New York, NY, 1984).

[154] J. Higle and S. Sen, "Statistical verification of optimality conditions for stochastic programs with recourse," *Annals of Operations Research* 30 (1991a) pp. 215–240.

[155] J. Higle and S. Sen, "Stochastic decomposition: an algorithm for two stage linear programs with recourse," *Mathematics of Operations Research* 16 (1991b) pp. 650–669.

[156] J.-B. Hiriart-Urruty, "Conditions nécessaires d'optimalité pour un programme stochastique avec recours," *SIAM Journal on Control and Optimization* 16 (1978) pp. 317–329.

[157] J.K. Ho and E. Loute, "A set of staircase linear programming test problems," *Mathematical Programming* 20 (1981) pp. 245–250.

[158] J.K. Ho and A.S. Manne, "Nested decomposition for dynamic models," *Mathematical Programming* 6 (1974) pp. 121–140.

[159] W. Hoeffding, "Probability inequalities for sums of bounded random variables," *Journal of the American Statistical Association* 58 (1963) pp. 13–30.

[160] A. Hogan, J. Morris, and H. Thompson, "Decision problems under risk and chance constrained programming: dilemmas in the transition," *Management Science* 27 (1981) pp. 698–716.

[161] A. Hogan, J. Morris, and H. Thompson, "Reply to Professors Charnes and Cooper concerning their response to 'Decision problems under risk and chance constrained programming: dilemmas in the transition'," *Management Science* 30 (1984) pp. 258–259.

[162] R.A. Howard, *Dynamic Programming and Markov Processes* (MIT Press, Cambridge, MA, 1960).

[163] C.C. Huang, W. Ziemba, and A. Ben-Tal, "Bounds on the expectation of a convex function of a random variable: with applications to stochastic programming," *Operations Research* 25 (1977) pp. 315–325.

[164] P.J. Huber, "The behavior of maximum likelihood estimates under nonstandard conditions," *Proceedings of the Fifth Berkeley Symposium on Mathematical Statistics and Probability*, (University of California, Berkeley, CA, 1967).

[165] J.C. Hull, *Options, Futures and Other Derivatives*, third edition, (Prentice-Hall, Upper Saddle River, NJ, 1997).

[166] G. Infanger, "Monte Carlo (importance) sampling within a Benders decomposition algorithm for stochastic linear programs; Extended version: including results of large-scale problems," Technical Report SOL 91-6, Systems Optimization Laboratory, Stanford University (Stanford, CA, 1991).

[167] G. Infanger, *Planning under Uncertainty: Solving Large-Scale Stochastic Linear Programs* (Boyd and Fraser, Danvers, MA, 1994).

[168] International Business Machines Corp., "Optimization Subroutine Library Guide and Reference, Release 2," document SC23-0519-02, International Business Machines Corp. (Armonk, NY, 1991).

[169] R. Jagganathan, "A minimax procedure for a class of linear programs under uncertainty," *Operations Research* 25 (1977) pp. 173–177.

[170] R. Jagganathan, "Use of sample information in stochastic recourse and chance-constrained programming models," *Management Science* 31 (1985) pp. 96–108.

[171] R. Jagganathan, "Linear programming with stochastic processes as parameters as applied to production planning," *Annals of Operations Research* 30 (1991) pp. 107–114.

[172] P. Jaillet, "A priori solution of a traveling salesman problem in which a random subset of the customers are visited," *Operations Research* 36 (1988) pp. 929–936.

[173] R.A. Jarrow and A. Rudd, *Option Pricing* (Irwin, Homewood, IL, 1983).

[174] J.L. Jensen, "Sur les fonctions convexes et les inégalités entre les valeurs moyennes," *Acta. Math.* 30 (1906) pp. 175–193.

[175] W.C. Jordan and S.C. Graves, "Principles on the benefits of manufacturing process flexibility," Technical Report GMR-7310, General Motors Research Laboratories, Warren, MI (1991).

[176] P. Kall, *Stochastic Linear Programming* (Springer-Verlag, Berlin, 1976).

[177] P. Kall, "Computational methods for solving two-stage stochastic linear programming problems," *Journal of Applied Mathematics and Physics* 30 (1979) pp. 261–271.

[178] P. Kall, "Stochastic programs with recourse: an upper bound and the related moment problem," *Zeitschrift für Operations Research* 31 (1987) pp. A119–A141.

[179] P. Kall, "An upper bound for stochastic linear programming using first and total second moments," *Annals of Operations Research* 30 (1991) pp. 267–276.

[180] P. Kall and J. Mayer, "SLP-IOR: an interactive model management system for stochastic linear programs," *Mathematical Programming* 75 (1996) pp. 221–240.

[181] P. Kall and D. Stoyan, "Solving stochastic programming problems with recourse including error bounds," *Math. Operationsforsch. Statist. Ser. Optim.* 13 (1982) pp. 431–447.

[182] P. Kall and S.W. Wallace, *Stochastic Programming* (John Wiley and Sons, Chichester, UK, 1994).

[183] J.G. Kallberg, R.W. White, and W.T. Ziemba, "Short term financial planning under uncertainty," *Management Science* 28 (1982) pp. 670–682.

[184] J.G. Kallberg and W.T. Ziemba, "Comparison of alternative utility functions in portfolio selection problems," *Management Science* 29 (1983) pp. 1257–1276.

[185] M. Kallio and E. Porteus, "Decomposition of arborescent linear programs," *Mathematical Programming* 13 (1977) pp. 348–356.

[186] R.E. Kalman , *Topics in Mathematical System Theory* (McGraw-Hill, New York, NY, 1969).

[187] E. Kao and M. Queyranne, "Budgeting costs of nursing in a hospital," *Management Science* 31 (1985) pp. 608–621.

[188] N. Karmarkar, "A new polynomial-time algorithm for linear programming," *Combinatorica* 4 (1984) pp. 373–395.

400   References

[189] A. Karr, "Extreme points of certain sets of probability measure, with applications," *Mathematics of Operations Research* 8 (1983) pp. 74–85.

[190] J. Kemperman, "The general moment problem, a geometric approach," *Annals of Mathematical Statistics* 39 (1968) pp. 93–122.

[191] A.I. Kibzun and Y.S. Kan, *Stochastic Programming Problems with Probability and Quantile Functions* (John Wiley Inc., Chichester, UK, 1996).

[192] A.I. Kibzun and V.Yu. Kurbakovskiy, "Guaranteeing approach to solving quantile optimization problems," *Annals of Operations Research* 30 (1991) pp. 81–93.

[193] A. King, "Finite generation method" in: Y. Ermoliev and R. Wets, Eds., *Numerical Techniques for Stochastic Optimization* (Springer-Verlag, Berlin, 1988a) pp. 295–312.

[194] A. King, "Stochastic programming problems:Examples from the literature" in: Y. Ermoliev and R. Wets, Eds., *Numerical Techniques for Stochastic Optimization* (Springer-Verlag, Berlin, 1988b) pp. 543–567.

[195] A. King and R.T. Rockafellar, "Asymptotic theory for solutions in generalized M-estimation and stochastic programming," *Mathematics of Operations Research* 18 (1993) pp. 148–162.

[196] A.J. King and R.J-B Wets, "Epiconsistency of convex stochastic programs," *Stochastics and Stochastics Reports* 34 (1991) pp. 83–92.

[197] K.C. Kiwiel, "An aggregate subgradient method for nonsmooth convex minimization," *Mathematical Programming* 27 (1983) pp. 320–341.

[198] W.K. Klein Haneveld, *Duality in Stochastic Linear and Dynamic Programming* (Lecture Notes in Economics and Mathematical Systems 274, Springer-Verlag, Berlin, 1985).

[199] W.K. Klein Haneveld, "Robustness against dependence in PERT: an application of duality and distributions with known marginals," *Mathematical Programming Study* 27 (1986) pp. 153–182.

[200] M.G. Krein and A.A. Nudel'man, *The Markov Moment Problem and Extremal Problems* (Translations of Mathematical Monographs 50, 1977).

[201] H. Kushner, *Introduction to Stochastic Control* (Holt, New York, NY, 1971).

[202] M. Kusy and W.T. Ziemba, "A bank asset and liability management model," *Operations Research* 34 (1986) pp. 356–376.

[203] B.J. Lageweg, J.K. Lenstra, A.H.G. Rinnooy Kan, and L. Stougie, "Stochastic integer programming by dynamic programming" in: Y. Ermoliev and R. Wets, Eds., *Numerical Techniques for Stochastic Optimization* (Springer-Verlag, Berlin, 1988) pp. 403–412.

[204] G. Laporte and F.V. Louveaux, "The integer L-shaped method for stochastic integer programs with complete recourse," *Operations Research Letters* 13 (1993) pp. 133–142.

[205] G. Laporte, F.V. Louveaux, and H. Mercure, "Models and exact solutions for a class of stochastic location-routing problems," *European Journal of Operational Research* 39 (1989) pp. 71–78.

[206] G. Laporte, F.V. Louveaux, and H. Mercure, "An exact solution for the a priori optimization of the probabilistic traveling salesman problem," *Operations Research* 42 (1994) pp. 543–549.

[207] G. Laporte, F.V. Louveaux, and L. Van Hamme, "Exact solution to a location problem with stochastic demands," *Transportation Science* 28 (1994) pp. 95–103.

[208] L. Lasdon, *Optimization Theory for Large Systems* (Macmillan, New York, NY, 1970).

[209] C. Lemaréchal, "Bundle methods in nonsmooth optimization" in: *Nonsmooth optimization (Proc. IIASA Workshop)* (Pergamon, Oxford-Elmsford, New York, NY, 1978) pp. 79–102.

[210] J.K. Lenstra, A.H.G. Rinnooy Kan, and L. Stougie, "A framework for the probabilistic analysis of hierarchical planning systems," *Annals of Operation Research* 1 (1984) pp. 23–42.

[211] F.V. Louveaux, "Piecewise convex programs," *Mathematical Programming* 15 (1978) pp. 53–62.

[212] F.V. Louveaux, "A solution method for multistage stochastic programs with recourse with application to an energy investment problem," *Operations Research* 28 (1980) pp. 889–902.

[213] F.V. Louveaux, "Multistage stochastic programs with block-separable recourse," *Mathematical Programming Study* 28 (1986) pp. 48–62.

[214] F.V. Louveaux and D. Peeters, "A dual-based procedure for stochastic facility location," *Operations Research* 40 (1992) pp. 564–573.

[215] F.V. Louveaux and Y. Smeers, "Optimal investments for electricity generation:A stochastic model and a test-problem" in: *Numerical Techniques for Stochastic Optimization* (Springer-Verlag, Berlin, 1988) pp. 33–64.

[216] F.V. Louveaux and Y. Smeers, "Stochastic optimization for the introduction of a new energy technology," *Stochastics (to appear)* (1997).

[217] F.V. Louveaux and M. van der Vlerk, "Stochastic programming with simple integer recourse," *Mathematical Programming* 61 (1993) pp. 301–325.

[218] I.J. Lustig, R.E. Marsten, and D.F. Shanno, "Computational experience with a primal-dual interior point method for linear programming," *Linear Algebra and Its Application* 152 (1991) pp. 191–222.

[219] A. Madansky, "Bounds on the expectation of a convex function of a multivariate random variable," *Annals of Mathematical Statistics* 30 (1959) pp. 743–746.

[220] A. Madansky, "Inequalities for stochastic linear programming problems," *Management Science* 6 (1960) pp. 197–204.

[221] M. Maddox and J.R. Birge, "Bounds on the distribution of tardiness in a PERT network," Technical Report, Department of Industrial and Operations Engineering, University of Michigan (Ann Arbor, MI, 1991).

[222] O. Mangasarian and J.B. Rosen, "Inequalities for stochastic nonlinear programming problems," *Operations Research* 12 (1964) pp. 143–154.

[223] A.S. Manne, "Waiting for the breeder" in: *Review of Economic Studies Symposium* (1974) pp. 47–65.

[224] A.S. Manne and R. Richels, *Buying Greenhouse Insurance—The Economic Costs of Carbon Dioxide Emission Limits* (MIT Press, Cambridge, MA, 1992).

[225] H.M. Markowitz, *Portfolio Selection; Efficient Diversification of Investments* (John Wiley, Inc., New York, NY, 1959).

[226] K. Marti, "Approximationen von Entscheidungsproblemen mit linearer Ergebnisfunktion und positiv homogener, subadditiver Verlusfunktion," *Zeitschrift für Wahrscheinlichkeitstheorie und Verwandte Gebiete* 31 (1975) pp. 203–233.

[227] K. Marti, *Descent Directions and Efficient Solutions in Discretely Distributed Stochastic Programs*, (Lecture Notes in Economics and Mathematical Systems 299, Springer-Verlag, Berlin, 1988).

[228] L. McKenzie, "Turnpike theory," *Econometrica* 44 (1976) pp. 841–864.

[229] P. Michel and J.-P. Penot, "Calcul sous-différentiel pour des fonctions lipschitziennes et non lipschitziennes," *Comptes Rendus des Seances de l'Académie des Sciences Paris. Serie 1. Mathématique* 298 (1984) pp. 269–272.

[230] J. Miller and H. Wagner, "Chance-constrained programming with joint chance constraints," *Operations Research* 12 (1965) pp. 930–945.

[231] G.J. Minty, "On the maximal domain of a 'monotone' function," *Michigan Mathematics Journal* 8 (1961) pp. 135–137.

[232] F. Mirzoachmedov and S. Uriasiev, "Adaptive step-size control for stochastic optimization algorithm," *Zhurnal vicisl. mat. i mat. fiz.* 6 (1983) pp. 1314–1325 (in Russian).

[233] B. Mordukhovich, "Approximation methods and extremum conditions in nonsmooth control systems," *Soviet Mathematics Doklady* 36 (1988) pp. 164–168.

[234] D.P. Morton, "An enhanced decomposition algorithm for multistage stochastic hydroelectric scheduling," Technical Report NPSOR-94-001, Department of Operations Research, Naval Postgraduate School (Monterey, CA, 1994).

[235] J.M. Mulvey and A. Ruszczyński, "A new scenario decomposition method for large scale stochastic optimization," *Operations Research* 43 (1995) pp. 477–490.

[236] J.M. Mulvey and H. Vladimirou, "Stochastic network optimization models for investment planning," *Annals of Operations Research* 20 (1989) pp. 187–217.

[237] J.M. Mulvey and H. Vladimirou, "Applying the progressive hedging algorithm to stochastic generalized networks," *Annals of Operations Research* 31 (1991a) pp. 399–424.

[238] J.M. Mulvey and H. Vladimirou, "Solving multistage stochastic networks: an application of scenario aggregation," *Networks* 21 (1991b) pp. 619–643.

[239] J.M. Mulvey and H. Vladimirou, "Stochastic network programming for financial planning problems," *Management Science* 38 (1992) pp. 1642–1664.

[240] B.A. Murtagh and M.A. Saunders, "MINOS 5.0 User's Guide," Technical Report SOL 83-20, Systems Optimization Laboratory, Stanford University (Stanford, CA, 1983).

[241] K.G. Murty, "Linear programming under uncertainty: a basic property of the optimal solution," *Z. Wahrscheinlichkeitstheorie und Verw. Gebiete* 10 (1968) pp. 284–288.

[242] K.G. Murty, *Linear Programming* (John Wiley, Inc., New York, NY, 1983).

[243] S.C. Myers, "Finance theory and financial strategy," *Interfaces* 14:1 (1984) pp. 126–137.

[244] J.L. Nazareth and R.J-B Wets, "Algorithms for stochastic programs: the case of nonstochastic tenders," *Mathematical Programming Study* 28 (1986) pp. 1–28.

[245] G.L. Nemhauser and L.A. Wolsey, *Integer and Combinatorial Optimization* (Wiley-Interscience, New York, NY, 1988).

[246] H. Niederreiter, "Quasi–Monte Carlo methods and pseudorandom numbers," *Bulletin of the American Mathematical Society* 84 (1978) pp. 957–1041.

[247] S.S. Nielsen and S.A. Zenios, "A massively parallel algorithm for nonlinear stochastic network problems," *Operations Research* 41 (1993a) pp. 319–337.

[248] S.S. Nielsen and S.A. Zenios, "Proximal minimizations with D-functions and the massively parallel solution of linear stochastic network programs," *International Journal of Supercomputing and Applications* 7 (1993b) pp. 349–364.

[249] M.-C. Noël and Y. Smeers, "Nested decomposition of multistage nonlinear programs with recourse," *Mathematical Programming* 37 (1987) pp. 131–152.

[250] V.I. Norkin, Y.M. Ermoliev, and A. Ruszczyński, "On optimal allocation of indivisibles under uncertainty," *Operations Research* 46 (1998) pp. 381–395.

[251] S. Parikh, *Lecture Notes on Stochastic Programming* (University of California, Berkeley, CA, 1968).

[252] M.V.F. Pereira and L.M.V.G. Pinto, "Stochastic optimization of a multireservoir hydroelectric system—A decomposition approach," *Water Resources Research* 21 (1985) pp. 779–792.

[253] G.Ch. Pflug, "Stepsize rules, stopping times and their implementation in stochastic quasigradient algorithms" in: Y. Ermoliev and R. Wets, Eds., *Numerical Techniques for Stochastic Optimization* (Springer-Verlag, Berlin, 1988) pp. 353–372.

[254] J. Pintér, "Deterministic approximations of probability inequalities," *ZOR—Methods and Models of Operations Research, Series Theory* 33 (1989) pp. 219–239.

[255] E.L. Plambeck, B-R. Fu, S.M. Robinson, and R. Suri, "Sample-path optimization of convex stochastic performance functions," *Mathematical Programming* 75 (1996) pp. 137–176.

[256] W.B. Powell, "A comparative review of alternative algorithms for the dynamic vehicle allocation program" in: B. Golden and A. Assad, Eds., *Vehicle Routing: Methods and Studies* (North-Holland, Amsterdam, 1988).

[257] A. Prékopa, "Logarithmic concave measures with application to stochastic programming," *Acta. Sci. Math. (Szeged)* 32 (1971) pp. 301–316.

[258] A. Prékopa, "Contributions to the theory of stochastic programs," *Mathematical Programming* 4 (1973) pp. 202–221.

[259] A. Prékopa, "Programming under probabilistic constraints with a random technology matrix," *Mathematische Operationsforschung und Statistik* 5 (1974) pp. 109–116.

[260] A. Prékopa, "Logarithmically concave measures and related topics" in: M.A.H. Dempster, Ed., *Stochastic Programming* (Academic Press, New York, NY, 1980).

[261] A. Prékopa, "Boole-Bonferroni inequalities and linear programming," *Operations Research* 36 (1988) pp. 145–162.

[262] A. Prékopa, *Stochastic Programming* (Kluwer Academic Publishers, Dordrecht, Netherlands, 1995).

[263] A. Prékopa and T. Szántai, "On optimal regulation of a storage level with application to the water level regulation of a lake," Survey of Mathematical Programming (Proc. Ninth Internat. Math. Programming Sympos., Budapest, 1976), Vol. 2 (North-Holland, Amsterdam, 1976).

[264] H.N. Psaraftis, "On the practical importance of asymptotic optimality in certain heuristic algorithms," *Networks* (1984) pp. 587–596.

[265] H.N. Psaraftis, G.G. Tharakan, and A. Ceder, "Optimal response to oil spills: the strategic decision case," *Operations Research* 34 (1986) pp. 203–217.

[266] L. Qi, "Forest iteration method for stochastic transportation problem," *Mathematical Programming Study* (1985) pp. 142–163.

[267] L. Qi, "An alternating method for stochastic linear programming with simple recourse," *Stochastic Processes and Their Applications* 841 (1986) pp. 183–190.

[268] H. Raiffa, *Decision Analysis* (Addison-Wesley, Reading, MA, 1968).

[269] H. Raiffa and R. Schlaifer, *Applied Statistical Decision Theory* (Harvard University, Boston, MA, 1961).

[270] A.H.G. Rinnooy Kan and L. Stougie, "Stochastic integer programming" in: Y. Ermoliev and R. Wets, Eds., *Numerical Techniques for Stochastic Optimization* (Springer-Verlag, Berlin, 1988) pp. 201–213.

[271] H. Robbins and S. Monro, "A stochastic approximation method," *Annals of Mathematical Statistics* 22 (1951) pp. 400–407.

[272] S.M. Robinson and R.J-B Wets, "Stability in two-stage stochastic programming," *SIAM Journal on Control and Optimization* 25 (1987) pp. 1409–1416.

[273] R.T. Rockafellar, *Convex Analysis* (Princeton University Press, Princeton, NJ, 1969).

[274] R.T. Rockafellar, *Conjugate Duality and Optimization* (Society for Industrial and Applied Mathematics, Philadelphia, PA, 1974).

[275] R.T. Rockafellar, "Monotone operators and the proximal point algorithm," *SIAM Journal on Control and Optimization* 14 (1976a) pp. 877–898.

[276] R.T. Rockafellar, *Integral Functionals, Normal Integrands and Measurable Selections* (Lecture Notes in Mathematics 543, 1976b).

[277] R.T. Rockafellar and R.J-B Wets, "Stochastic convex programming: basic duality," *Pacific Journal of Mathematics* 63 (1976a) pp. 173–195.

[278] R.T. Rockafellar and R.J-B Wets, "Stochastic convex programming, relatively complete recourse and induced feasibility," *SIAM Journal on Control and Optimization* 14 (1976b) pp. 574–589.

[279] R.T. Rockafellar and R.J-B Wets, "A Lagrangian finite generation technique for solving linear-quadratic problems in stochastic programming," *Mathematical Programming Study* 28 (1986) pp. 63–93.

[280] R.T. Rockafellar and R.J-B Wets, "Scenarios and policy aggregation in optimization under uncertainty," *Mathematics of Operations Research* 16 (1991) pp. 119–147.

[281] W. Römisch and R. Schultz, "Distribution sensitivity in stochastic programming," *Mathematical Programming* 50 (1991a) pp. 197–226.

[282] W. Römisch and R. Schultz, "Stability analysis for stochastic programs," *Annals of Operations Research* 31 (1991b) pp. 241–266.

[283] S.M. Ross, *Introduction to Stochastic Dynamic Programming* (Academic Press, New York, London, 1983).

[284] H.L. Royden, *Real Analysis* (Macmillan, London, NY, 1968).

[285] R.Y. Rubinstein, *Simulation and the Monte Carlo Method* (John Wiley Inc., New York, NY, 1981).

[286] A. Ruszczyński, "A regularized decomposition for minimizing a sum of polyhedral functions," *Mathematical Programming* 35 (1986) pp. 309–333.

[287] A. Ruszczyński, "Parallel decomposition of multistage stochastic programming problems," *Mathematical Programming* 58 (1993a) pp. 201–228.

[288] A. Ruszczyński, "Regularized decomposition of stochastic programs: algorithmic techniques and numerical results," Working Paper WP-93-21, International Institute for Applied Systems Analysis, Laxenburg, Austria (1993b).

[289] G. Salinetti, "Approximations for chance constrained programming problems," *Stochastics* 10 (1983) pp. 157–169.

[290] Y.S. Sathe, M. Pradhan, and S.P. Shah, "Inequalities for the probability of the occurrence of at least m out of n events," *Journal of Applied Probability* 17 (1980) pp. 1127–1132.

[291] H. Scarf, "A minimax solution of an inventory problem" in: K.J. Arrow, S. Karlin, and H. Scarf, Eds., *Studies in the Mathematical Theory of Inventory and Production* (Stanford University Press, Stanford, CA, 1958).

[292] R. Schultz, "Continuity properties of expectation functionals in stochastic integer programming," *Mathematics of Operations Research* 18 (1993) pp. 578–589.

[293] A. Shapiro, "Asymptotic analysis of stochastic programs," *Annals of Operations Research* 30 (1991) pp. 169–186.

[294] W.F. Sharpe, "Capital asset prices: a theory of market equilibrium under conditions of risk," *Journal of Finance* 19 (1964) pp. 425–442.

[295] D. Simchi-Levi, "Hierarchical planning for probabilistic distribution systems in the Euclidean spaces," *Management Science* 38 (1992) pp. 198–211.

[296] L. Somlyódi and R.J-B Wets, "Stochastic optimization models for lake eutrophication management," *Operations Research* 36 (1988) pp. 660–681.

[297] L. Stougie, *Design and Analysis of Algorithms for Stochastic Integer Programming* (Centrum voor Wiskunde en Informatica, Amsterdam, 1987).

[298] B. Strazicky, "Some results concerning an algorithm for the discrete recourse problem," in: M.A.H. Dempster, Ed., *Stochastic Programming* (Academic Press, New York, NY, 1980).

[299] A.H. Stroud, *Approximate Calculation of Multiple Integrals* (Prentice-Hall, Inc., Englewood Cliffs, NJ, 1971).

[300] J. Sun, L. Qi, and K-H. Tsai, "Solving stochastic transshipment problems as network piecewise linear programs," Technical Report, School of Mathematics, The University of New South Wales (Kensington, UNSW, Australia, 1990).

[301] G.H. Symonds, "Chance-constrained equivalents of stochastic programming problems," *Operations Research* 16 (1968) pp. 1152–1159.

[302] T. Szántai, "Evaluation of a special multivariate gamma distribution function," *Mathematical Programming Study* 27 (1986) pp. 1–16.

[303] G. Taguchi, *Introduction to Quality Engineering* (Asian Productivity Center, Tokyo, Japan, 1986).

[304] G. Taguchi, E.A. Alsayed, and T. Hsiang, *Quality Engineering in Production Systems* (McGraw-Hill Inc., New York, NY, 1989).

[305] H.A. Taha, *Operations Research: An Introduction*, Fifth edition (Macmillan, New York, NY, 1992).

[306] S. Takriti, "On-line solution of linear programs with varying right-hand sides," Ph.D. Dissertation, Department of Industrial and Operations Engineering, University of Michigan (Ann Arbor, MI, 1994).

[307] M.J. Todd and B.P. Burrell, "An extension of Karmarkar's algorithm for linear programming using dual variables," *Algorithmica* 1 (1986) pp. 409–424.

[308] D.M. Topkis and A.F. Veinott, Jr., "On the convergence of some feasible Eddirection algorithms for nonlinear programming," *SIAM Journal on Control* 5 (1967) pp. 268–279.

[309] C. Toregas, R. Swain, C. Revelle, and L. Bergmann, "The location of emergency service facilities," *Operations Research* 19 (1971) pp. 1363–1373.

[310] S. Uriasiev, "Adaptive stochastic quasigradient methods" in: Y. Ermoliev and R. Wets, Eds., *Numerical Techniques for Stochastic Optimization* (Springer-Verlag, Berlin, 1988) pp. 373–384.

[311] F.A. Valentine, *Convex Sets* (McGraw-Hill Inc., New York, NY, 1964).

[312] R. Van Slyke and R.J-B Wets, "L-shaped linear programs with application to optimal control and stochastic programming," *SIAM Journal on Applied Mathematics* 17 (1969) pp. 638–663.

[313] R.J. Vanderbei, M.S. Meketon, and B.A. Freedman, "A modification of Karmarkar's linear programming algorithm," *Algorithmica* 1 (1986) pp. 395–407.

[314] P. Varaiya and R.J-B Wets, "Stochastic dynamic optimization approaches and computation" in: M. Iri and K. Tanabe, Eds., *Mathematical Programming: Recent Developments and Applications* (Kluwer, Dordrecht, Netherlands, 1989) pp. 309–332.

[315] J.A. Ventura and D.W. Hearn, "Restricted simplicial decomposition for convex constrained problems," *Mathematical Programming* 59 (1993) pp. 71–85.

[316] J. Von Neumann and O. Morgenstern, *Theory of Games and Economic Behavior* (Princeton University Press, Princeton, NJ, 1944).

[317] A. Wald, *Statistical Decision Functions* (John Wiley, Inc. New York, NY, 1950).

[318] D. Walkup and R.J-B Wets, "Stochastic programs with recourse," *SIAM Journal on Applied Mathematics* 15 (1967) pp. 1299–1314.

[319] D. Walkup and R.J-B Wets, "Stochastic programs with recourse II: on the continuity of the objective," *SIAM Journal on Applied Mathematics* 17 (1969) pp. 98–103.

[320] S.W. Wallace, "Decomposing the requirement space of a transportation problem into polyhedral cones," *Mathematical Programming Study* 28 (1986a) pp. 29–47.

[321] S.W. Wallace, "Solving stochastic programs with network recourse," *Networks* 16 (1986b) pp. 295–317.

[322] S.W. Wallace, "A piecewise linear upper bound on the network recourse function," *Networks* 17 (1987) pp. 87–103.

[323] S.W. Wallace and R.J-B Wets, "Preprocessing in stochastic programming: the case of linear programs," *ORSA Journal on Computing* 4 (1992) pp. 45–59.

[324] S.W. Wallace and T.C. Yan, "Bounding multi-stage stochastic programs from above," *Mathematical Programming* 61 (1993) pp. 111–129.

[325] R.J-B Wets, "Programming under uncertainty: the equivalent convex program," *SIAM Journal on Applied Mathematics* 14 (1966) pp. 89–105.

[326] R.J-B Wets, "Characterization theorems for stochastic programs," *Mathematical Programming* 2 (1972) pp. 166–175.

[327] R.J-B Wets, "Stochastic programs with fixed recourse: the equivalent deterministic problem," *SIAM Review* 16 (1974) pp. 309–339.

[328] R.J-B Wets, "Convergence of convex functions, variational inequalities and convex optimization problems" in: R.W. Cottle, F. Giannessi and J.-L. Lions, Eds., *Variational Inequalities and Complementarity Problems* (John Wiley, Inc., New York, NY, 1980a) pp. 375–404.

[329] R.J-B Wets, "Stochastic multipliers, induced feasibility and nonanticipativity in stochastic programming" in: M.A.H. Dempster, Ed., *Stochastic Programming* (Academic Press, New York, NY, 1980b).

[330] R.J-B Wets, "Solving stochastic programs with simple recourse," *Stochastics* 10 (1983a) pp. 219–242.

[331] R.J-B Wets, "Stochastic programming: solution techniques and approximation schemes" in: A. Bachem, M. Grötschel, and B. Korte, Eds., *Mathematical Programming: State-of-the-Art 1982* (Springer-Verlag, Berlin, 1983b) pp. 560–603.

[332] R.J-B Wets, "Large-scale linear programming techniques in stochastic programming" in: Y. Ermoliev and R. Wets, Eds., *Numerical Techniques for Stochastic Optimization* (Springer-Verlag, Berlin, 1988).

[333] R.J-B Wets, "Stochastic programming" in: G.L. Nemhauser, A.H.G. Rinnooy Kan, and M.J. Todd, Eds., *Optimization* (Handbooks in Operations Research and Management Science; Vol. 1, North–Holland, Amsterdam, Netherlands, 1990).

[334] R.J-B Wets and C. Witzgall, "Algorithms for frames and lineality spaces of cones," *Journal of Research of the National Bureau of Standards Section B* 71B (1967) pp. 1–7.

[335] A.C. Williams, "A stochastic transportation problem," *Operations Research* 11 (1963) pp. 759–770.

[336] A.C. Williams, "Approximation for stochastic linear programming," *SIAM Journal on Applied Mathematics* 14 (1966) pp. 668.

[337] R.J. Wittrock, "Advances in a nested decomposition algorithm for solving staircase linear programs," Technical Report SOL 83-2, Systems Optimization Laboratory, Stanford University (Stanford, CA, 1983).

[338] R. Wollmer, "Two stage linear programming under uncertainty with 0-1 integer first stage variables," *Mathematical Programming* 19 (1980) pp. 279–288.

[339] H. Woźniakowski, "Average-case complexity of multivariate integration," *Bulletin of the American Mathematical Society (new series)* 24 (1991) pp. 185–194.

[340] S.E. Wright, "Primal-dual aggregation and disaggregation for stochastic linear programs," *Mathematics of Operations Research* 19 (1994) pp. 893–908.

[341] D. Yang and S.A. Zenios, "A scalable parallel interior point algorithm for stochastic linear programming and robust optimization," Report 95-07, Department of Public and Business Administration, University of Cyprus (Nicosia, Cyprus, 1995).

[342] Y. Ye, "Karmarkar's algorithm and the ellipsoid method," *Operations Research Letters* 6 (1987) pp. 177–182.

[343] Y. Ye and M. Kojima, "Recovering optimal dual solutions in karmarkar's polynomial algorithm for linear programming," *Mathematical Programming* 39 (1987) pp. 305–317.

[344] J. Žáčková, "On minimax solutions of stochastic linear programming problems," *Časopis pro Pěstování Matematiky* 91 (1966) pp. 423–430.

[345] S.A. Zenios, *Financial Optimization* (Cambridge University Press, Cambridge, UK, 1992).

[346] W.T. Ziemba, "Computational algorithms for convex stochastic programs with simple recourse," *Operations Research* 18 (1970) pp. 414–431.

[347] W.T. Ziemba and R.G. Vickson, *Stochastic Optimization Models in Finance* (Academic Press, New York, NY, 1975).

[348] P. Zipkin, "Bounds for row-aggregation in linear programming," *Operations Research* 28 (1980a) pp. 903–916.

[349] P. Zipkin, "Bounds on the effect of aggregating variables in linear programs," *Operations Research* 28 (1980b) pp. 403–418.

Author Index

Subject Index